D0365443

ONLY THE NAILS REMAIN

Books by Christopher Merrill

Workbook (poems)
Outcroppings: John McPhee in the West (editor)
Fevers & Tides (poems)
The Forgotten Language: Contemporary Poets and Nature (editor)
From the Faraway Nearby: Georgia O'Keeffe as Icon (editor, with Ellen Bradbury)
The Grass of Another Country: A Journey Through the World of Soccer (nonfiction)
Anxious Moments, prose poems by Aleš Debeljak (translator, with the author)
Watch Fire (poems)
What Will Suffice: Contemporary American Poets on the Art of Poetry (editor, with Christopher Buckley)
The Old Bridge: The Third Balkan War and the Age of the Refugee (nonfiction)
The Four Questions of Melancholy: New and Selected Poems of Tomaž Šalamun (editor)
The Forest of Speaking Trees: An Essay on Poetry
Your Final Pleasure: An Essay on Reading
The Way to the Salt Marsh: A John Hay Reader (editor)
The City and the Child, poems by Aleš Debeljak (translator, with the author)

ONLY THE NAILS REMAIN

Scenes from the Balkan Wars

CHRISTOPHER MERRILL

ROWMAN & LITTLEFIELD PUBLISHERS, INC.
Lanham • Boulder • New York • Oxford

ROWMAN & LITTLEFIELD PUBLISHERS, INC.

Published in the United States of America
by Rowman & Littlefield Publishers, Inc.
4720 Boston Way, Lanham, Maryland 20706
http://www.rowmanlittlefield.com

12 Hid's Copse Road
Cumnor Hill, Oxford OX2 9JJ, England

British Library Cataloging in Publication Information Available

Library of Congress Cataloging-in-Publication Data

Merrill, Christopher.
Only the nails remain : scenes from the Balkan wars / Christopher Merrill.
p. cm.
Includes bibliograhical references and index.
ISBN 0-8476-9820-3 (alk. paper)
1. Merrill, Christopher. 2. Yugoslav War, 1991–1995—Personal narratives, American. 3. Former Yugoslav republics. I. Title.
DR1313.8.M48 1999
949.703—dc21 99-16783
CIP

Printed in the United States of America

∞™ The paper used in this publication meets the minimum requirements of American National Standard for Information Sciences—Permanence of Paper for Printed Library Materials, ANSI Z.39.48–1992.

for Frederick Turner

Contents

Acknowledgments

I would like to thank the following individuals without whose help and inspiration this book could not have been completed: Amy Singleton Adams, Shirley Adams, Marjorie Agosin, Stephen Ainlay, Agha Shahid Ali, Dick and Gabrielle Anderson, Kemal Bakaršić, Ellen Bradbury, Emilie Buchwald, Tracy Cabanis, Margaret Collins, Robert Cording, Patrick Cox, Frederick Cuny, Mary Jo Cuny, Jadranka and Predgrag Čicovački, Savo Cvetanovski, Mike Davis, Aleš and Erica Debeljak, Christopher Dustin, Ferida Duraković, Peter Eisenhauer, Jane Fox, Robert Fox, Mark Freeman, Tom Gavin, Sara Gillespie, Leslie Golden, Clifford and Rosellen Gowdy, Elizabeth Grossman, Marijan Gubic, Sue Halpern, David Harland, Emily Hiestand, Susan Holahan, Rebecca Hoogs, Bujar and Susan Hudhri, Mark Irwin, Matthew Kaufman, James Kee, Edmund Keeley, Metka Krašovec, Elaine Lamattina, Tom Lawler, Dennis Maloney, Richard Matlak, Jeanne McCulloch, Charles and Suzanne Merrill, Lisa Gowdy-Merrill, Paula and William Merwin, Kent Morris, David O'Brien, Ron Padgett, Lynne Partington, Jadranka and Marina Pintarić, Zvonimir and Ivanka Radeljković, Pat Reed, Ed Reid, Peter Hadji-Ristic, Phil Alden Robinson, Tomaž Šalamun, Nicholas Samaras, Barry Sanders, Jim Schley, Mark Shelton, Amela and Goran Simić, Greg Simon, Mojca Šoštarko, David St. John, Sarah Stanbury, William Strachan, Veno Taufer, Mike Tharp, James Thomas, Eva Toth, Darcy Tromanhauser, Frederick Turner, Andrew Wachtel, William Wadsworth, Brooke and Terry Tempest Williams, John Wilson, and Krista Witanowski.

I am very grateful to the Open Society Institute, which provided the funds and impetus for two of my journeys, and especially to Heather Iliff, Katya Koncz, and Ammar Mirascija, who coordinated my stay in Sarajevo in March 1995. My thanks as well to Domenick DiPasquale and Sandy Rouse for providing me with a U.S. Information Agency travel grant to Slovenia in September 1997, and to the Dean's Office at the College of the Holy Cross for covering my travel costs to the UNESCO Barcelona Conference on Higher Education in Bosnia and Herzegovina.

Parts of this book have appeared, in somewhat different form, in the following publications: *The American Poetry Review*, *The Antioch Review*, *DoubleTake*,

Krasnogruda, Left Bank, Mānoa, Middlebury Magazine, Notre Dame Review, The Old Bridge: The Third Balkan War and the Age of the Refugee, by Christopher Merrill (Minneapolis, MN: Milkweed Editions, 1995), *Orion Magazine, Sarajevo: An Anthology for Bosnian Relief*, edited by John Babbitt, Carolyn Feucht, and Andie Stabler (Elgin, IL: Elgin Community College, 1993), *Tin House*, and *The True Subject: Writers on Life and Craft*, edited by Kurt Brown (Saint Paul, MN: Graywolf Press, 1993).

Grateful acknowledgment is made to New Rivers Press for permission to reprint eight lines from "A Bard from an Ancient War" from *The Slavs beneath Parnassus: Selected Poems* by Miodrag Pavlović, translated by Bernard Johnson, © 1985 by Miodrag Pavlović; to Ecco Press for permission to reprint "Eclipse II" from *The Selected Poems of Tomaž Šalamun*, edited by Charles Simic, © 1988 by Tomaž Šalamun, and Elliot Anderson, Anselm Hollo, Deborah Kohloss, Sonja Kravanja-Gross, Bob Perelman, Michael Scammell, Charles Simic, Veno Taufer, Michael Waltuch; to Swallow Press/Ohio University Press for permission to reprint ten lines from "The Downfall of the Kingdom of Serbia" from *The Battle of Kosovo*, translated by John Matthias and Vladeta Vučković, © 1987 by John Matthias and Vladeta Vučković; to Harcourt Brace & Company for permission to reprint fourteen lines from "The End and the Beginning" from *Poems New and Collected, 1957–1997*, by Wisława Szymborska, translated by Stanisław Barańczak and Clare Cavenagh, copyright © 1998 by Wisława Szymborska; and to Bloodaxe Books for permission to reprint "Kolo of Sorrow" by Mak Dizdar, translated by Francis R. Jones, from *Scar on the Stone: Contemporary Poetry From Bosnia*, edited by Chris Agee, copyright © 1998 by Chris Agee.

A Note on Pronunciation

c is pronounced "ts"
č is "ch"
ç is "ch"
ć is "tu" as in "fu*tu*re"
đ is "j"
j is "y"
lj is "liyuh"
nj is "ni" as in "mi*ni*on"
š is "sh"

SWEDEN
LATVIA
Riga
RUSSIA
DENMARK
Copenhagen
LITHUANIA
Kaliningrad
RUSSIA
Vilnius
Gdansk
Minsk
Hamburg
BELARUS
Berlin
Warsaw
GERMANY
POLAND
Kiev
Prague
Cracow
Lvov
UKRAINE
CZECH REP.
SLOVAKIA
Munich
Vienna
Bratislava
MOLDAVIA
Kishenev
AUSTRIA
Budapest
HUNGARY
Ljubljana
SLOVENIA
Zagreb
Venice
VOJVODINA
Novi Sad
ROMANIA
CROATIA
YUGOSLAVIA
Belgrade
Bucharest
Constanta
BOSNIA
Sarajevo
BLACK SEA
ITALY
Adriatic Sea
Split
Varna
BULGARIA
MONTENEGRO
Priština
Sofia
Dubrovnik
KOSOVO
Podgorica
Skopje
Rome
ALBANIA
Tirana
MACEDONIA
Istanbul
Naples
GREECE
Thessalonika
TURKEY
Tyrrhenian Sea
Aegean Sea
Ionian Sea
Izmir
Athens
MALTA
MEDITERRANEAN SEA
0
100
200
300
miles

CZECH REP.
Morava
SLOVAKIA
UKRAINE
Inn
Donau
Vienna
Bratislava
Salzburg
AUSTRIA
Budapest
Graz
HUNGARY
CARINTHIA
Klagenfurt
Čahorce
Mt. Triglav
Bled
Maribor
L. Bohinj
SLOVENIA
Celje
Drava
Ljubljana
Danube
Szeged
Arad
Subotica
Tisza
Mures
Sava
Trieste
Sežana
Zagreb
ROMANIA
Piran
CROATIA
Drava
VOJVODINA
Rijeka
Pakrac
Vukovar
Novi Sad
Jasenovac
Cetingrad
KRAJINA
Otoka
Bosanski Novi
Županje
Bihać
Banja Luka
Bosna
Belgrade
Danube
Bosanska Krupa
Sanski Most
YUGOSLAVIA
DINARIC ALPS
Tuzla
Zvornik
Drina
SERBIA
Zadar
Knin
BOSNIA
Srebrenica
Žepa
Livno
Sarajevo
Pale
Ancona
Goražde
Morava
Split
Mostar
Adriatic Sea
Ñis
Sofia
MONTENEGRO
Dubrovnik
Cetinje
Prištinë
BULGARIA
Herceg-Novi
KOSOVO
Podgorica
Prizren
L. Scutari
Shkodër
Skopje
Vardar
ITALY
ALBANIA
Tiranë
MACEDONIA
Durrës
Bari
L. Ohrid
Elbasan
L. Prespa
Naples
GREECE
Thessaloniki
Brindisi
Taranto
Vlorë
Gjirokastra
Aegean Sea
0
100
200
miles
Ionian Sea

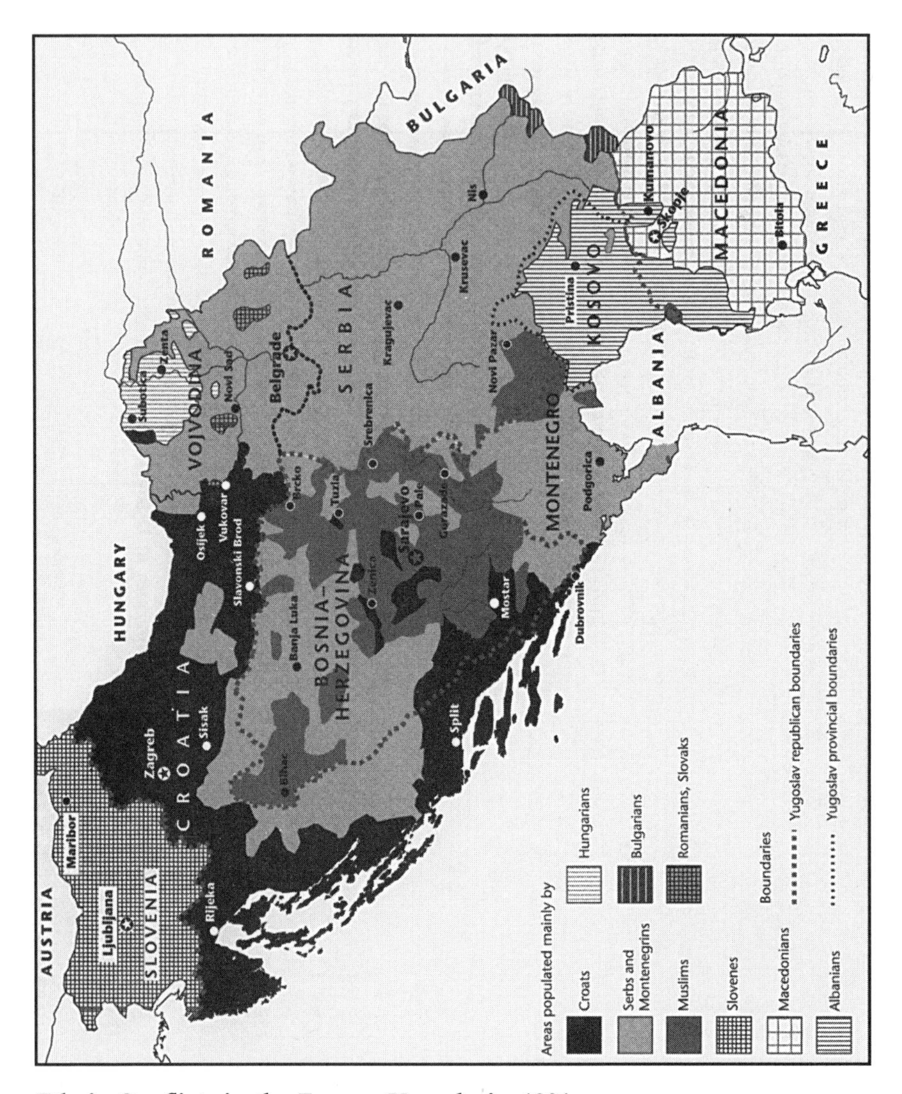

Ethnic Conflicts in the Former Yugoslavia, 1991

In war you lose your sense of the definite, hence your sense of truth itself, and therefore it's safe to say that in a true war story nothing is ever absolutely true.

—Tim O'Brien

I will take nails,
long nails
and hammer them into my body.
Very very gently,
very very slowly,
so it will last longer.
I will draw up a precise plan.
I will upholster myself every day
say two square inches for instance.

Then I will set fire to everything.
It will burn for a long time.
It will burn for seven days.
Only the nails will remain,
all welded together and rusty.
So I will remain.
So I will survive everything.

—Tomaž Šalamun

Was bleibet aber, stiften die Dichter.

—Friedrich Hölderlin

Prologue

Just before noon a shell lands in the top floor apartment of the next building over, severing the leg of an old woman who lost her husband to a sniper's bullet last month.

Ćiki appears in the doorway, clutching a five-pound piece of shrapnel from the shell, which left a hole in the façade of his in-laws' house. Ćiki is a Serbian architect married to a Muslim; at the start of the siege they moved in with her parents, and when she fled Sarajevo he stayed with them, though they despise him. No one trusts Ćiki. Rumor has it he is calling in coordinates for strikes in this neighborhood. He chalks the date on the shrapnel. Muha props it on the table next to the flowers.

"It looks like the devil's tooth," says John.

Now there are twelve of us in the basement—humanitarians and engineers and soldiers; a housewife, an architect, a tennis coach: Australian and American, British and French, Bosnians of all stripes—Serb, Croat, and Muslim.

"How many years do you think living with this much adrenalin in your system takes off your life?" says Pat, who traded in twenty-five years of journalism to work in disaster relief.

"If it keeps up like this?" says Vic. "About fifty."

With one hand he is curling a dumbbell, in the other he holds a Serbo-Croatian dictionary. Vic has a gift for languages, and he is careful to describe his course of study as Bosnian lessons.

Mirna is delighted with his progress. "Listen to his accent," she says. "Like a native."

"Because we are all old in Bosnia," says Slaven.

"What we do in Sarajevo is wait," Vic says. "Wait for an order to be filled, for something to begin, for a missile to strike—"

"Civilians," John interrupts.

"The Četniks aren't the only ones doing that," says Vinnie.

"Izetbegović is a sheep with no teeth," Vic mutters, holding the dumbbell out at arm's length. "He should just hand over control of the government to the military."

"When the first shell struck here this morning what I saw in my mind's eye was the Bread Line Massacre," Mirna snaps. "Who killed those people?"

Ćiki edges toward the garage. Mirna gives him an icy look. Vic tells him to stay where he is.

"In war the goal is to win," says Vinnie, "and you have to have the support of the people. It's very easy to kill your own people to rev them up to fight. I've seen children killed by a sniper in a Bosnian position. There are no rules in war."

"Is that why I stopped caring about the shelling last winter?" Mirna says sarcastically. "All I cared about was staying warm. I'd lie in bed as long as I could, then get up and cook as much as possible and put out the fire, because there wasn't any fuel, except the trees we cut down."

"Everyone was schlepping water," John remembers, "and water weighs eight pounds a gallon. If it's fifteen degrees, and you're spilling it all over yourself, you're in misery, especially if you're old and have to climb ten flights of stairs to your apartment."

"But standing in line for water when snipers were working—that was hell," says Mirna.

"On days like this," says John, "when the Serbs want to take someone out, they just open up the telephone book, lock his address into their coordinates, and fire. Boom: another mailbox gone."

Ćiki rubs his jaw. Muha offers him a chair. The architect stands still. "Here," Muha insists. "Sit." Ćiki shakes his head.

Mirna turns to me. "Muha has changed so much since the war began," she whispers. "You have no idea."

Vic puts down his dumbbell. "And now for a shower," he says, spraying his armpits with antiperspirant.

"If a Bosnian fires a mortar," Vinnie says, "the Serbs track his round, make an adjustment, and shoot right back. There are spies all over the city to confirm where the targets are."

Mirna glares at Ćiki. He lets out a nervous laugh.

But Vinnie does not believe the Serbs will storm Sarajevo. "The street to street fighting would cost them too many troops. What they'll do is keep up the terror. Shelling. Sniping. Killing children. Two, three, ten a day: it wears people down."

"But do you see how well-dressed everyone is?" says John. "Sarajevans still have pride."

Mirna rolls her eyes. "That's how we pretend things aren't as bad as they are. Do you know what the contestants at the Miss Besieged Sarajevo contest said yesterday when they were asked what they hope for in the future? 'What future? We have no future. We could be killed at any time.'"

"The way the momentum is going now the Bosnians will lose," Vinnie says. "Arms are getting in, but the Serbs and Croats are getting all of them. I try not to be pessimistic about it."

Mirna goes into the bedroom. "How dark it is," she exclaims—and then there is another explosion.

Pat opens a book called *The Destruction of Yugoslavia* but immediately closes it. "What should you read during an artillery attack?" she asks. I brought my copy of St.-John Perse's *Collected Poems* into the basement. Vic has an anthology of erotica in his backpack. His is the better choice, everyone agrees, though in fact we are too scared to read at all. But Mirna has the best answer to Pat's question: "Marx and Lenin," she calls out from the bedroom. "They burn forever."

This was not where I expected to find myself when in 1991 a Slovenian friend invited me to join him on a hike across his homeland. Ours was to be a literary excursion through the mountains of Yugoslavia's northernmost republic: we would drink wine, listen to folk songs and stories, revel in the wild. I wanted to explore a place where literature is more than a decorative art. Nor did I foresee Yugoslavia's violent breakup, though perhaps a closer reading of the poetry and fiction of the South Slavs would have prepared me—and the world—for the coming horrors. Once enmeshed in the Third Balkan War, however, I began to wonder how people shaped by literature would respond to the central tragedy of the post–Cold War era. What blood, what loss, what poetry speaks through this pain? Ezra Pound called literature "news that stays news." Notwithstanding the inconsistencies, bewilderment, and occasional wrong-headedness of the writers and artists I met in my travels I found their experience and witness of challenging circumstances not only compelling but also current in a manner distinct from the reportage of the front-line journalists whose courage inspired me.

"The history of the Slavs is, for example, utterly miserable," a contemporary poet has written. "Some small nations biting each other, because the weather is bad." That history had turned especially nasty when I undertook my first Balkan journey, in August 1992. Slovenia had won its independence, and an uneasy truce was holding in Croatia. But journalists had just discovered the existence of Serbian-run concentration camps in Bosnia-Herzegovina. Images of starving men behind barbed wire, broadcast around the world, raised the specter of genocide, a form of evil which many in the West had imagined belonged to the past. The so-called new world order suddenly appeared to be anything but orderly. It was then I decided to travel farther along some of history's most sensitive fault lines, and what follows is a record of my journeys, ten in all over the next five years, through the Balkan peninsula, a strange and tragic place in which nothing is ever as it seems.

Take war. The Yugoslav wars of succession were waged in large measure against civilians. (The joke in Sarajevo was that if you were in the military you had a better chance of surviving the siege than anyone else.) This, then, is a book about civilians, writers and artists in particular, caught up in a decisive historical moment, when their worlds shifted underfoot. I traveled to all the provinces and republics of the former Yugoslavia—Bosnia-Herzegovina (hereafter Bosnia), Croatia, Kosovo, Macedonia, Montenegro, Serbia, Slovenia, and Vojvodina—as well as to Albania, Austria, Bulgaria, Greece, Hungary, Italy, and Turkey, to meet with people on every side of the Third Balkan War. And what continually impressed me was the triumph of the imagination over the most terrible reality.

"How is it that one day you were climbing a mountain in Slovenia and the next thing you knew you were getting shelled in a basement in Sarajevo?" a friend asked.

"You could say I was dragged into this story," I replied.

"I understand that," my friend said after a moment. "It's the same with marriage."

My love affair with the Balkans was indeed turbulent. *Only the Nails Remain* thus chronicles a passion, by turns cultural, historical, religious, and political, for the people and poetry of those small nations given to biting each other whenever the skies cloud over.

Part I

August–September 1992

1

Journey

"Can you imagine?" she said. "The phone rang in the middle of the war: it was a Serbian writer calling to ask if we had read his new novel yet. 'Look,' I told him. 'We haven't time. We're stuck in our basement, dodging your bombs.' 'Well,' he said, 'that's a good place to read my book.'"

She shook her head in disbelief. Metka Krašovec was a slender, middle-aged woman with a bob of dark, curly hair and a melodious voice. Dressed casually in a floral print skirt and a white blouse, she looked as if she had just come from her studio, where she was at work on a series of portraits of women. Two paintings in particular would haunt a viewer, offering contrasting visions of the same troubled model. In the first the young woman's head, lined with shadows, jutted out of a turbulent sea; embedded in the second, unfinished work titled "Ana Happy" were shards of broken mirrors. Metka wrapped a black cape around her shoulders.

"It's freezing," she said. "I wish we could build a fire."

We were drinking coffee in the living room of her house in Bled, a resort near the Austrian border. Autumn was in the air. A cold front had moved in overnight, the sky was overcast, and rain had fallen, on and off, since early morning. And it was almost as chilly inside as out. This three-story log house—a replica of a dacha—was built by a White Russian who had fled the October Revolution. Metka's father, a Yugoslav diplomat, used his connections to acquire it, and now the house was her summer residence—the rest of the year she lived in Ljubljana, the Slovenian capital, teaching at the Academy of Fine Arts. The change of weather had caught her by surprise.

"The problem is," she said, "we don't know how to work the chimney. Whenever we light a fire, the house fills with smoke."

The mention of the word *smoke* made her husband, the poet Tomaž Šalamun, wince. A soft-spoken man, more studied in appearance than Metka, Tomaž was wearing a sea green Italian shirt, white pants, and leather sandals. His hair was greying around his ears; the lenses in his wire-rimmed glasses were thick. He was very nervous. Jailed briefly in the 1960s as a dissident, he had emerged as a cultural hero; by the late 1980s he had become Slovenia's most renowned literary figure, the author of more than twenty books of poetry. But he had not written a word since before his country's Ten-Day War of secession from Yugoslavia, the so-called Weekend or Television War waged one year earlier. "Poetry may ruin you for anything else," he liked to say, explaining his silence; though if Metka overheard him, he would quickly correct himself: "Maybe I'm just lazy."

During Slovenia's first year of independence in more than twelve centuries, Tomaž had turned his attention to making money, spending one day a week in Trieste—a two-hour drive away—trading stocks. He had discovered a measure of truth in Wallace Stevens's maxim: "Money is a kind of poetry." Yet after an initial grace period in which his gains had been substantial enough to earn him the respect of some Italian brokers, the poet had seen his luck change. His mounting losses were a source of constant anxiety. Worse, four days earlier, as a gift to Metka (and also to improve his luck), he had given up smoking. More than once this afternoon he had said how much he missed his habit. He studied the fireplace, as if willing to risk catastrophe to inhale a little wood smoke.

"There is no purer joy than smoking," he murmured, a beatific smile spreading across his face.

I reminded him of an entry in Cesare Pavese's journal: "Life without smoking is like the smoke without the barbecue."

"And that was four days before he committed suicide!" Tomaž laughed. He rose to his feet. "Time to go."

"Will you be able to drive?" Metka teased.

He considered her question seriously. "I don't trust myself these days," he told me.

I was traveling with Aleš Debeljak, who had remained quiet during our visit, suffering from what Metka said was a common ailment in this country of poets: melancholy. If there are three Slovenians in a room, odds are all three are poets, was the national joke, melancholy the national disease. Aleš's sadness grew out of his ambition to make a name for himself in literary circles; he sometimes talked as if he had not achieved anything, though at thirty he was regarded as Slovenia's preeminent young writer, having already published several books of poems and essays. He taught the social history of literature at the university; worked as an editor for an Austrian publisher; compiled anthologies; wrote articles and reviews; gave poetry readings around the world. He had even appeared on television with Slovenia's president to discuss the nature of friendship. Impelled by the war, during which he had worked as a translator for CNN, now he took advantage of the attention focused on the Balkan crisis to promote his country's literature.

"We need to strike while the iron's hot," he was fond of telling editors.

Clean-shaven, Aleš had a narrow, innocent regard, which accentuated his deep blue eyes; when he let his beard grow for more than a day, he took on a hungry look. Like many Slovenians, he was both frugal and shrewd. Once he had walked the streets of Paris in search of a prostitute who would give him a student discount for her services—and he had found one. He was by turns romantic and rakish, a libertine determined to marry. Similar contradictions marked his literary work. His poetry was lyrical and vaguely surreal, his prose logical, incisive. He held a doctorate from Syracuse, yet he criticized his peers for seeking inspiration in American poetry. Slovenian poets must rediscover their Central European roots, he argued, even as he lined up American publishers for his books. Rilke, Trakl, Celan—these were his masters, not the poets of the New York School.

He was, in fact, the first writer of his generation not to imitate Tomaž Šalamun, which did not prevent him from becoming the older poet's closest friend. He was as sure of his ideas as Tomaž was tentative, and as we drove out of Bled—"A place dipped in green," was how Metka described the resort—to a party in southern Austria, Aleš started speaking in the authoritative manner he adopted whenever the war came up in conversation. It was Metka's story of the Serbian writer's telephone call that riled him. The man is a brilliant novelist, said Aleš, but his politics are appalling.

"The Serbs want to be the Russians of the Balkans, imposing their language and culture on us. What arrogance! In Belgrade I can pass as a Serb living outside their borders, because I speak their language fluently. But if you think they'll speak our language when they come here you're gravely mistaken. And," he added, leaning forward into the space between the front seats to drape his arms over Tomaž's and Metka's headrests, "what do they know about our literature? I know theirs intimately. I know their poems by heart. We all do."

He gazed out the rain-streaked window at a roadside chapel commemorating plague victims. "It was only when the war began that I discovered who my real friends were," he said. "In our basement, with war planes flying overhead, only two Serbs called to see how I was. Croats, Bosnians called, but only two Serbs, who happened to be translators of Slovenian writers. The man who wrote a heartfelt introduction to the Serbo-Croatian translation of my *Dictionary of Silence* didn't even call. How many of your Serbian friends did you hear from?" he asked Tomaž and Metka.

"Two, maybe three people," said Metka, who had spent part of her childhood in Belgrade and had often exhibited her work there.

Tomaž mentioned a prominent American poet of Serbian descent.

"He doesn't count," Metka giggled. "Name someone from Belgrade."

Tomaž shrugged.

"I thought this community of writers in Yugoslavia was proof we could live together," Aleš sighed, settling back into his seat. "I was wrong. The Serbs have a streak of irrationality they suckle in their mothers' blood."

We sped by a baroque church with a pale-yellow stucco facade and a black tower, an onion-shaped dome tapering into a spire ("There's a Catholic church on every hill," Aleš said), then past cornfields, woodlots, and a tidy village. In a softer voice he said to me, "Remind you of Vermont?"

Everywhere I looked was green: thick grass and gardens surrounded two-story alpine houses; hayfields covered the Sava River basin; the road up the steep Karavanke range separating Slovenia from the Austrian province of Carinthia was lined with trees. Half of this mountainous country is forested—"*Balkan* comes from the Turkish word for a chain of wooded mountains," historian Barbara Jelavich writes—and in some regions, as in parts of New England, the woods are reclaiming farmland. Slovenia is indeed scarcely bigger than Vermont in total square miles; not for the last time was I reminded of the lush landscape around Lake Champlain, where I had spent my college years.

The UN had accepted Slovenia's application for permanent membership in May (along with those of Bosnia and Croatia), and a giddiness lingered in the air, as if the whole country was caught up in freshman orientation week at college. After a kind of prolonged adolescence, Slovenia had revolted and joined the international community. Here was a country testing its wings: anything seemed possible to many of its new citizens, who until recently had perhaps never imagined feeling that way.

Slovenia has a population of two million, half a million less than that of the American Colonies during the Revolutionary War. Slovenians proudly told me that in drafting the American Constitution Thomas Jefferson was inspired by what one writer calls a "colorful Slovene ceremony": namely, the installation of dukes. In the Slovene Principality, centered in Austrian Carinthia, the ancient Slovenians, the western Slavs then settling the Balkans, were liberated in 623 A.D. from the Avars to establish an independent realm stretching from the Elbe River to the Adriatic Sea. For almost a hundred years they maintained a democracy of sorts—their only brush with freedom. In the eighth century the Slovenians came under Frankish rule, yet they continued convening a general assembly to elect their leader. He was then installed—with special rites—by a peasant, "the embodiment of the people," according to a Communist-era guidebook. And the ceremony persisted until 1414, early in the Habsburg reign. Thomas Jefferson underlined a reference to it in his copy of Jean Bodin's *Six Livres de la République*, a small gesture destined to assume new meaning in the Slovenian imagination.

Here was proof, the guidebook concluded, "that democracy flourished [in the Slovenian lands] centuries before the adoption of the Magna Carta in 1215"—a questionable assertion that Slovenians accepted as historical truth. This was a central Slovenian myth, vital to a people making the transition from what an exile returning from America labeled "a sort of dictatorship"—the self-managed, nonaligned or nationalistic form of socialism created by Marshal [Josip Broz] Tito—to "a sort of democracy." Like all South Slavs, the Slovenians were accustomed to living under the yoke of one power or another. (The word *slave* has roots in both *Slav* and *Slovene*.) It had taken them more than a millennium to reenter the turbulent waters of independence. Their ship was a rickety version of nation-building, the blueprints of which had been drawn up in the glory days of 1848 when, as Czesław Miłosz points out, a certain manifesto beginning with the words, "A specter is haunting Europe, the specter of Communism," might have read, "A specter is haunting Europe, the specter of nationalism."

In progressive Slovenian circles that specter took the shape of a Program for a United Slovenia, which at the end of the "springtime of nations" was pushed aside. But the demand for a unified country, with its own parliament and its own language, kept growing, spurred by the writings (in German) of France Prešeren (1800–1849), the national poet. Prešeren dreamed, in true Romantic fashion, of greater freedom for Slovenians and declared "God's blessings on all nations who long and work for that bright day"—a theme amplified by Slovenia's finest prose writer, Ivan Cankar (1876–1918), who articulated the need for this "people of serfs" to be transformed into "a people who will write their own judgment." Cankar also spoke for the "Yugoslav idea," the plan to forge a union with Serbs and Croats, which had arisen during the revolution in Vienna two weeks after the publication of *The Communist Manifesto*. In the wake of the Balkan Wars (1912 and 1913) that idea again fueled public debate; and the demise of the Ottoman Empire, coupled with the downfall of the Dual Monarchy at the conclusion of World War One, set the stage for the creation, in 1918, of the Kingdom of Serbs, Croats, and Slovenes—Yugoslavia's first incarnation, which owed its charter to Woodrow Wilson's enunciation of the Fourteen Points and his leadership at Versailles.

The Slovenians' hopes for more freedom in a union of South Slavs were soon dashed. At the start of Serbian King Alexander's reign they founded their university in Ljubljana, built two national theaters, and set up Radio Slovenia. But Alexander's experiment in democracy ended in dictatorship; in 1929, when he rechristened his country Yugoslavia (literally, Land of the South Slavs), Slovenia was abolished in name and made into a province. (Which may explain why Slovenia receives scant mention in Rebecca West's *Black Lamb and Grey Falcon*, her celebrated book about Yugoslavia. Her travels, inspired in part by Alexander's assassination in 1934, took her first to Zagreb, her train passing through the Drava province at night, and thus Slovenians do not figure in her narrative. Her longest description of them—"a sensible and unexcitable people who had better opportunities than their compatriots to live at peace"—may seem, in a book of 1,158 pages, parsimonious.)

Not until the final days of World War Two did Slovenians restore their national government. And although they prospered in Yugoslavia's second attempt to enter the community of nations, under Tito they did not "write their own judgment," partly because the rapid industrialization of the post-war years did not allow their mainly peasant population to exchange its servile habits for the liberal spirit found in the growing urban classes; partly because the single-party government's abridgement of individual rights accentuated the symptoms of what some observers call the "Slovene Syndrome": a people subjected for too long to the whims of a foreign power, like a criminal imprisoned beyond a reasonable length of time, loses its ability to function without shackles; and partly because the Serbs, Slovenians kept telling me, doomed any union of South Slavs.

"The Slovenians in Carinthia knew what the Serbs were all about," Aleš muttered as we approached the border. "That's why these mountains divide us from our historic lands north of the Drava River. In 1920 they held a plebiscite in Carinthia, and the Slovenians voted to remain in Austria. They could see what was coming. The butchery in Bosnia is nothing new for the Serbs."

Tomaž shuffled my passport in with the Slovenian passports (the same shade of blue as mine) and, bluffing like a cardsharp, flashed them at the guard manning the checkpoint. The guard waved us through. What might once have seemed impossible—Slovenian passports and a border guard with no interest in examining their contents—was now commonplace. I considered the practical difficulties of creating a new country, the enormous scale of the Slovenian undertaking. They needed a constitution, legislative and legal reforms, security forces, diplomats, embassies, a foreign policy. They had to design a new flag, military uniforms, and their own currency; rename streets and businesses, schools and political parties; reroute communications systems; rethink their whole world. Ezra Pound's injunction to his fellow poets—"Make it new"—applied to a broad spectrum of Slovenian life. Little was free from the pressure to reinvent itself: a daunting prospect. It was one thing to secede from Yugoslavia, quite another to build a nation.

On 25 June 1991, six months after a plebiscite in which nearly 90 percent of Slovenian voters had favored the right to secede, Slovenia—Yugoslavia's only ethnically homogeneous republic—joined neighboring Croatia in declaring independence, ignoring diplomatic signals from the United States and the European Community (EC) to remain in Yugoslavia. The next day, at checkpoints along Slovenia's 420-mile frontier, guards raised their country's new flag, even as the Yugoslav National Army (JNA) moved in to retake control of the border posts. War erupted; and while the JNA—one of Europe's largest standing armies—far outnumbered and outgunned Slovenia's forces, it was not prepared for the resistance it encountered. Slovenian units of the Territorial Defense, the paramilitary organization Tito had established after the Soviet Union's suppression of the 1968 "Prague Spring," defended their homeland guerrilla-style; their casualties were light (a dozen killed, less than 150 wounded), their victories impressive: more than 3,200 Yugoslav conscripts surrendered to the Slovenians, who in only ten days of fighting had won their freedom. The JNA Chief of Staff accused the seceding republic of waging "a dirty and underhanded war." Indeed Slovenian Minister of Defense Janez Janša, whose criticism of the JNA had once landed him in prison on a treason charge, had over the previous year secretly marshaled his forces to fight such a war. Now at the age of thirty-three Janša was his country's first military hero. This summer his memoirs were a best-seller.

A nation's development, de Tocqueville suggested, is conditioned by the circumstances of its birth. "The entire man is, so to speak, to be seen in the cradle of the child," he wrote, though only America could be studied this way, since no other nation possessed a beginning as clearly marked as ours. Slovenia's origin is as murky as that of any European country, but the circumstances of its birth as a nation-state may reveal "the primal cause of the prejudices, the habits, the ruling passions, and, in short, all that constitutes what is called the national character," as de Tocqueville proposed to do for America. If so, then the salient feature of Slovenia's birth was that after a long incubation its all but bloodless entry into the community of nations precipitated the Third Balkan War. A troubling legacy for the people who made up only 8 percent of Yugoslavia's population? Hard to say.

What *was* clear was that their signing of the Brioni Accord in July 1991 (an EC-mediated agreement requiring the JNA to withdraw its forces from Slovenia, and Slovenia and Croatia to suspend their independence efforts for three months), was the prelude to barbarism on a scale not seen in Europe since World War Two. The journalist Misha Glenny observes that with this accord "the European Community embarked on a policy of localized solutions in the Balkans which have neither stopped the violence nor resolved the underlying causes of that violence"—a point borne out that August, when the JNA attacked Croatia, vowing to defend the more than half-million Serbs living there.

The fighting in Croatia was unimaginably vicious: 10,000 killed, tens of thousands wounded, hundreds of thousands displaced from their homes, countless atrocities; the uneasy truce negotiated late in the fall had given Serbian authorities time to concentrate on Bosnia, where since April of this year war had raged. The shelling of cities like Vukovar, Dubrovnik, and Sarajevo, Aleš insisted, demonstrated the primitive nature of Serbian warcraft. And as we drove down the winding road into Austria, passing religious shrines and alpine houses indistinguishable from those on the Slovenian side of the mountains, he resumed his diatribe against his former countrymen.

"What military significance does Dubrovnik have?" he cried. "It's a cultural monument."

"If they can't have it, they'll destroy it," said Metka. "They're just like little boys."

"What I'm afraid of is that they're setting an example to the tyrants of the world," said Aleš. "And they're getting away with it. If no one stops them, ten years from now another demagogue will come along and do the same thing somewhere else."

Tomaž studied the invitation to the party, balancing it on the steering wheel for a moment before proclaiming its engraving of an apple tree "beautiful, but useless." The address and directions mystified him, and when he turned onto a road that ended at a fence propped on a hill, he threw up his hands. He squinted at the farmhouse beyond the fence, worrying aloud that we had forgotten our manners by neglecting to bring food and wine. A hayfield stretched across the valley below us—a peaceful scene that on another day might have made him thankful to have put another border between us and the fighting in Bosnia. But the talk of war had rattled Tomaž, and Metka's assurances that there would be plenty to eat and drink at the party hardly calmed him. He was superstitious enough to imagine his confusion was an omen: we would be lucky if nothing worse befell us than that we showed up late. Metka examined the invitation and, pronouncing it hopeless, gently steered her husband toward the village of Cahorce, where the party had already begun.

To distract Tomaž, Metka and Aleš took turns telling a story about his brief career as a door-to-door salesman. In the early 1970s, forbidden for political reasons to find employment as either a teacher or a curator (his own conceptual works had been exhibited around Yugoslavia), he tried peddling how-to books—the worst possible occupation for someone as unassuming as Tomaž. He rarely managed even to show potential customers the books he despised. One day, though, he summoned the courage to ask a woman impatient with his pitch to tell him which writers she read.

"Kafka, Proust, and Šalamun," she replied.

"*I* am Šalamun," he said, amazed.

Presently we arrived at the party hosted by the owner of Wieser Verlag, a small Austrian publishing house specializing in writers from Eastern and Central Europe. The clouds were so low that by five o'clock dusk was falling; rain was imminent. But the backyard of the publisher's spacious country home was flush with poets, fiction writers, journalists, and a television crew. Two large canopies draped with colored lights and a gazebo surrounded by apple trees completed a square, its fourth side the stucco wall of the house. Framed and lighted illustrations of Wieser's book covers hung beneath the windows, like flags; new titles were displayed on a table tucked under a staircase near the cellar entrance, where bartenders served beer, white wine, and schnapps.

Writers from all over the former Yugoslavia had gathered here, and when Metka and I were separated from Tomaž and Aleš the first person we met was the Serbian novelist who had telephoned her during the war. Though he immediately recognized her, he could not remember her name. She gave him no hints. He shifted his weight from foot to foot, embarrassed, unable to place her until, finally, another writer hailed her. The novelist's face turned bright red. Of course he knew who she was. Hadn't he and his family stayed at her house in Bled? He hurried to the cellar to get a drink.

"What did we tell you?" said Metka.

Near the food tent a radio journalist was interviewing Aleš about Slovenia's independence; before the night was over Aleš would offer opinions on his favorite subjects—the writer's role in these changing times, the inability of current political and poetic language to address the "new world order," the moral and cultural abyss opening in Europe as a result of the war in Bosnia—to newspapermen and television journalists, too. He was Wieser's Slovenian editor, and literary business dominated his conversations. I said this must be a good place for him to make connections.

"It's not what you think," he shot back. "This isn't literary politics, like you have in America."

Tomaž, meanwhile, was trapped in the food line with Wieser's Austrian editor. Ludwig was a bald, affable man with a round belly and bright blue eyes. He loved Tomaž's work, he said, and recited—in Slovenian—an early poem, declaring it had prophesied the atrocities in Bosnia. Tomaž was unnerved. His was a mystical view: a single verse might unleash terrific forces. "Every true poet is a monster," he wrote. "He destroys people and their speech." He had seen (or foreseen?) enough: one reason he gave for not writing anymore. On his plate he piled thick slices of ham, potato salad, and rolls; he ladled rich meat stew into a bowl. He sighed with relief when Ludwig wandered away.

"Eat," he said to me. "Look at how much food there is!"

One writer conspicuously absent was Peter Handke, the Austrian novelist and playwright who had collaborated with Wim Wenders on the screenplay for *Wings of Desire*. Handke was a fixture at international literary events, and Carinthia was his homeland, in fact and fiction. What was more, Wieser was about to publish a collection of interviews with him. But Handke knew better than to attend this party, said my Slovenian friends. The author of *Offending the Audience* had alienated his favorite readers by insisting that Slovenia should have stayed in Yugoslavia.

"I lost my homeland when Slovenia seceded," Handke said in an interview. "I would never wish each republic to be separate. Perhaps as a child you want your village to be a kingdom, but not when you're an adult."

He did not agree with the widespread Slovenian belief that their language and traditions had been threatened by the Serbs; it pained him to know that diversity, which in his mind had been Yugoslavia's cardinal virtue, was the first casualty of the war.

"The Slovenian nation was a fact, not an illusion," he declared. "One could say dream instead of illusion."

Handke's idyllic dream of a workers' paradise no longer interested Slovenians. Nor did they heed his warning about the ways in which nationalism diminishes a people. On one level his work investigated borders and languages, German and Slovenian (his books were sprinkled with Slovenian words), and he liked to give his protagonists some of his Slovenian mother's characteristics. He identified profoundly with her people. In his novel *Repetition* he writes of the Slovenians: "Like them, I was gaunt, bony, awkward, with rough-hewn features and arms that dangled inelegantly, and my nature like theirs was compliant, willing, undemanding, the nature of a people who had been kingless and stateless down through the centuries, a people of journeymen and hired hands (not a noble, not a master among them)—and yet we children of darkness were radiant with beauty, self-reliant, bold, rebellious, independent, each man of us the next man's hero."

Sentimental claptrap, Slovenian writers believed.

In the middle of the lawn, on uneven ground, a card table tilted; on it a stack of books was arranged next to a microphone. Journalists crowded around the table and guests drifted closer. To celebrate the fiftieth birthday of Ali Podrimja, an Albanian poet from Kosovo, Wieser had printed a special edition of his work—fifty copies of fifty poems, in Albanian and German. Ludwig gave an impassioned speech supporting Albanian independence. Kosovars, he argued, made up an old Balkan country that deserved to be free of Serbian dominance. His evidence? Podrimja himself—the youngest grandfather in Albanian poetry, he chuckled. And a prolific writer, too. Podrimja went on to read a selection of the poems that had made him a hero in Kosovo. No one at the party understood him. He was slouched at the table, a short dark-haired man in a grey windbreaker, reading softly, as if to himself. Ludwig's more forceful reading of the German version of these poems may also have meant nothing to some guests, but the symbolism of the occasion was not lost on anyone.

Yugoslavia's poorest province, Kosovo was rich in minerals and history—the site of a famous battle between the Serbs and Ottoman Turks. On Saint Vitus Day in 1389, the Serbian kingdom, which over the previous two centuries had become the strongest power in the Balkans, suffered its worst defeat. Legend has it that Blackbird's Field, as the killing ground came to be known, was where Prince Lazar, the nobleman leading the Serbian armies, sacrificed his earthly crown for a heavenly kingdom. He led "his men into battle knowing what the tragic outcome was to be," writes poet-scholar John Matthias, "as one might lead a host of martyrs consciously into a conflagration."

The result was five hundred years of Turkish occupation, during which the Battle of Kosovo was commemorated in heroic ballads and folk poems. "Serbs are possibly unique among peoples in that in their national epic poetry they celebrate defeat," poet Charles Simic notes. "Other people sing of the triumphs of their conquering heroes while the Serbs sing of the tragic sense of life." Their belief in "a great nation strangled at birth" is rooted in this oral tradition. And that loss remains a primary subject for Serbian poets. The late Vasko Popa devoted a cycle of poems to Blackbird's Field, where the white peonies are supposedly stained red with the blood of fallen Serbs: "A field like none other/ Heaven above it/ Heaven below," he wrote in *Earth Erect*. And the closing stanzas of Miodrag Pavlović's "A Bard from an Ancient War" give a sense of the collective Serbian grief:

> Now the open graves have no need of words
> and the pines turn away their heads from man,
> only in the homes of the hanged
> do they ask me to speak of the rope.
>
> Who now in the fields will await
> the morning summons of the sun's arising,
> who will find a new beginning for the song
> and a better ending?

Recently, a Communist apparatchik, Slobodan Milošević, had discovered "a new beginning" in Kosovo, and a bloody song it was. Tito's death in 1980 left a power vacuum in Yugoslav politics, and, in 1987, on a trip to the Kosovo capital of Prištinë, Milošević recognized a potent political force in Serbian fury at their diminished standing in the autonomous province. The 1974 Yugoslav Constitution had granted Kosovo a status equivalent in most ways to that of the republics—a bitter blow to Serbs who could not reconcile themselves to the idea that in this Albanian-dominated land they were no longer "lords of the blue fields/ And the ore-rich mountains with no foothills," as Popa called them: the Serbian Jerusalem, which contained their most important religious shrines and monasteries. That the new constitution ratified changing demographics—Kosovo was now 90 percent Albanian, thousands of Serbs having emigrated in search of better economic opportunities—only aggravated the wound. No one should dare to beat you, Milošević told Serbs in Prištinë. And on 28 June 1989—Saint Vitus Day—he gave a stirring speech to several hundred thousand Serbs gathered at Blackbird's Field. "After six centuries, we are engaged in battles and quarrels," he said. "They are not armed battles, but this cannot be excluded yet." What was once a poetic subject became his call to arms.

The imposition of martial law in the province enabled the Serbian minority to take control of the media, schools, hospitals, businesses, and security forces; the federal decision to revoke Kosovo's autonomy (a campaign spearheaded by Milošević) fueled an Albanian separatist movement, which brought from the authorities so many reprisals that Peter Handke, among others, believed that Albanians, not Slovenians or Croats, faced the biggest threat of all. Once the war ended in Bosnia, conventional wisdom ran, Milošević would set his sights on Kosovo, and no one would stop him. Hence the bitterness of the poem Ali Podrimja read about the Berlin Wall coming down: "When I wanted to go through the Albanian wall/ my feet my head were soaked in blood."

The audience was hushed.

When Podrimja closed his book, a quartet consisting of Ludwig, the publisher, and two elderly women sang Slovenian folk songs a cappella, their voices high and clear: a celebration of a culture vanishing on this side of the border. Up the road was Klagenfurt, the ancient Slovenian capital; home of the first large Slovenian publishing house and newspapers, the city was Germanized over the centuries. Since the fall of the Habsburgs and the fateful 1920 plebiscite, the Slovenian population in Carinthia had shrunk from 80,000 to less than 18,000, thanks to assimilation, a subject Handke explores in *Repetition*. "[W]e had been a family of hirelings, of itinerant workers, homeless and condemned to remain so," Filip Kobal, the narrator, explains. Exiled from Slovenia, his family lives in Carinthia, where "[t]he only right we retained, in which we could find brief moments of peace, was the right to gamble. And when my father gambled, even as an old man, he hadn't his equal in the village. As he saw it, another aspect of the sentence of banishment was that in his home he was obliged not only to give another language precedence over Slovenian, which had after all been the language of his ancestors, but to ban the use of Slovenian altogether. As he regularly showed when talking to himself, often very loudly, in his workshop, he himself spoke it in his innermost consciousness, but he felt forbidden to let it out or pass it on to his children." But the familial attachment to the *idea* of Slovenia does not die: Filip Kobal is drawn to the land south of the mountains, where "invention and freedom were one." There he passes himself off as a Slovenian, "playing with conviction the role of a man who had returned to his home country after a long absence." Telling stories in Slovenian, he discovers, is one way to create an imaginary homeland: his only refuge. What sanctuary Handke had found in writing about his mother's people—"With my play of words I can create an ideal picture of the landscape," he said of his Slovenian settings—had perhaps disappeared with Yugoslavia's dissolution, a death not unlike his mother's suicide.

He said of the alpine republic, "it's so disappointing that all of this is only Slovenia."

A Slovenian folksinger stepped up to the microphone and tuned his guitar. He usually performed with a band of old men, none under the age of eighty. When they went on tour, he had to put them to bed after each concert and feed them their medicine before he could go out. But this night he was singing alone. His music was haunting—mournful ballads in a minor key, an exercise in nostalgia. This polyglot, pluralistic party was a vintage Austro-Hungarian affair. It took a publisher to unite Franz Joseph's disparate peoples, I was thinking, when the Slovenian journalist who had interviewed Aleš plucked an apple from the tree I was leaning against and gave me a smile.

"This is the same tree on the invitation," said Nina Zagoričnik, a striking woman with shoulder-length auburn hair and sharp features. I remembered Tomaž's description of the engraving on the invitation—"beautiful, but useless"—when Nina offered me the apple. "Like Eve," she said in a thick accent. "Everything will happen under the apple tree."

"And what, exactly, is everything, Eve?" I asked.

She changed the subject. "When I saw Aleš in Ljubljana during the war, I decided not to leave. I had a ticket to Munich to go to the film festival. But I thought, if Aleš can stay, so can I."

I started laughing when the singer covered Lou Reed's "Take a Walk on the Wild Side."

"What's funny," said Nina.

"The music," I said. "What a strange juxtaposition: these beautiful folk songs and Lou Reed."

"Why?" she said. "We love Lou Reed."

So it seemed, for the next song was his "Satellite of Love."

"Slovenia," Nina confided, "is possible only as long as one thinks constantly of Italy. I got *this*," she added, pronouncing the last word so that it rhymed with *lease* and smoothing her hands down the front of her body, "in Trieste. You like?"

"Very nice," I murmured—and then she was gone.

The singer packed up his guitar. The rain had held off, and the air felt warmer even before Aleš brought me a glass of schnapps. He had just met a Serbian actor from Croatia, a poet-singer whose work had encouraged the romantic entanglements of an entire generation in Yugoslavia. One of Aleš's prized possessions was a record album of the actor's love songs, which he had played for many women. "My whole erotic history passed before my eyes," he grinned. But the actor was now a man without a country: in Belgrade he was denounced as a Croat, in Zagreb as a Serb. He had applied for asylum in Slovenia. Aleš was pleased to think that soon they might be countrymen.

"Drink up," he said. "By the way, Nina loves poets. She named her daughter after one."

"Cheers," I said, raising my glass.

"*Na zdravje!*" he corrected me. "Your health. And look in my eyes when we clink glasses, or you'll have bad luck. You don't need that in the Balkans."

Metka and Tomaž waited in the car while Aleš made a last sweep of the party. After saying goodbye to Ludwig, he turned to me and said, "Can you imagine? The man loves Slovenia so much he's learning our language, even though he was arrested once in Ljubljana for helping a drunken Slovenian poet. Ludwig was also drunk, of course. And while he was in custody the police beat him so badly he was deaf for six months. That kind of thing was one of Communism's dirty little secrets."

2

Pohorje

Our plan was idyllic: two poets walking in the mountains, traversing Slovenia. Aleš had proposed the journey in the winter of 1991. We were translating his new book, *Anxious Moments*, and the prospect of joining him that summer for a hike across the land described in his prose poems thrilled me. True, in his work were premonitions of impending disaster. He wrote of military convoys, destroyed villages, "phosphorous lighting the passion in soldiers' eyes"—portents I dismissed as poetic images until Slovenia's declaration of independence in June led to war. The week I was to fly to Ljubljana, the JNA bombed Brnk Airport, wrecking an Adria Airways Airbus; two Austrian photographers driving a jeep down the runway were blown up. We put off our trip for a year.

Now we stood at the base of a ski run in the Pohorje Mountains, adjusting our backpacks, impatient to begin our first trek. The bus from Maribor, Slovenia's second city, an industrial center in the northeast, had let us out moments after the tram had left for the summit. I suggested we climb the steep trail to the ridge we would hike along for the next three days instead of waiting an hour for another tram: a foolish decision. On the hottest day of the century in Slovenia, at the edge of the Styrian valley, it was sweltering. For relief we talked about downhill skiing.

"We'll know the end is here," said Aleš, crossing the slope to head into the woods, "when Slovenians stop skiing in France, Austria, and Switzerland."

A popular saying. Slovenians are a mountaineering people; ski racers are admired almost as much as poets, no small matter in a country that builds monuments to literary figures instead of military heroes. In the war's aftermath, however, that saying had acquired more currency. Slovenia was Yugoslavia's most prosperous republic, its social product roughly five times as great as that of Kosovo. But independence had cost the new state its traditional Yugoslav sources of raw materials and markets for its finished goods, which amounted to nearly 60 percent of its exports. Efforts to reorient the economy from a command structure to market conditions worsened the depression, lowering the standard of living. And the burden of caring for refugees from Croatia and Bosnia only added to the hardship. At least 100,000 Slovenians were out of work, though the government had yet to pass a privatization law and make deep cuts in state-owned industries. Thus Aleš's father was in danger of losing his job building laser sights for, among others, the Serbian tanks leveling Sarajevo. I could not imagine the damage those gunners might inflict once they stopped aiming.

We came to a clearing, sweating heavily. A ragtag unit of the new army jogged downhill.

"Are you hiking up?" cried a young soldier.

"If we can make it," said Aleš.

"There are two meters of snow up there," the soldier joked. "Take me with you!"

There was no snow this late in the summer. Like the Appalachians, the Pohorje are the eroded remnants of a much larger chain—in this case the Central Alps. Fifty kilometers long and rising only 1,500 meters above sea level, the crescent-shaped Pohorje range lies on the banks of the Drava River, a natural passage from the Carinthian basin to the fertile Pannonian plain, the country's agricultural heart. In 1960, half of the Slovenian population was agrarian; thirty years later, industrialization had reduced agriculture to 5 percent of GNP. But farming instincts die slowly. Aleš's parents, for example, like other villagers who moved to the city to find work, kept a garden near their apartment. In the countryside mixed households combined farm and factory work.

On the bus ride from Ljubljana I had observed that every available acre of land seemed to be under cultivation. Slovenians, said Aleš, are Catholic by religion, Protestant by work ethic. And private ownership of the land—the rule rather than the exception under the Communists—increased their diligence. Large white stuccoed houses with red-tiled roofs looked out over cornfields and hops in rows of tall green tents like a bivouac: the "green gold" exported to American breweries. Near the ancient city of Celje were vineyards—almost half of Maribor lies above a vast wine cellar—and the lower slopes of the Pohorje produce a Riesling nicknamed *ritoznojčan*, "the sweat of your ass." Aleš remembered another saying: *Only olive-growing regions are civilized*—a slur against Serbia. The Serbs wage war, he said, because they need access to the sea and civilization.

The soldiers zigzagged down the slope, like ski racers on a dry-land slalom course, and disappeared around the bend. Brilliant—that was how Aleš described Tito's decision to establish the Territorial Defense in which these soldiers learned their warcraft. "What a demonic genius! First he held us together all those years, then he gave us a guerrilla army to fight our way out."

Into a stand of firs we trudged then up another hill to a restaurant on the ridge, where we refilled our water bottles and drank cartons of apricot juice. We would follow the Partisan courier route through the Pohorje, the first part of an elaborate trail-and-hut system crisscrossing the country, once used by peasants to escort Partisans through German-occupied territory. And it was here in the Pohorje that the first Americans to infiltrate the Third Reich coordinated the Partisan demolition of railroad tracks and tunnels leading to the strategically vital Ljubljana Gap, through which armies and traders since antiquity had traveled from the Mediterranean to Central Europe.

Exhausted, we hung our wet shirts to dry from the straps of our backpacks, laced up our boots, and headed west, past the first of many plaques commemorating Slovenian Partisans. Within minutes we had lumbered into a patch of red raspberries, and as the sweetness of the slightly overripe fruit exploded on my tongue, Aleš recalled the German minority in Maribor, merchants whose families had settled there in the sixteenth century. Hitler said, "Make this land German again" (though it had never been German), and Germanized the Slovenian place and family names. It was the same strategy he used in the Sudeten, where the local Germans asked the Third Reich to "liberate" them, and ended just as miserably: with the Germans' expulsion—a lesson lost on Milošević, at whose prompting the Serbian minorities in Bosnia, Croatia, and Kosovo now asked Belgrade to come to their defense.

The trail meandered through the woods along the ridge. In the shade we picked up our pace, and soon we came to an inn Partisans had burned down to keep the Germans from occupying it, one of more than a hundred operations undertaken in the Pohorje. The Axis aggressors, surprised at the ease with which they conquered the Royalist forces in April 1941—routing the army in eight days, putting to flight Serbian King Peter and his government—divided up the country, with Germany controlling much of Slovenia and Serbia. Italy's holdings included Istria and Ljubljana, the Dalmatian coast, Montenegro, Albania, and Kosovo; Hungary retook its pre-1919 territories of Bačka and Baranja (both part of present-day Vojvodina, an autonomous province with a significant Hungarian minority in northern Serbia); and Bulgaria ruled over most of Macedonia. Yugoslavia's sovereignty, like the peace in Europe, had lasted less than a generation.

Of critical importance was the establishment of an independent Croatian state, *Nezavisna Država Hrvatska* (NDH), within the German sphere of influence. The ultranationalist Ante Pavelić, installed as head of a puppet government in Zagreb, and his fanatical supporters in the Ustaša movement attempted to annihilate Croatia's large Serbian population. Their formula: "Kill a third, convert a third [of the Serbian Orthodox Christians to Catholicism], expel a third." The Ustaše did not succeed in "ethnically cleansing" their land, though they put to death hundreds of thousands of Serbs, Jews, Gypsies, and other Croats. Their primitive methods embarrassed the Nazis—and inspired civil war throughout the occupied territories.

Historian Barbara Jelavich, noting the "importance of the Balkan tradition of resistance and the romanticization of the role of the guerrilla fighter and the bandit," suggests that "[f]rom the first days of the Axis occupation conditions in Yugoslavia were almost ideal for the formation of an armed resistance. The very swiftness of the war and the surrender meant that the country was never completely occupied, nor was it possible for the conquerors to collect and destroy all of the available arms. As a result, after the armistice [of 17 April 1941], armed men without leaders were in ready supply." In Serbia they formed Četnik units loyal to the government in exile; elsewhere they became Partisans under the command of Tito, general secretary of the Yugoslav Communist Party. The Allies recognized King Peter's London-based government, but gave military assistance to Tito's forces, which emerged as the strongest and best organized. The home armies fought a guerrilla war against the occupiers—and each other; Croats and Muslims often teamed up against the Serbs.

Partisan success, particularly in the mountains of Bosnia and Montenegro, where they tied up several German divisions, convinced the Allies to support Tito. The Četniks under Serbian nationalist Draža Mihailović, who feared the Communists even more than the Nazis, turned from resisting the Nazis to fighting the Partisans; because Tito's best troops were Serbs who had escaped the Ustaša terror, Serbs were soon pitted against Serbs. Axis aggression thus provoked, and Allied strategy intensified, Yugoslavia's civil war, in which over a million people lost their lives, most of whom "must be reckoned active or passive victims of the policy of Partisan warfare," military historian John Keegan writes. Indeed Tito rose to power, according to one biographer, "because of the carnage among the Yugoslavs." After a show trial in 1946, Mihailović was executed for treason.

Reprisals from the occupying armies were common—and even courted, since the fear of atrocities served the resistance. "In war we must not be frightened of the destruction of whole villages," said Edvard Kardelj, Tito's deputy (and author of *The Slovene Nationality Question*). "Terror will bring about armed action." There was good reason to be scared. In Croatia and Bosnia the Partisans launched offensives designed to draw retaliation from the Germans, while in Serbia the Nazis' efforts to discourage guerrilla action could be summed up in a stark formula the Ustaše must have admired: one hundred hostages executed for each German killed, fifty for each wounded. No region was spared the occupiers' wrath. (In the Pohorje, for instance, in January 1945 the Nazis hanged a hundred Slovenians in an apple orchard, a place known thereafter as the Hostages' Avenue of Apple Trees.) In every part of Yugoslavia villiagers fled into the woods, as they had throughout their history, then struck at armies unprepared for irregular warfare, as well as at other villages.

Slovenian Partisan units, named after national literary figures like France Prešeren and Ivan Cankar, waged sporadic war against the Germans, Italians, and collaborators, betrayal being a popular form of expression in the Balkans. The Pohorje Battalion's downfall was a famous story of treachery. Organized in October 1942, this unit attacked the Germans and set fire to scores of alpine huts. After the New Year, the battalion was betrayed; twenty thousand Nazis surrounded seventy Partisans and killed them all, the last two shot out of trees, in the words of one writer, "like good-for-nothing squirrels." Thus the myth of the Pohorje as "the home of Slovenian freedom and resistance."

Nowhere is the relationship between history and mythology more unsettling than in the Balkans. A culture constructs its sense of itself according to stories rooted in history and transformed by artistic, psychological, and political necessities. In its poetry, fiction, and folklore a society creates a vision of itself distinct from the historical record. But what happens when mythology replaces history, child of the Enlightenment? The myth that every Partisan was a Communist, for example, was now being revised downward to suggest that fewer than 10 percent had pledged allegiance to Moscow. What was the truth? Difficult to say. And Aleš was not alone in blaming the Third Balkan War on Serbia's collective belief in dangerous myths articulated by its literary community.

"I don't know how to address the Serbian writers who used to be my friends," he said. "They were defeated at Kosovo. *Six hundred years ago!* So what do they do? Turn that loss into a national symbol, then use it to justify war. Who else would do that? A nation has to face up to its evil side, or else it just reproduces it. Serbian writers—with very few exceptions—avoid dealing with what their government's doing in Bosnia. I certainly don't publish in Serbia anymore.

"That's why their war crimes must be prosecuted. Nuremberg gave the German people a chance to be de-Nazified. Serbs need their own Nuremberg, their own cleansing process."

Across the meadow was a church with a fresh coat of white paint bright as sunlight on snow, and I had to shield my eyes until we were past it. Through the dry grass and haze of high summer we marched, pausing regularly to change shirts and sip water until our water bottles were empty. Waning sunlight filtered through the trees. Bits of frog's silver, the local fool's gold, were scattered in the dirt. A blind snake wriggled across the path and, when it came to a rock, curled its ocher length into the shape of a sickle blade. Three crones hauled buckets of raspberries past us. Catholic shrines and Partisan memorials, wooden crucifixes and stone plaques engraved with red stars, all blurred together.

By dusk the heat had let up, so we hiked down a steep hillside to see a towering waterfall: another bad decision. Hand over hand, I inched along the cables bolted into the rock wall, struggling to balance my heavy pack, my legs trembling. I was thirsty, hungry, worn out. And a mountain hut in which to spend the night was an hour away. But Aleš had found a reserve of energy. He slipped off his pack and scrambled across the fallen logs surrounding the pool at the base of the waterfall.

"You see why we're so proud of our land?" he cried, standing in the spray.

Night was falling fast when we regained the trail, a white gravel logging road that gave way to wheel ruts cutting across a meadow. Before we reentered the forest we passed a solitary homestead: two barns in a fenced pasture, stacks of firewood, a garden full of cabbages. The small house showed no sign of life. The wheel ruts became a muddy footpath; the woods were pitch-dark. We returned to the house, hoping to camp nearby.

The house was still dark when Aleš knocked on the door. Minutes passed before a young woodcutter in olive-green overalls came out. He was blond, blue-eyed, thin; his scraggly beard gave him a haggard look. Yes, we could spend the night. He led us to a pine grove above his pasture. On the soft needles we unrolled our sleeping bags, the woodcutter watching us closely. Aleš went with him to fill our water bottles at his spring. It was a long time before he returned.

"The poor guy has no one to talk to except his mother," said Aleš, climbing into his sleeping bag. "He wouldn't let me go."

A warm wind washed over us. A cowbell clanked in the meadow. We could dimly make out a woman leading a cow into the pasture. "That must be the man's sister," said Aleš. "She probably comes by every evening. Not much of a social life."

Ten years ago, eight families had lived in the area; only the woodcutter's remained. This homestead belonged to a time when it was possible to live off the woods—logging, hunting, grazing, cultivating. But people from Maribor were buying these places for vacation houses. The woodcutter and his sister ventured down to the valley on a Moped only to buy olive oil and seasonings.

Where would they meet people their own age? I wondered.

"It's almost impossible," said Aleš. "Who wants to move to the mountains? If his sister's lucky, she's married to another logger."

Geography is a better guide to destiny than character in the Balkans, where the vagaries of foreign occupation and rule have shaped the lives of ordinary men and women. The woodcutter's fate was tied to the Pohorje; if his proved to be the last generation to live here, his family's migration would be only the latest in a history of departures. Once inhabited by Celts (near Maribor is an archaeological excavation of an extensive Celtic settlement), this land is marked by those who believed the universe was governed by demonic powers. Trees are said to be sacred to Slovenians, as they were to the Druids who performed their ceremonies in groves like this one, where I was rolling over and over, trying to find a place to sleep among the rocks and roots.

What arboreal spirits the Slovenians inherited from their pagan predecessors now reside in their folk expressions and literature—in their language, that is. Peter Handke, a firm believer in literature as "the center of a people," once said that "what is left behind is a history of words, in which words contain that which was and that which might have been." Thus Aleš's mother wanted him to marry "a girl born under the linden tree"—i.e. Slovenian—and in one of his poems a woman sang "sotto voce, splitting the trunks of pines, softening the shuddering hides of deer." Both seemed possible.

There is nothing like a long hike in hot weather to sharpen your imagination. Stars filled the sky, and as I drifted in and out of sleep I saw dark shapes moving through the trees. Ancient spirits? No. Small deer roaming toward a lean-to hunters stocked with hay, the herd's winter provisions. A line from Wordsworth ran through my head: "there is a Spirit in the woods"—an idea perhaps occasioned by eighteenth-century excavations of Celtic settlements. Interest in things Celtic nourished Romantic poets; like the folklore collected by the Brothers Grimm (much of it based on Celtic lore), curiosity about Druidic practices fostered the return of magic into literature, unleashing new poetic energies in Wordsworth and Coleridge, Goethe and Novalis. And it was this underground tradition, coursing from medieval alchemists to the French Surrealists, that inspired me as a writer: poetic explorations of the dark and marvelous, that terra incognita mapped most brilliantly at the dawn of modernism.

But these poetic cartographers, Wordsworth in particular, are better known for their praise of childhood, love of nature, and fascination with the French Revolution. Two hundred years ago, Wordsworth traveled on foot across France, rejoicing in that country's liberation from monarchical bonds. The poet gave himself over to what one critic calls "the movement of the spirit of a whole people." He saw the best and worst of the revolutionary urge fated to become the religion of the modern age, and then he wrote his masterpiece, *The Prelude.* His epic work on the growth of a poet's mind is at once a celebration of freedom and a cautionary tale about human folly; only his faith in nature and recollection of rapturous moments from childhood saved him from despair at the mounting evidence of the Revolution's failures—the Terror, war between England and France, famine, suspension of civil liberties, the Pope crowning Napoleon Emperor, which Wordsworth memorialized in a startling fashion: "This last opprobrium, when we see a people/ ...take a lesson from the dog/ Returning to his vomit." Indeed "the degradation of the era" drove the radical poet into the Tory camp and dried up his inspiration, two facts that haunted me.

The cowbell clanked in the pasture. I rolled over onto my back and gazed at the stars. The warm wind reminded me of a sea breeze. I remembered bodysurfing as a child—the blue swirl of the Atlantic, the taste of sand and salt as I tumbled in the surf—and how our family house was swept out to sea in a hurricane. That was the year of the Cuban missile crisis. And now that the Cold War was over we faced another historical watershed: the end of the modern age. In his *Prison Notebooks* Antonio Gramsci wrote: "The crisis consists precisely in the fact that the old is dying and the new cannot be born; in this interregnum a great variety of morbid symptoms appears."

The new? For Aleš it meant addressing the Croatian poet Branko Miljković's question—"Will freedom know how to sing in the same way the slaves sang about it?" *Anxious Moments*, a product of his stay in America, offered a provisional answer. He believed his cycle of prose poems, completed more than a year before the JNA attacked Slovenia, was in some respects prophetic. "Don't you think there's a certain beauty," one poem concludes, "in the way the distant glow of the fire—consuming villages in the south—trembles above the rolling hills, far from the town neither you nor I want to leave?"

"My writing," Aleš once said, "is still very much affected by the fact that I come from a country that was constantly torn apart, divided between many foreign masters, that never had the tradition of statehood. It was the language, the words, the writing, that allowed for the development of national as well as personal identity." While American free verse and prose poetry pointed him toward his current work, he thought Central European poetry could be "a witness to current events, not by reflecting them in a straightforward manner, but by reshaping them, digesting them, and incorporating them into the perennial dilemma of humankind."

That tradition fascinated me. Steeped in *The Prelude*, I imagined hiking across Slovenia as a scaled-down version of Wordsworth's adventures in France. When I first met Aleš, the day after a crowd of East Germans had demolished the Berlin Wall, I discovered that we shared not only poetic interests but also a love of the wild, a passion kindled during our subsequent translation of *Anxious Moments.* Ours would be an excursion in nature, in the aftermath of another revolution. My friend liked to say that in Slovenia man and nature have lived in harmony for more than a millennium. What does that mean? I wondered, trying in vain to identify the constellations overhead. As I dozed off I decided to interview the woodcutter, but when we awoke in the morning he was gone.

Once we started hiking, though, it felt as if we were tracking him, because at regular intervals along the trail we came upon neat piles of logs. A sizable part of Slovenia's economy involves woodworking enterprises, yet I saw no evidence of clear-cutting, in stark contrast to the forests of the American Northwest, where I was living at the time.

"We can't just cut down all the trees and move on," said Aleš. "We have to live here."

Slovenia was not spared the ecological consequences of Communist industrial policy: the air in its cities is polluted, its rivers and streams are fouled, its farmlands suffer from overuse of chemical fertilizers and pesticides, and acid rain endangers half of its forests. Yet in these woods I found signs of harmonious interaction with the earth: well-kept farms, hunters' lean-tos and the neighboring tree houses in which they wait for their prey, protected springs, holy shrines. High in the mountains there was even a stone church covered with scaffolding. The nave looked as if it had been dynamited; four masons were restoring the damaged walls. The government was working with the Church to refurbish crumbling parishes, hoping to bolster tourism. The Communists' anticlerical campaign had come to nothing: the faithful were returning to the fold, Christian politicians were gaining power, and the Church was quietly pressing its claim to more than half of the public lands. Had the last fifty years been an exercise in skepticism, an aberration for what might again become a religious country?

The Slovenians' conversion to Christianity in the eighth century (their golden age of "democracy") was instrumental in the development of their language, which one writer describes as "a weapon and tool of their national existence and survival." Pagan rebels were no match for German might and missionaries who translated the Lord's Prayer into Slovenian. A Protestant writer, Primož Trubar (1508–1586), inaugurated the Slovenian literary tradition with the publication in 1550 of *The Catechism*; thirty years later, another Protestant, Jurij Dalmatin (1547–1589), published the first Slovenian Bible, which had the same impact on Slovenian literature that the King James Bible had on English writing, providing a language and context in which to work out one's destiny. Slovenians might turn to the Catholic Church again for instruction, but their language remained "a depository of all the care and fervor of the proponents of Reformation," as a nineteenth-century poet wrote.

By mid-afternoon we were slogging through an alpine bog. Among a score of pools and ponds were scrub pines, heather, water lilies, dragonflies called snake shepherds, a family gathering blueberries. From there we hiked through a meadow and out onto a logging road, where an old cowherd in a battered Yugo drove up to ask—in one of the more obscure of some forty Slovenian dialects—if we had seen his cattle. Barely able to understand him, Aleš shook his head. The cowherd drove away.

Up a last steep hill, through a copse of fir, we climbed to the tall Partisan memorial on top of Lake Peak, from which we saw more farms, meadows, mountains, and forests.

"Our Partisans were the last to join Tito's army," said my friend, changing his T-shirt for our final trek to the hut we would stay in that night. "And we were the first to leave his country."

The hut turned out to be a ski lodge with private rooms, an attic lined with bunk beds, and a restaurant with a bar. Over the sound system came a German version of Kenny Rogers' "Lucille"; the television in the bar was tuned to the government station, and the evening news was devoted to the war in Bosnia. The crowded room was silent. We ate a hearty beef and vegetable soup, washed down chunks of dark rye bread with beer, and viewed footage from the Serbian-run concentration camps—emaciated men behind a barbed-wire fence. The fighting in the Balkans, the commentator noted, had displaced two million people, 70,000 of whom were refugees in Slovenia.

"The West wants to turn Slovenia into a *cordon sanitaire*," Aleš said when the news ended. "They'll give money, but they won't open their borders."

The television was switched off, and two middle-aged couples at the bar began to sing. They had hiked over from another hut an hour away because, they said, it was too quiet there. They sang folk songs until well after dark, pausing to drink rounds of schnapps or to argue over the exact origin of a particular song. Did it come from this mountain village or that one?

Aleš glanced at the singers. "Their arguments are more serious than you might imagine."

Minstrels, bards, and poets—they sing a country into being. In the eighteenth century the legends and tales, folk songs and poems passed down by Slovenians were collected—another Romantic project—into a book containing more than 8,000 texts. The oral tradition helped preserve the Slovenian character in the face of foreign occupation and Germanization, a task continued in the literary world. (No wonder Prešeren inspired the Partisans.) In *The Songlines*, Bruce Chatwin's brilliant book about the Australian Aboriginals' ancient "Dreaming-tracks," a character remarks that "'[m]usic...is a memory bank for finding one's way about the world.'" And Chatwin wrote, "wherever men have trodden they have left a trail of song (of which we may, now and then, catch an echo)." Here was a Slovenian version of that trail. At nine o'clock the singers knocked back another round of drinks, then left, laughing and accusing one another of having forgotten their flashlight at the other hut.

That night we slept in the common room, surrounded by a dozen hikers, a novel experience for me. Not for Aleš. "You won't believe what I've heard in these places," he whispered. "Couples making love. Drunks throwing up! Little did my parents know what kind of education my sister and I would get when they first took us hiking." In my narrow bunk I remembered something he said at dinner: "It's impossible to understand this country without wandering around the mountains."

Balkan history is the history of the woods. It was in another forest farther south—Kočevski Rog, where ethnic Germans (brought into the area by feudal lords in the fourteenth century) lived until Hitler ordered their emigration in

1941—that the Partisans convened an assembly of delegates in 1943: Slovenia's first national act. Kočevski Rog takes pride of place in Slovenian mythology, for good and ill. Headquarters of the Partisans, after the war it was also the site of terrible atrocities, when Tito ordered the execution of thousands of Slovenian Home Guards (anti-Communists who sided with the Germans). They were some of the 250,000 Yugoslavs—Slovenian collaborators, Ustaše, Četniks—the Communists killed in mass shootings, death marches, and concentration camps.

Their bodies were dumped into caves in Kočevski Rog, a fact kept secret during the Tito era because the forest was an important base for the JNA. In the Ten-Day War the Slovenian militia waged a major battle there; now the same forest houses units of the Territorial Defense. The history of Kočevski Rog, like that of the Pohorje and other Balkan forests, is by turns heroic and horrible. In these places the Balkan peoples discovered what sociologist Liah Greenfeld might call their "style of thought," the cultural expressions integral to their identity—song and story, religion and language, politics and war. Or so I imagined as I drifted off to sleep. In the morning, after breakfasting on fried eggs and sweet mountain tea, we set out for Slovenj Gradec, our final destination.

Down a winding trail we hiked to a clearing, where Aleš paused to listen to something. Two shirtless men off to our right were sharpening scythes.

"A sound I remember from my childhood," he said. "These hills are too steep to cut with tractors, and no one can afford them anyway. The folk songs we heard last night are preserved by these kinds of laborers. Ten men will work the same field, and the strongest one will set the tempo with his singing. If you can't keep up, you get ridiculed."

This morning the peasants might have reserved their ridicule for us. We walked slowly through the meadow, through knee-high grass that had turned golden-brown. Aleš's back was hurting, and my boots had bruised my ankles. I wondered if we would make it to Slovenj Gradec. Walt Whitman thought the body's decay was poetry's only subject—a theme I meditated on at length. The heat had not let up, though dark clouds were gathering on the horizon. We passed several families without uttering more than greetings. I took off my boots and walked in my socks. Aleš carried his backpack as if it were a bag of groceries. We rarely spoke, and then only to complain.

Next to a weather station was a mountain hut, a stone structure used by the Partisans, and this was where we stopped for lunch. The hut was filled with refugees from Bosnia: the military barracks in the valley, converted into a camp for the men, women, and children fleeing the fighting and "ethnic cleansing," were overcrowded. A handsome family was playing cards at the picnic table next to ours. A plump woman in a bright red sweater and floral print skirt was picking wildflowers. An old woman pushed a baby in a stroller. We changed shirts, drank water, ate nothing.

"I don't have the heart to ask them any questions," said Aleš.

We hiked in silence down an abrupt wooded incline, rested at a crossroads, next to a crucifix, then hobbled up a precipitous hillside. I could not think of a good reason to walk across Slovenia.

Eventually we came to the edge of another clearing, a steep alpine meadow more than a kilometer across, a green-and-golden field occupying an entire mountainside. Here were long straight lines of hay cut by hand, a richly ribboned pattern of light and dark extending from the woods below us to the trees straddling the ridge. The sky was blue, the wind was picking up. It would rain before we reached Slovenj Gradec, a prosperous town where UNESCO holds an annual conference on the transfer of knowledge. The wind cooled us off, and we stumbled up to the grove of oak, alder, and mountain ash leading to Kremzar Peak. In a notch down to our left was a solitary farmhouse; the clearings on the Austrian side looked like islands in an ocean of trees. I had never seen such beauty, beauty borne of a people's respect for the natural order.

"I can see why you would fight for this land," I said.

Aleš had hiked here many times, yet he seemed moved. "Do you see how they left that carpet of purple?" he said, pointing at a patch of wildflowers. "There's no need to take it all." He dropped his backpack, stretched out on the ground, and gazed at the sky. "When I was translating for CNN," he said, "we asked a young soldier how he could fight against his friends. He replied, 'Yes, I have Serbian friends, but the minute they attacked my country, they were no longer my friends.' That encapsulates the truth about this beautiful, crazy part of the world. The rest is just philosophy."

3

Ljubljana I

"Why did I return?" said Boštjan Zupančič. "I ask myself that all the time."

We were sitting outside Bistro Romeo, a pizzeria in the old part of the capital. Our table overlooked the Ljubljanica River, a foul tributary of the Sava, and on this hot afternoon the smell of algae and sewage was overwhelming. I pushed aside the Neapolitan pizza Boštjan had ordered and sipped my beer, shielding my eyes from the sun. A canoeist paddled up the green trough that loops around the city. I wondered if he had lost his mind.

Boštjan, a tall, middle-aged man with a beard, had spent most of his adult life studying and teaching in the United States. Only in 1986 had he resettled in Slovenia, accepting a professorship in constitutional law at the University of Ljubljana. He was still trying to find his bearings. In the first democratic elections he had run for office and lost; a consulting contract to the government of Papua New Guinea had evaporated upon his arrival in the South Pacific last month; and this morning a commercial venture had gone awry, because he had neglected a basic rule of business in Slovenia—the need, in his words, to "massage the wife" of any potential partner or client.

"Everything here is great," he said with little conviction, "except the work. You can't undertake any constructive project without encountering obstructions, for no rational reason. It's human nature: part of us always wants to put sand in the machine. But after all the battles you fight for irrational reasons, you have no time left for rationality."

His conversation was peppered with references to psychology, and more than once he urged me to read a book called *The Evolving Self.* Boštjan was a student of Slovenia; his methodology came from the couch. His countrymen, he was convinced, were stuck at the lowest level of moral development.

"This place is still in a Freudian period," he said. "People have anxieties, unmotivated outbursts, neurosis. I get allergies and flu here I never got in the States. It's because of our long history as a subaltern nation. You see, when a man walks out of the house he becomes a boy. He's castrated. The woman's the superego for the son. She's the model."

No wonder the youth of his girlfriend figured in so many conversations among his friends. His marriage had broken up in America, and I was given to understand that he treated his new flame, an undergraduate, as if she were his daughter. I had an impish desire to ask him if his attraction to a girl half his age was a function of Austro-Hungarian politics, but I thought better of it.

The forced migration of peasants to the city, he said, destroyed the urban mentality. "You'd get a country girl for a housemaid, then lose her to an office job. Before World War One there were no Slovenian inscriptions on the buildings—everything was in German. The countryside was 100 percent Slovenian. But now peasant values dominate: patriarchal, authoritarian, insular, inert."

He pronounced the last four words as if he were reading them from a cue card. "For urbanites change is everything," he said, "especially in the Balkans, where the division between city and country is so sharp. This war, for example. They call it tribal. I think it marks the resurgence of primitivism."

"So why did you return?" I repeated.

Boštjan thought for a moment. "I spent an entire weekend putting reasons on a computer: Lotus 1-2-3. Do you know it? I came out with about two thousand points on either side and a difference of only six. In the end I guess I came back because of Plečnik."

Jože Plečnik (1872–1957) was the architect responsible for reconstructing Ljubljana after an earthquake in 1895 destroyed much of the city center. He had a taste for grandeur, though his sources of inspiration were divided between Austria and Italy: Viennese Secession and Venetian Baroque. Best known for designing the castle in Prague in which Czech President Václav Havel lived, in his home town Plečnik built bridges, promenades, columns and colonnades, the covered market arcades, the National Library, and the municipal cemetery. A few steps away was Shoemakers' Bridge, which he rebuilt in 1930; lined with pillars and lampposts, it crosses the narrow river and leads to what was once the Jewish quarter, now a single block of cobblers' shops. The Jews left the city in the sixteenth century, the joke goes, because Slovenians were so tight-fisted. In truth, they were banished from the Habsburg lands during a wave of anti-Semitism; those who returned suffered continuing persecution until World War Two, when the Nazis strung barbed wire around Ljubljana, turning the city into a detention camp; the remaining Jews were among the 6,000 residents sent to concentration camps.

If architecture had inspired Boštjan's return, it was not enough to satisfy him. Slovenians have a saying about exile—*He who is away for three years no longer has a home*—which the professor's friends applied to him. Indeed he took pride in the distance he kept from the literati, saying they had been coddled by the Communists. Yet twice he bragged of being the best-paid writer in Slovenia.

His feelings for America were just as complicated. He despised its commercialism—and bought his shoes in New York. His advanced degrees came from Harvard, which he loved and hated equally. He wanted to hire Laurence Tribe, the professor who gave him his lowest grade, a B– in Evidence, to draft Slovenia's constitution, and when he asked him what he was paid per hour, Tribe said, "I'm paid per thought." Boštjan found that hilarious.

But writing a constitution was a serious matter. Boštjan's countryman, Edvard Kardelj, was responsible for the 1974 Yugoslav constitution, which granted autonomy to Kosovo and Vojvodina, devolved power to the republics, and set the stage for political paralysis in the 1980s. When Milošević created a voting bloc consisting of Serbia, Montenegro, Vojvodina, and Kosovo, the other republics responded in kind; economic and political reforms ground to a halt. The unwieldy constitution thus pitted Serbian efforts to centralize authority in Belgrade against accelerating centrifugal forces in Slovenia and Croatia, hastening the destruction of Yugoslavia.

"And what does a constitution cost these days?" I said.

"A hundred thousand dollars," said Boštjan. "But we turned to the Germans for help. The independence process here was very anti-American, because you were too slow to recognize us."

I gazed at Shoemakers' Bridge. "And where do you fit in?"

"I play the role of gadfly," he said. "But I will be displaced. I lived too long in the West."

In Ljubljana I stayed down the street from a refugee center, in the spare bedroom of a drab apartment belonging to a woman named Mojca, whom I rarely saw. The five-story buildings in the complex were functional only in appearance; the hot water worked intermittently, the grass in the courtyard was overgrown. Graffiti in English: **NO ONE IS INNOCENT. WHO CARES? SERBIA DIE HARD. VOTE DEMO. MORE DOPE.** Swastikas surrounded the inscriptions. A Slovenian version of John Denver's "Country Roads" played at all hours; church bells rang through the night. The smell of manure mingled with garlic, the first spread on the gardens behind each house in the neighborhood, the second a staple of Slovenian cooking. Mojca's small kitchen was cluttered with wine bottles filled with water, in case the war resumed.

On my second night in her flat I was awakened from a deep sleep by the ringing telephone. A drunken man was on the line.

"Where's Mojca?" he demanded in slurred English.

"Away for the weekend," I said.

"Why?" he said.

"She's on holiday," I said.

He muttered something in Slovenian. I told him I did not understand his language.

"Why didn't you learn?" he shouted.

"Because I'm not from here," I explained.

He muttered incoherently until I hung up. He called back immediately. I heard music in the background, clinking glasses, loud voices. "Get out of my life!" he said, and hung up.

The next day Mojca returned from a week of meditation in the forest, under the tutelage of an American Buddhist. With long dark hair and oversized glasses, Mojca sat at her kitchen table, chain-smoking and drinking herbal tea. She was in no mood to go back to work. The founder of Slovenia's first hotline for battered women was thinking about translating German feminist writers.

"Are there more calls since independence?" I asked.

"Attacks are up," she said—and more violent. But reports were down, because of the war in Bosnia. "The women think their problems are minor next to what's going on down there."

I told her about the telephone call I had received.

She giggled in delight. "That's perfect," she said. "He won't bother me anymore, not if he thinks an American's staying here."

Tomaž Brate was Slovenia's busiest architect, and he had not designed anything in years. He was managing editor of *Piranesi*, a new architectural journal. I assumed his glossy, bilingual magazine was named after the eighteenth-century artist whose etchings of prisons have inspired so many poets and revolutionaries. But when I leafed through the inaugural issue I found no sign of the Venetian's desperate ladders, staircases, catwalks, gangways, hanging ropes and chains. *Piranesi* grew out of a series of meetings held in the Adriatic resort town of Piran, where Tomaž and his colleagues explored how Central Europe was "miniaturized" in Slovenian culture, style, and geography.

Tomaž had promised me an architectural tour of Ljubljana, and one drizzly morning we wandered through the old part of town, searching for rain gear. In a clothing shop he asked the clerk if we could borrow a poncho. "Of course," she said, then realized she had none left. We walked back out into the rain, where we debated the merits of repairing to Drama, his favorite bar (also known as his office). In the end we decided to climb the narrow winding lane to the castle overlooking the city: there we could get a good overview of Ljubljana—and a drink in the castle's new bar.

Tomaž was not yet thirty, but he had already suffered a major heart attack. He chain-smoked, drank large quantities of coffee, wine, and beer, and worried constantly about two things: his health and funding for *Piranesi*. The cost of printing 3,000 copies was $100,000, almost all of which came from the ministries of culture and environmental protection. "It's strange to meet with these ministers who until a year ago were my drinking buddies," he said as we started up Castle Hill. "When they get serious in our meetings, I want to tell them, 'Talk like you used to.'"

The lane gave way to a steep gravel path; on one side, through acacias and hornbeams, was the center of Ljubljana, a sea of red-tiled roofs; on the hillside was a stone wall from which trailed moss, ferns, and hanging plants. Tomaž stopped frequently to catch his breath, light a cigarette, and tell me about the city he had moved to at the age of thirteen.

The Mediterranean meets the Alps in Ljubljana, a shallow basin into which water flows from several mountain ranges and rivers. It was fitting to walk in the rain: the first inhabitants of this area, at the end of the Neolithic era, were known as "marsh folk." The large bárje, or marsh, south of here governed the settling of Ljubljana from the arrival of the Romans, at about the time of Christ's crucifixion, until the last decades of the eighteenth century, when a canal was built behind the castle to drain the marsh. Only in 1931, when Ljubljana's walls and bridges were reinforced with concrete, was the city saved from the periodic flooding of the Ljubljanica. It was difficult to imagine the sluggish river ever causing problems, but during the Middle Ages, Tomaž assured me, city residents sometimes stepped directly from their first-floor windows into boats. Prešeren even wrote a ballad about a feckless woman dancing with a water troll, who leads her into the river, where she drowns.

The Ljubljanica's source is in twelve springs near the marsh, which lies atop a mysterious world of caves, tunnels, subterranean precipices, and natural bridges. If a walk in the mountains is basic to understanding the Balkans, then a more particular kind of knowledge may be found in the Slovenian Karst, which extends almost to Trieste. This landscape of porous limestone is home to the marvelous: fissures and sinkholes, underground streams and fountains, stalactites and stalagmites. Southwest of Ljubljana, for example, is Cerkniško Jezero, a 26-kilometer expanse that is a lake half of the year, a meadow the other half; when the water returns, so do the fish, which have fed the local population for centuries. And the Karst begins at the city's edge.

Ljubljana grew concentrically, each district encircling the Roman core, like a tree ring; destroyed by earthquakes, in 1511 and 1895, the city was rebuilt both times according to this circular logic, over which was laid another idea—connecting Castle Hill to Rožnik Hill, two kilometers to the north. This visual bridge, or "Ljubljana Gate," connects the old city to the new. The farther you go from the center the Romans called Emona, the less interesting the architecture, until on the outskirts of town there are only monotonous apartments built in the International style: a lifeless counterweight to Plečnik's pillars and obelisks, his efforts to resurrect classical Rome in his version of Ljubljana, which since his death has become home to some 340,000 people.

"In Yugoslavia all the people had to be workers," said Tomaž. "So peasants were moved into the city, and apartment buildings were needed. Unfortunately, third-rate architects designed them."

I was surprised, when we reached the hilltop, to see the castle renovations taking a modern form. The original plans were gone. Fortified at the time of the Celts and Illyrians, the first castle was built by Carinthian Dukes in the ninth century; the Habsburgs acquired it in 1335, refurbished it in the fifteenth century, and their provincial rulers occupied it for more than two hundred years. But it fell into disrepair, serving at one point as a garrison and prison. What remained were the excavations, a scattering of walls, and the tower. A cafeteria and bar had been added; even in its unfinished state, the castle—Ljubljana's cultural acropolis, Plečnik said—was a popular wedding site.

"Let's get out of the rain," said Tomaž.

In the bar he told me that although he came from a village his thinking was urban. Yet he had passed up a chance to finish his studies in Vienna, believing it had lost its Central European context and importance after World War Two.

"Twenty years ago," he began, "its architects said Vienna was a finished city. It didn't need new buildings. It was just a monument to Central Europe, a monument without life. But with the borders opening there's a radical change—competitions, new buildings going up, another line of the underground. Vienna's coming back to life, a reincarnation of the metropolis of Middle Europe. But because it lost forty years of cultural touch with the rest of us it's still a city without connections."

Venice was altogether different. "It's not a city," he said with feeling. "It's something cosmic. The water and canals: it's always moving, from nowhere to nowhere. The city can be full of tourists and I don't mind. The marriage of architecture and water transcends the physical facts of tourists. Venice is the site of ritualistic meetings. What do I meet when I go to Venice? Water and light."

He ordered another carafe of wine. "Vienna and Venice are Slovenia's two poles: we lean either toward the German or Roman cultures."

At the turn of the century Viennese architectural elements reigned in Slovenia. Then came the International style, befitting the Yugoslav idea. Since the early 1980s Italy had become for Slovenians the most beautiful country.

"A group of architects from Karst tried to give expression to the whole area—buildings, people, nature, poetry," said Tomaž. "They wanted to express the idea of small stone houses in something big. They looked for a simpler vocabulary, a stripped-down style. Regionalism as opposed to Internationalism. But in the last year you see a Viennese-Venetian mix. Vienna's changing status is already starting to have its effect on us. Where will that lead?"

He refilled our glasses. "Of course there's no money to build anything. So I'll write a book about Central European monuments." He smiled. "And we'll have another drink."

One morning the poet Uroš Zupan took me to the public market. Slovenian literati marked decades in poetic terms: Tomaž Šalamun was the poet of the '70s, Aleš Debeljak embodied the '80s, and in the '90s Uroš defined the prevailing spirit. He was a tall, lanky man, with a scraggly beard and deep blue eyes, in gym shorts and a muscle T-shirt. He was oblivious to his surroundings.

"Unlike everyone else," he said, "I know nothing about the political and economic situation. I prefer to read and write poetry, to drink and walk around."

But Uroš had lately given up drinking, certain he was headed for madness—or death. He did not want to suffer the same fate as his favorite writer, Cesare Pavese, who shot himself in a hotel room, despairing over his love life. Uroš, grieving over a failed affair, cut a very romantic figure.

In the sunlit market were fresh tomatoes from Macedonia. How did they get here? I asked. Uroš shrugged. He was busy gathering onions and new potatoes, which an old woman measured out on a scale, adding lead weights until the pans balanced. I recalled a scene from a film about Pablo Neruda: how the poet caressed the fruits and vegetables in the market, storing up sensations. Neruda translated his passion for the things of the earth into his "Elemental Odes," celebrating "life forces wherever they may be"—salt, a lemon, an artichoke, a large tuna, like "a bullet/ from the ocean/ depths." Uroš had the same love for the market. Here were wax beans gathered in sheets tied together into pouches; melons, apples, red and yellow peppers in wooden crates; herbs in sacks of white cloth rolled up like turbans; buckets and bunches of fresh flowers—gladioli, chrysanthemums, white daisies, sprays of baby's breath; stalls of new and used clothes, tables on which shoes and sandals were piled, young men hawking woven baskets and leather goods, old women in red aprons calling to us.

Plečnik's unfinished arcades along the river housed the butchers and fishmongers, where Uroš bought calamari. From there we wandered to a suite of tables displaying magazines, a car which was the grand prize in a lottery, a street musician whose guitar case was filled with tolars, though he could barely sing—all new since the war. Three soldiers were waiting for a bus; on the wall behind them someone had scrawled, **FUCK JNA**, an accurate description of Slovenia's military objectives during the war, when its militia besieged several JNA barracks, threatening to starve the conscripts.

"Our military," Uroš snickered.

Then to the apartment he shared with a choreographer on tour in Belgium. Like many young artists, Uroš was supported by his father, who wanted him to finish school. But he preferred to write and translate. His main influences were American rock-and-roll and Beat poetry, not Slovenian literature and folklore. Aleš had thus dedicated a new essay to him, urging him to return to his roots, to which Uroš replied, "Blah, blah, blah."

He prepared calamari stuffed with garlic and cheese to music from a small tape deck. The Grateful Dead. Jackson Browne. Paul Winter. Then Albinoni's "Adagio in G Minor," a current hit in the Balkans. Vedran Smailović, a cellist in Sarajevo, had captured the world's attention with his performances of this haunting music, after a Serb-launched mortar shell in May had killed twenty-two Sarajevans lined up outside a bakery. Smailović played the Albinoni in front of the bakery for twenty-two days straight, risking snipers' bullets and artillery to honor the dead. It pleased Uroš to think that "Adagio in G Minor" was based on a single sheet of music discovered on a street in Dresden after the Allied firebombing of the city. He remembered the first day of the war in Slovenia.

"I left a bar at five in the morning, and I was pretty stoned when I came to a barricade put up by the JNA. It was a complete surprise. I was so scared," he said with a laugh, "that when I got home I started reading Neruda's love sonnets. I only slept a few hours before my roommate called to tell me to buy bread. But every store was sold out. And everyone was in a panic."

Ironically, later that day, a hundred meters from Uroš's apartment, a JNA helicopter thought to be carrying weapons was shot down. Loaves of bread spilled out of the burning wreck. The pilot killed in the crash was Slovenian. Indeed the JNA suffered the bulk of the casualties.

Presently Aleš arrived. Uroš opened a bottle of wine for us, and we ate lunch directly from the salad bowl, "peasant style." Aleš described his recent trip to Argentina to read to the Slovenian émigré community, which was divided into two camps, pre– and post–World War Two exiles, a common feature of the Yugoslav diaspora. The first wave of émigrés, Istrian Communists forced out by Italian Fascists, had assimilated into Argentine society; the second, Nazi collaborators and anti-Communists, were frozen in time. "The third generation still speaks Slovenian," said Aleš. "They have their own doctors, dentists, and grocery stores, like a Slovenian Disneyland. They're isolated in the past."

Yet they had inspired Slovenian nationalists, just as some of their Ustaša neighbors in Argentina, including Ante Pavelić, had encouraged independence-minded Croats. Trapped in sentimentalism, the post-war émigrés fueled efforts to maintain their national identity. "Dawn breaks over the snows of Argentina," Aleš himself wrote. "You lie still, on the edge of the world." On the banks of the Río de la Plata Slovenian literature flourished during Communism: unable to acquire books from their homeland, the exiles wrote, translated, and published their own works, including a Slovenian *Divine Comedy*.

The literary junket closed a circle for the exile community. The two sides had grown closer since independence, and post-war émigrés were returning to Slovenia, some to visit, others to live. What inspiration they offered now took the form of capital investments. Literature was no longer the sole repository of value. Nor did Aleš mourn the fact that the best and brightest were now going into business and politics instead of the arts and humanities. They've become legitimate occupations, he said. And when we went out in the early evening, he pointed to a Partisan monument about to be dismantled: those who had fought the Communists during World War Two (supplied with German money, food, and arms) were reasserting themselves (with émigré support). We stopped on a bridge over the river. Uroš and Aleš were amazed to see fish feeding in the polluted algae.

"Look how big they are!" they exclaimed together.

To tease Uroš, Aleš told one more story: how, on his flight home from Argentina, he had picked up the woman next to him with what another writer called "devastating quickness." When the plane developed mechanical problems and made an emergency landing in São Paolo, Aleš and the woman were perhaps the only passengers happy to learn they would have to spend the night in a hotel. Uroš groaned: he had not slept with a woman since he had stopped drinking.

"Uroš is writing a book called *Seven Thousand Nights Without a Woman*," said Aleš, playing on Karlo Štajner's *Seven Thousand Days in Siberia*, the chronicle of a Yugoslav diplomat who spent twenty years in Soviet prisons and camps.

"It's already written," Uroš sighed.

Aleš had other plans, so we left him on the bridge and walked downtown. Uroš was known among his friends as "a field worker" (someone who cruises bars and cafés), and it was growing dark by the time we reached his proving ground, a long cobblestone street in the old part of the city. The bars and cafés were overflowing, men and women of all ages strolled down the street, young couples sprawled on the steps around a monument, teenagers bicycled past.

Our first stop was Uroš's bar, distinguished from its neighbors by oversized eye charts on either side of its entrance. I was puzzled by these signboards until Uroš mimed a drunk trying to read the small print. He slapped his hand over one eye, charged blindly into the bar, and returned with beer for me, cranberry juice for him. A couple in their early thirties sauntered up.

He was a printer and sculptor who refused to show his work in Slovenia. "I could work here for ten years and where would it get me?" he said. His sights were set on galleries in several European capitals. But when I asked him if he had lined up any exhibitions yet, he started talking with Uroš.

"Do you know Slovenian?" his wife asked me.

"Not a word," I said.

"You better learn," she said, nodding at Uroš. "They're gossiping about you."

"Is it any good?" I said.

"Slovenians are the most critical people in the world," she said.

She was the mother of twin boys born during the war, six weeks early. "The most awful period of my life," she said, and steered our conversation toward a happier time, when she had studied English at the university. After writing a thesis on Margaret Atwood's novels, she lived in London for six months, and there she fell in love with Kenneth Patchen's poetry.

"A strange combination," I said.

"What?" she said.

"The writers you like," I said. "A Canadian feminist and a surrealist from Ohio."

She thought for a moment. "I can't talk about this anymore, or I'll become a bad mother."

Her husband took her arm and asked me to resolve a dispute with Uroš. "Do you say 'put out' or 'shed' a few tears?"

"Shed," I replied.

"Very well," he said with a laugh, throwing his arms around Uroš's neck. "Put out a few tears!" Then he left with his wife.

Uroš and I walked up and down the street, now ducking into a bar, now talking to someone sitting outside a café. He waved or nodded at dozens of people. I thought: who needs secret police when you know everyone? Certainly this was an orderly crowd, which broke up before midnight, when the buses stopped running. Waiting for our bus, Uroš gazed at a young woman. "I need to learn how to approach women sober," he blurted out, "and it better happen soon!"

Street cleaners wheeled past in yellow trucks. Church bells rang across the city.

In Prešeren Square, where the medieval town meets the new city, two Mormon missionaries were proselytizing by the Franciscan church. They stood next to a new Renault, grand prize in this week's lottery, and tried in vain to catch the attention of passersby. Prohibited from evangelizing door-to-door, the young men were reduced to stopping Slovenians in the street, a strategy that rarely worked.

"They're usually going somewhere in a hurry," one told me.

The Mormons had arrived six months before the war. Now there were even sister missionaries to talk to the women who, in their words, "had America in their eyes." And they had converted some Slovenians. It was easier to work here than in Italy, where the Church was stronger.

Scaffolding draped the façade of the Baroque church. At noon it was dark and musty inside. Only a handful of people sat in the pews. An old woman was dusting the altars. The frescoes covering the vault needed restoration. A minor drama was in the offing: two Gypsies, a father and son, selling matchbook-sized pictures of Christ, caught a priest's attention. He chased them away, screaming at the top of his lungs. I followed them outside, where they vanished into the crowd. The Mormons were talking to a young man, who was examining the Renault. He was not listening to them.

It was raining the day Aleš took me to a Serbian restaurant for lunch. On the covered terrace we sat in plastic chairs, surrounded by palms. Glen Campbell's "Gentle on My Mind" wafted out of the sound system. Like other Serbian establishments in the capital, this restaurant was on the verge of closing, and there was an air of desperation about the place. The owner addressed his waiters in Serbian, one of whom took our order in broken Slovenian, which upset my friend.

"Imagine living here for fifteen years and not learning the local language," he sneered. "That's like going to England and not learning English. The Serbs have a saying: 'Speak Serbian, and the whole world will understand you.' What delusions of grandeur. And we couldn't speak our language in public until the end of World War Two. Slovenian was the language of privacy before then."

The waiter brought us a bottle of red wine from Vojvodina, the ethnically mixed region—Magyars, Serbs, Croats, Romanians, and Slovaks live there—that, like Kosovo, lost its autonomy in the 1990 Serbian constitution. Growing numbers of Hungarians viewed their brethren in the south with the same mixture of irredentism and compassion that Albanians felt for the Kosovars.

"Milošević was testing the waters in Kosovo and Vojvodina," Aleš said, pouring the wine. "He sent in his supporters to stage rallies and demand reunion with Serbia, as if the idea came from the local ethnic populations. Then the bureaucrats were forced to say, 'This is the will of the people.' And the other republics just stood by and watched."

The waiter delivered our meal—fried paprika; a salad of tomatoes, onions, cucumbers, and goat cheese; round loaves of bread; and *pleškavica*, a meat patty with cheese and onions cooked into it. We ate greedily, and when we were finished Aleš said, "Catalans, Scots, Basques, Slovenians—they're of a piece. We need a new language to address that. The old model won't hold. The minute you equate murderer and victim, as the West has done in Bosnia, you become a murderer. Unlike poetry, there's no limit to the imagination when it comes to killing, as the Serbs have shown."

Our waiter was talking with the owner. Aleš signaled for the check. "Maybe in a year or two there will be a backlash against those who speak only Serbo-Croatian in public," he said.

"It was a great time for metaphors," said the journalist. "That was the only way to escape the censor."

It was in the concert ticket line at the National Gallery that Bernard Nežmah told me about working for *Mladina*, the Socialist Youth Alliance weekly which in the 1980s was known for its hard-hitting journalism; its editorials criticized the JNA and supported the Albanian cause in Kosovo; its writers were among Yugoslavia's leading dissidents. In 1988, *Mladina* denounced the military for its alleged involvement in arms sales to the Third World and its plans to aid police in a crackdown on dissidents. The authorities retaliated, arresting several journalists, charging them with possession of secret military documents: a crucial event in Yugoslavia's dissolution.

For despite public outrage and mass protests, the accused were prosecuted and convicted in closed trials in Ljubljana, receiving sentences ranging from five to eighteen months. One journalist, Janez Janša, was soon hatching plans destined to have international repercussions. On a prison leave in June 1989, the future defense minister attended the first congress of the Slovenian Democratic Alliance, where he called for the creation of "parallel armies" (rooted in the Territorial Defense Units) in each republic. *Mladina* kept up its attacks on the "Yugoslav Shogun Army authorities," and within two years Janša had helped to win Slovenia's independence, plunging the rest of Yugoslavia into civil war. *Mladina* was about to celebrate its fiftieth anniversary. Founded in World War Two as part of the Resistance, now it was just another journal with a modest circulation of 25,000.

Bernard Nežmah missed the excitement of writing on the cutting edge of history. While we waited, he recounted two stories about the city zoo: 1) A vulture escaped, only to be recaptured on the river bank. It could not fly away: its wings were too heavy, and Ljubljana lacks both the mountain heights from which a vulture can drop into flight and the thermals on which to rise. 2) A fox dug a hole around the raven's cage, and one raven flew away. Each day it returned to be fed by the other ravens, because it did not like the food it found outside.

"Are we speaking in metaphors?" I said.

The concert was sold out, but Bernard wangled press passes for us. I was sweating when I took my seat in the main gallery, under an oval Baroque painting of Saint Francis. The museum was packed (it is said that classical music draws more people in Ljubljana than soccer), there was no air conditioning, and television lights were burning. The flutist Irena Grafenhauer's red dress was stained with sweat. Yet she and a harpist gave a spirited performance of works by Blavet, Lauber, Ibert, and Fauré. During intermission I learned that if a fire broke out no one would escape (a cheery thought) and heard for the third time that Mahler had spent a year in Ljubljana. When I returned to my seat, an usher asked me for my ticket. Unable to find my press pass, I threw up my hands and said in exasperation, "American." "Ah, American," he laughed, waving me through. It was even hotter now, and sweat dripped from the musicians' brows. This was a homecoming for

Grafenauer, who was on the faculty of the Mozarteum in Salzburg. Cameras recorded her every move.

When the concert ended, I joined the crowd outside her dressing room. Slovenians should hear Native American music, Bernard thought, and he wanted Grafenauer to perform it. She said she would think about it. He smiled and left. I asked her if she could imagine living in Slovenia again.

"I was born for music," she said. "That is the language I express myself in. I am like a bird. I could live anywhere."

I went outside, into a cooling drizzle, and walked to the bus stop, where a toothless Bosnian cornered me. He claimed to have fought in Tripoli, Tunis, Kuwait, and Sarajevo; from time to time he mumbled what he said were Muslim prayers. He was blind drunk. A bus arrived, and just before I boarded it he grabbed my arm. "Germans, good," he slurred. I nodded. "English, good," he said, and I nodded again. The driver motioned me onto the bus. The Bosnian would not let me go. His eyes were bloodshot, his breath reeked. "Americans, bad," he muttered, and stumbled down the block.

I took a seat behind the other passengers and stared out the window. It was some time before I realized I was on the wrong bus. I kept hoping I would recognize a building or a street. Instead, we passed through several unfamiliar neighborhoods. There were fewer and fewer lights. One by one the passengers got off, and then the bus came to a stop, on the outskirts of the city. The driver turned off the engine and put on his jacket. His shift was over. I had a long walk ahead of me.

In Ljubljana I had the longest lunch of my life. My host was Tomaž Šalamun, the setting, a garden restaurant, and for eight hours carafes of white wine appeared at our table. Tomaž was in a bad state. He had lost $8000 on Lotto, he was nearing what he called the crucial nineteenth day without a cigarette, and he could only write, it seemed, at artists' colonies in New England.

"American poets offered me a way out of Slovenia's strictures," he said. "Your country's spaces open up my cells. But I don't know if I can handle that now."

He refilled our glasses. "When I was in America, I wrote two to three poems a day. Images just flew toward me. My poems always begin with the visual. I have to see it, then I write. And I must hurt myself to write. But I can no longer hurt myself in America. I'm too far away in case I go mad. I must learn to hurt myself here."

I told him I had been lying awake at night, bombarded by frightening images.

"That's good," he said. "That means you're about to write."

I said I thought it meant I was going mad.

"No," he said. "It's a gift from the gods."

His mystical view of language was not uncommon in Slovenia. Poets, according to the myth, sang the Slovenian nation into being, held it together for a thousand years, and foretold its liberation. Even so, Tomaž was terrified to think that Ludwig, the excitable editor at the Wieser party, saw in one of his very first poems, "Eclipse," a vision of the war in Bosnia:

I will take nails,
long nails
and hammer them into my body.
Very very gently,
very very slowly,
so it will last longer.
I will draw up a precise plan.
I will upholster myself every day
say two square inches for instance.

Then I will set fire to everything.
It will burn for a long time.
It will burn for seven days.
Only the nails will remain,
all welded together and rusty.
So I will remain.
So I will survive everything.

Tomaž was haunted by this poem from thirty years before.

"Is that what landed you in jail?" I said.

He thought for a moment. "Poetry has rendered me unfit for everything," he said cryptically.

What I remember of the rest of the evening is hazy. It was dark when we left the restaurant. Tomaž drove up and down a street barricaded at one end. "Tomaž," I said, "this isn't working." A look of bewilderment covered his face—and then he tried again. At last he turned around, and nearly backed into the river. Somehow we arrived at KUD, the Prešeren Cultural Center, where poets from the former Yugoslavia had gathered to give a reading protesting the war in Bosnia. Four hundred people drifted around the outdoor stage, drinking beer and smoking. It felt like a fraternity party.

Tomaž introduced me to Boris Novak, President of the Slovenian PEN Club. I knew he was the author of a handbook of poetic forms titled *Forms of the World*, and all I could think to say was, "What ethical system will Slovenians find to replace socialism?"

He took my question seriously. "Nothing," he said. "The Church is not strong, and the people distrust the government, which they have been in the habit of cheating. That's why so much New Age stuff is flooding into the country."

I nodded—and almost passed out. I hurried inside to get something to drink. I stood at the bar swilling orange juice.

The reading began two hours late. On the stage were Aleš, Uroš Zupan, a Macedonian poet, and a Serbian poet who read a statement about the siege of Sarajevo—Slovenians called it equivocal—the French philosopher Jacques Derrida had supposedly delivered in Belgrade. Few were listening.

But when it was Aleš's turn to speak, television cameras lit up the stage. A woman turned to me and said, in perfect English, "our next president." I stared at her.

The reading drew to a close. Tomaž offered to drive me home.

"You're too drunk," I said.

He frowned at me. "How did we pay the bill at the restaurant?" he said.

"It's a mystery," I said.

Eventually he decided to walk to his apartment, a decision he regretted in the morning, when it took him three hours to find his car. He just made his flight to Macedonia, where he was to receive a prize at the Struga Poetry Festival. Worse, when he checked into his hotel, he discovered that in his drunkenness he had packed—for four days—eight shirts and no pants!

One sweltering afternoon I climbed a circular staircase to a fourth-floor apartment overlooking the river. Across the hall was an open door in which an old barber worked; a cocker spaniel slept at his feet, a young man's shorn locks covering its paws and face.

The apartment belonged to a cultural affairs editor, who wanted me to meet her friends, Alan and Emma—Slovenia's most interesting fashion designers. They were a study in contrasts. Emma, a fiery redhead, was casually dressed in a red blouse and jeans. Alan, a Croat (and member of the Slovenian Cultural Commission, though he spoke no Slovenian), reminded me of the phantom of the opera. Emma said, "Alan only overdresses." This day he wore a white shirt, a black vest, a tie pinned with an elaborate brooch; he had jet black hair and a ghostly pallor. From time to time he hurried to the bathroom to reapply his powder, crying "Ooh, ooh" at the mirror; his makeup, he said, occupied him "every day, all day. I start at six in the morning, then again eight hours later, like in a factory. I'm afraid of water—the sea, not a bath. I hate mountains and saunas; it's too hard to bring a large mirror to a mountaintop, and in the sauna the makeup runs away. In this heat I am always repairing."

Emma explained that his mother, an odd woman, used to buy him makeup when he was a boy, and during the war, in the basement of his apartment, while others brought food, medicine, and books, Alan packed only his makeup kit. He soon decided not to stay in the basement.

"I could not stand to be among people who do not wash their hair," he said.

His work was no less eccentric. He had just had his first show in Milan, and going through the catalogue he giggled at his extravagant designs: "Death in

Venice," "Russian Butterfly," and "Cock-Hunting," the last featuring a woman in a swatch of chartreuse, a hoop skirt that resembled a table on which she laid her hands; a cloth nightingale spread its wings over her dark beret. Her scarf, gloves, and boots were orange; there were ornamental buttons everywhere.

Alan's friends called him an innocent in sexual matters—one critic suggests his "deepest cut into the subconscious is his attempt to obliterate the distinctions between conventionally male and female garb." Thus he dressed a woman in black, gave her a cane, and called the design "Oscar Wilde." What part of *Death in Venice* had he named his design after? Only the title, he said, and his love of Venice; he had not read the book. History inspires me, he insisted. No one influences me.

"Where will you be in ten years?" I said. "Ljubljana?"

"Not at all," he replied, touching his cheeks. "Maybe Sarajevo." His makeup was running again. He ran from the room.

Emma smiled. She was as practical as he was outlandish. Like Alan, she designed costumes for theater and dance companies to pay the bills, but her real work lay in a series of tight-fitting suits spotted and colored like leopards. The first issue of a new magazine included some of her designs, a spread of photographs of athletic women crouching and stretching like cats.

"I work only with animal skins," she said. "Fighting for animal rights is absurd when they are killing people all the time."

Equally absurd was the convention of models and mannequins displaying clothes, so she used actors and dancers for her happenings, which she staged in caves or abandoned power stations. Only guerrilla tactics would wake her fellow Slovenians. Her destiny lay in Ljubljana.

"I never had the wish to go outside," she said. "But I would like my country to look to the East for inspiration. The West is not the answer. In the factories the young designers imitate the West. That's a pity. MTV-style is not for us."

By now Alan had returned, his face pale again, and he wanted me to admire his brooch, bought in an antique store for only two dollars. "You could have one, too," he said, raising an eyebrow at my jeans and work shirt.

"Think of that," said Emma.

An exhibition, at KUD, titled *Sarajevo*, with photographs and texts by the American photojournalist Jana Schneider and Ivo Štandeker, a reporter for *Mladina*. In June a mortar shell launched from a Serbian position in the hills around Sarajevo had wounded Schneider and killed Štandeker, her lover. What he had written during the Ten-Day War now seemed prophetic:

> The sight of burning trucks, the glass, the mud. In front, a woman is dragging a body, the tanks behind her are motionless. Avoid the explosion, cross the meadow. It was

full of water and near the lead truck were the mutilated dead and wounded. What could one do?

Rumor had it, the gunner had aimed at a reflection off Schneider's camera lens, journalists being a favorite Serbian target. On opening night, the cultural center was packed with silent men and women studying Schneider's work. She was not present. People said she was lucky to be alive.

I had spent the day with Aleš at Lake Bled. We rowed out to the church on the island, hiked up to the castle overlooking the lake, later swam in the clear, cool water. In a restaurant near a long gorge we lunched on ham omelets and white wine. I was in a fine mood, but my friend grew restless on the drive back to Ljubljana. His small Fiat had belonged to his best friend, a suicide at twenty-one, and Aleš was thinking about him because he was planning to buy a new car. Over the radio came the news of the deaths of twenty-five people in Sarajevo.

"This is how our life is," he said. "A blissful afternoon, and then the abyss."

In KUD's bright white exhibition space, I was horrified at the destruction Schneider had documented in Sarajevo. Here was a man walking through rubble, the walls of the building next to him pocked with shrapnel, two destroyed cars blocking the street. Here were bodies stretched out on sidewalks, burning houses, a **SNIPER WARNING** sign painted on a street corner, frightened men, women, and children, a cat in its death throes. And here was the story of Štandeker's death, a man who pushed on against his better judgment until he landed in the sights of the Serbs.

"Drink," said Tomaž Šalamun, handing me a glass of white wine.

"Difficult work," said a high school history teacher.

"My people are from Lebanon, and this war is like Beirut," said a South American woman working on a documentary about Bosnia.

"I was twenty years drinking coffee with my neighbor," said a refugee from Sarajevo. "How could he start shooting at me?"

"The war's the current running under everything," said Aleš, offering me a piece of cake. "In all other respects life goes on. Films are shot, books are published, we go swimming. Yet there's a difference: the abyss is yawning under our feet as we walk, opening its dark jaws."

"Sarajevans didn't fortify their basements or arm themselves," said the teacher, "because they couldn't believe war was coming, not with their mix of cultures. It was self-induced myopia. Everyone knew the Serbs were fortifying their gun emplacements around Sarajevo; they were preparing to fight for a long time. The Muslims thought it was just a military exercise."

"When the shelling began, a Serbian writer called a friend in Sarajevo," said the film maker. "The Bosnian cried, 'They're shelling Sarajevo!' 'Don't believe foreign propaganda,' said the Serb."

"This is the end of the history of something," said the teacher. "The Serbs are burning all the books and archives in Sarajevo."

"Have you heard about the Weekend Četniks?" said the refugee. "They come from Serbia and fight only on the weekends; the rest of the time they work. For them it's a turkey shoot."

"I'll tell you what the West's response means," said Aleš. "If you don't give water to a person dying of thirst and he dies, then you're a killer, even if you don't kill him with your own hands."

"In their rifle sights the Serbs make the sign of the cross over the Muslims before they kill them," said the teacher. "They brag about their shooting sprees."

"Many people are getting rich off this war," said Tomaž. "It will go on for years."

"In Sarajevo they're throwing snipers out the windows," said the refugee. "We're taking no prisoners now. Two months ago I wouldn't have thought that possible."

I needed fresh air. My head was spinning, and I was sweating heavily. "Are you all right?" said Tomaž. I shook my head. "It must be the photographs," I said. "I feel very strange."

"Man, you're high!" said Aleš, smiling.

"What do you mean?" I said.

"The cake," he said with a laugh. "It was cooked in hashish! You're stoned out of your mind."

"Oh," I said, and stumbled out into the courtyard.

Aleš followed me out to introduce me to a Bosnian Serb who had just arrived in Ljubljana. Goran Janković was sitting at a table, smoking and drinking. A thin, sharp-featured man in grey pin-striped shorts and a white T-shirt. He had short dark hair and hooded eyes, which made me think of a hawk. He spoke in a convoluted manner, or perhaps that was how I heard him.

"I left because of the craziness," he said. "I left because this is not my war. This is a war for criminals. And what you see in Bosnia are all the forms of hatred that will play themselves out in the ex-Soviet republics."

He went on to describe a harrowing escape from Bosnia, and it was well after midnight before I left. The buses had stopped running, and on my walk home the streets were dark and quiet. In Jana Schneider's work and the refugee's story I imagined I had glimpsed a hellish version of the future. The river was on my right. On my left were buildings I did not recognize. I wondered if I would find my way back to Mojca's apartment. A man dressed in white was sitting on a stool outside an open door. I stared at him for some time before I realized he was a baker. The air was filled with the aroma of fresh bread.

I had no idea where I was.

I had an 11 a.m. appointment at Radio Slovenia. Nina Zagoričnik, the willowy journalist from the Wieser party, wanted to interview an American writer for her cultural affairs program, but when she called to set up a time to meet she seemed interested only in my musical tastes.

"We will play Lou Reed in the background," she said breathlessly.

"Maybe you should just play music," I said.

"Why?" she said, and hung up.

I waited outside her office for twenty minutes. Aleš and I were leaving that afternoon to climb Mount Triglav. Twice already we had postponed our trek; it might snow any day now in the Alps, ending the climbing season. Don't be late, my friend had said. Just when I was about to give up, Nina and her little daughter Niké arrived on bicycles. I am in a rush, she said, because I knew you would be prompt. Then she disappeared into the building—only to return minutes later with another bicycle.

"Let's go," she said.

Through the busy streets I followed Nina and Niké, dodging cars and pedestrians. Niké kept swerving into the traffic, and I was relieved when we came to a stop in an alley behind an old apartment building.

"What's going on?" I said.

Nina called to a white-haired woman hanging clothes from her balcony. Niké darted into the building. The aging woman scrutinized me in what I had come to regard as characteristic Slovenian fashion—two parts suspicion, one part hostility. She said something to Nina—I had a feeling it was about me—and when we were out of earshot, Nina said simply, "my mother."

On the way back to the center of town, Nina adopted the air of a tour guide. "Look this," she kept saying, pointing at her grammar school, her high school, a playground. Here was the tunnel in which she had kissed her first boyfriend, and here was the home of her best friend. Naturally, we ended up at a café, drinking coffee and juice. Nina had said nothing about our interview.

"Time to go," she said suddenly, and off we went to the station.

When we took our seats in the studio, I said, "What are we going to talk about?" But the engineer in the control room had Nina's attention. "How long will this last?" I said. Nina did not answer. The interview began.

Nina asked me to name all the places I had lived—she even had me sketch a map of the United States—and then she pleaded for stories about famous American writers. "Can't you tell us something about Saul Bellow?" she cried. Sadly, I could not.

When we were done, Nina replayed the tape, stopping to write out the words she did not recognize, which would then be translated into Slovenian. A look of alarm flashed across her face.

"Excuse me," she said, turning the machine off. "I must go look at myself in the mirror." And she rushed out of the room.

When she returned some minutes later, she was twirling her brassiere around her index finger. "Why didn't you tell me I was wearing this?" she demanded.

I could not answer her.

She twisted around and pointed at her back, at the opening in her sun dress. "Why didn't you tell me?" she repeated.

"I guess I'm not in the habit of telling women I don't know very well that they're wearing underwear," I offered. "Besides, I didn't notice it."

"Then you must be blind," she laughed, tossing her brassiere into her handbag, and then she switched the tape back on.

4

Triglav

I inch along the rock wall, clutching the twisted iron cable anchored lengthwise across the granite face. Aleš and I started up the trail before sunrise; by midmorning we have gained more than a thousand vertical meters, and we are hours from the summit of Mount Triglav (Three Heads), the crown of the Julian Alps. The steep trail is crowded with men and women of all ages; helicopters swarm overhead. Slovenia's tallest peak (2,864 meters)—named after the highest deity of the ancient Slovenes, the three-headed god responsible for the sky, earth, and mysterious world of the Karst—annually draws thousands of climbers, none more important this summer than the nation's first president, Milan Kučan, and his minister of defense, Janez Janša. Tomorrow they will hold a "Summit on the Summit" to discuss Slovenia's—and their own political—future. You Slovenians are crazy, I tell my friend. Why? he says, tightening the straps on his backpack. You make a pilgrimage out of climbing a mountain no one in their right mind would attempt without ropes, I reply, and you make it look easy. That's because the mountains are a way of life for us, he says, not just a sport.

A helicopter ferrying supplies to the mountain hut near the summit flies past. A fine time for the Serbs to bomb Triglav, Aleš jokes, reminding me to maintain three-point contact with the rock—two feet and a hand, or two hands and a foot—which, with so many people on the trail, is not always possible. I consider the distance I will fall if I make a mistake—hundreds of meters down a sheer slope—and my head begins to spin. How foolish to spare my bruised ankles by wearing sneakers instead of hiking boots! I have already slipped several times on the rocks worn smooth by generations of alpinists. And it is easy to imagine pitching headlong down the slope and landing in the scree field far below. Behind me, an old man and his younger companions are waiting for us to continue. A familiar scene: we left the hut in the valley when they did this morning, but since Aleš and I set a fast pace through the trees and shade, every hour we have to rest and wring the sweat out of our shirts, at which point the old man, a steady, graceful climber, catches up with us. The tortoise and the hare, Aleš says to him by way of asking how often he has climbed this mountain. This is his first attempt, as it turns out: he had no interest in climbing Triglav when it was the tallest peak in Yugoslavia.

In the last decade, climbing Triglav has become a national enterprise for Slovenians. Groups are formed—one hundred women, one hundred plumbers—to mount regular assaults on the peak, and the attending publicity strengthens the claim that in the republic on "the sunny side of the Alps," where the tourist bureau places Slovenia, there is a mountaineering people distinct from the rest of the South Slavs. You're an aborted Slovenian if you haven't climbed Triglav, Aleš said on the ride here. But I am tired of things Slovenian. The bus let us off yesterday afternoon near a hay harp, a long narrow structure with a shingle roof, upright posts and, running parallel to the ground, wooden rails on which hay was drying: a distinctive feature of the Slovenian landscape. Then, walking through the village of Mojstrana, we passed a building with a display on its façade, thirty small paintings depicting beekeeping, yet another national activity. It was a relief to enter Slovenia's first, and only, national park. We walked for ten kilometers along a tributary of the Sava River, green with glacial melt, and before the sun set we stopped to drink the clear, cold water. At last we arrived at a hut named after Jakob Aljaž, the Catholic priest who in 1885 had bought Triglar's Summit for one forint (two cents) and erected a tower, determined to reverse the Germanization of the Julian Alps. In the dining room, a filmmaker completing a series of documentaries on Slovenia's mountains told us the president would be strapped into a rescue litter, then a helicopter would drag him, like a salami, up Triglav's dangerous north face. Anything to get reelected, the filmmaker quipped. Kučan is indeed one of the Balkans' most adept politicians, a reform-minded Communist, slow to rally to the secessionist cause, who nevertheless spearheaded his republic's independence drive. The secret deal he supposedly cut with Milošević—that Slovenia could leave Yugoslavia, with token resistance from the JNA, as long as it did not oppose the strongman's plans for Greater Serbia—is what remains unspoken in every conversation about Kučan. Slovenians admire their wily politician for waiting until the right moment to embrace the nationalist agenda, for guiding them through their war, even for deciding to skip this weekend's London Conference on the War in the Balkans to climb Triglav. A clever man, Aleš called him this morning when we left the hut. But when I asked him to explain why, he gestured instead toward a monument, in the form of a piton, to fallen alpinists and Partisans. You can still find trenches, barracks, foxholes, he said. If you dig a little, you'll even find bullets and missiles.

A land rich in human suffering—the site of peasant revolts, military campaigns between the Venetian republic and the Habsburgs, extensive fighting in World War One; Ernest Hemingway even saw action in these mountains. In the interwar period Triglav was a symbol of division for Slovenians in the Treaty of Rapallo (1920) the border between Italy and Yugoslavia ran through the summit, and the whole of Istria came under Italian rule. In World War Two the former Rapallo border marked the dividing line between the German and Italian invasions of Slovenia. Partisan units assembled in these mountains, including the Prešeren and Cankar brigades; because the fiercest battles were waged here in the struggle for Slovenia's liberation, Triglav became the nation's symbol of rebellion. (The Partisan caps were called *triglavka*.) And now the peak is Slovenia's symbol of independence.

A herd of sheep in the distance, their bells clanking in counterpoint to the sound of an accordion up the trail. Now *they're* crazy, Aleš says of the party ahead of us, the college students with guitars strapped to their backpacks who sing along with the accordionist. Out of the shade we climb into intense sunlight, which burns my arms, legs, and neck. The cliffs are blinding in their whiteness; above the timberline we enter a stark world of rock. Never have I felt so exposed, so vulnerable. I recall what the Polish poet Mieczyslaw Jastrun wrote during World War Two: "And far more hostile, more indifferent/ Than all that common and inhuman grave/ Is the beauty of the earth." Here is a single pasque flower surrounded by boulders, here a form of hawk's-beard found only on Triglav, and here a fissure lined with bellflowers. But that is all. I grip the cable until my hands are numb. I concentrate on my breathing, and at the sight of a large black spider weaving its web in a crevice I laugh out loud. God forbid if you should fall, says Aleš. But if you do, try to get your back to the cliff so your pack will break your fall a little and maybe slow you down. A lovely thought, I tell him.

By noon we have made it to the hut tucked in the saddle below the peak. My knees are bruised, my hands are bleeding. Helicopters are delivering supplies, clouds are building on the horizon, and the trail to the summit is a long line of climbers. Every table in the dining room is full of families drinking beer, and when we find places to sit I am too nervous to eat. Not Aleš. He is dipping a rind of black bread in his soup when I doze off, only to wake with a start from a dream that leaves me with an ominous feeling. Time to go, says Aleš.

Outside, climbers are sitting at picnic tables, smoking and watching the helicopters. A storm is coming in; hot as it is in the sun, by nightfall it might be snowing. A family emerges from the chapel next to the hut. Two little boys eagerly attach to their belts a rope wrapped around their father's waist and follow him up the trail, their mother close behind. They insisted we take them to the top, she beams. From the chapel, which has just been rebuilt, comes another family. This is how the battle escalates between the government and the Church, according to Aleš. Some people say the president is climbing Triglav to spite the Church. The clergy won't climb this mountain anymore. Two Italians from Trieste, equipped with expensive climbing gear, including ropes, crampons, ice axes, and helmets, watch us adjust our backpacks. We didn't know what to bring, one says sheepishly, staring at my sneakers. They don't need all that stuff, Aleš sneers when they leave.

What I would give to have their ropes! On the trail to the summit all that keeps me from a thousand-meter fall are spikes driven into the polished rock and, here and there, a short cable. I lose my balance time and again, slipping on the slick rock. Below us is the only glacier in the former Yugoslavia, and I am afraid I will end up buried in it; the trail is lined with plaques commemorating fallen alpinists. Victims of the mountains, Aleš intones. Two hundred meters from the peak, at Little Triglav, I give up. While Aleš scrambles up the last steep section, I rest beside a monument to Valentin Vodnik, a poet, priest, and naturalist from a nearby village. Scores of people amble by, including the old man climbing Triglav for the first time. He asks with a wicked smile if Vodnik's plaque is for me.

I gaze at the jagged peaks on the horizon, thinking of Goran Janković, the Bosnian Serb I met at the *Sarajevo* exhibition. For three years, while he worked as a journalist and published fiction in literary journals, his wife wanted him to move their family away from Bosnia, fearing the rising tide of Serbian nationalism. Only when he saw the Jews packing their bags did he make his escape plans. When they leave, he said, you know it's time to go. Like all males between the ages of fifteen and fifty, however, Janković was subject to conscription and thus forbidden to leave the country. So he sent his wife and three-year-old daughter to Croatia. A Croatian friend fighting with a Croatian Defense Council (HVO) unit allied with the Bosnians offered to drive him to safety, and Janković, armed with forged documents which gave him a Muslim name, left everything behind. The Serbs own my house now, he said. How could a house in Bosnian territory fall into Serbian hands? I said. He shrugged.

On the black market they bought five liters of gasoline, at ten times the normal price, then drove to another city closer to the front, where the price doubled again. A firefight broke out near the gas station—between Bosnian and Croatian forces, to the surprise of three hundred Bosnians and Croats crowding into the same shelter—and then Janković and his friend pushed on, taking turns driving through the Bosnian and Croatian checkpoints. At the border he paid a huge sum to leave the country, bribing guards from different armies at each of thirty checkpoints in the space of only ten kilometers. Everyone was collecting tolls. Those without enough money were sent back to Bosnia. The two men reached the Croatian town where Janković's wife and daughter were staying. He brought them to Zagreb and from there he traveled on alone to northeastern Croatia. One night, on foot, he stole across the high plains along the Slovenian border. By the next afternoon he was in Ljubljana, where his family joined him. I will never forget, he said, that I, a Serb, was rescued by Croats.

Scores of climbers pass through Little Triglav, eyeing me with suspicion. I try not to feel foolish about sitting on the side of the narrow trail. Multiply Janković's story by hundreds of thousands: these are the dimensions of the refugee crisis of the Third Balkan War, which has generated more than three million refugees and displaced people. And those who cannot leave? Or decide not to go? Why do some stay when others flee? Janković is a young man who might have joined the makeshift Bosnian Army. At the same age, in World War Two, the poet René Char was a leader of the French Resistance, at work on *Leaves of Hypnos*, a remarkable collection of journal entries and aphorisms. "I write briefly," Char confessed. "I can scarcely *be absent* for long." In the mountains of southern France he discovered that "Lucidity is the wound nearest the sun," recognizing the intimate connection between light and loss. One entry concerned the execution of a friend, which Char could have prevented, except that sparing his friend would have meant the destruction of an entire village. That day "The June sun slipped a polar chill into my bones," said the poet. And Goran Janković? If he told the story of his escape with verve, he also betrayed a certain faint-heartedness I recognized. I suspected he would tell his story often in the coming years (he claimed to be writing a novel), always in the feverish tone of someone hoping for absolution.

I hear a helicopter in the distance. (The joke: Slovenia's air space is too small for an air force. There is not a single private helicopter in the country.) In London, at the Conference on the War in the Balkans, diplomats from the UN, the United States, and the EC are making resolutions: a no-fly zone over Bosnia, tighter sanctions on the Serbs, Lord David Owen to replace Lord Carrington as EC negotiator. British Prime Minister John Major believes the pledges he has secured from the Serbian leadership to lift the siege of Sarajevo and let UN peacekeepers monitor their heavy weapons are sincere, despite this weekend's Serbian shelling of the National Library in Sarajevo. Fire breaks out—the gunners use phosphorous shells—and more than a million books burn, including 155,000 rare texts and manuscripts. By the end of the London Conference Bosnia's historical records will have gone up in flames. Meantime, Helsinki Watch is publishing a report with plenty of evidence to suggest that the very men John Major thinks he can trust—Slobodan Milošević, Radovan Karadžić, and General Ratko Mladić, commander of the Bosnian Serb forces—should be investigated as war criminals.

Aleš returns before long, buoyed by his ascent to the peak, where the college students with the guitar sold beer and the old man smoked a cigarette. I regret missing such an intimate party—only four people can stand on the summit at once—but not for long: the hike down is just as terrifying as the climb. There is nothing quite like hanging from a spike anchored in a cliff, hundreds of meters from the ground, while an overweight climber shuffles by—between you and the rock face.

Late in the afternoon, arriving at a hut nestled between two ridges, we go directly to the crowded bar and sit at a table with two middle-aged Croatian couples. Their first mountain excursion has left them in worse shape than me. Chain-smoking, downing fever pills with each round of beers, they press their wrists to their foreheads, wondering why they are still sweating. They came to the mountains because the hotels on the coast are filled with refugees, and they do not like paying the foreigner rate (three times what Slovenians are charged) for their stay in this hut. *Mountain Croats*—that is what some Croats call Slovenians. Denying the national and religious identities of your neighbors is indeed a favorite Balkan pastime. Croatian President Franjo Tuđman says Bosnian Muslims are Croats who converted to Islam, Milošević calls them Serbs. How else are they to justify carving a "Greater Croatia" or a "Greater Serbia" out of Yugoslavia? The Croats act like heirs contesting a will in Bosnia and along our borders, a professor in Ljubljana told me. They're blackmailing us, because they're still at war, their economy's wrecked, and they believe their media, which is as one-sided as Serbia's. No doubt the Croats at our table will tell their countrymen to let Slovenia keep its mountains. Aleš repeats a popular story about a doctor diagnosed with terminal cancer, who took up mountain climbing in the hope that the adrenaline rush he would experience hanging from a cliff would cure him—and it did. Even his arthritis and migraines vanished. The Croats are not impressed.

The college students with the guitars sit by the wood stove, singing folk songs—John Denver's "Country Roads" in Slovenian, Bob Dylan's "Knockin' on Heaven's Door" in English. They are not musicians, but no one seems to mind. We could be in a roadhouse, except that we are miles and mountain passes away from the nearest road. The old man toasts Aleš. People climb over tables to share their forks and Swiss Army knives. I am still too tense to eat; and when Aleš and I take the last two bunks in the common room, I sleep fitfully, dreaming I am clutching a cable—and falling.

In the morning clouds roll in, rain falls in spurts, fog covers Triglav's summit. You see the power of prayer? Aleš jokes. The Church will find a way to keep the president off the mountain! We walk through scree and gravel up a steep pass and over a barren stretch of ground that resembles a lunar landscape, then down into the valley of the Seven Triglav Lakes. On the ridge to our left are two chamois, and here, amid clumps of grass and wildflowers, is a kidney-shaped pond. This is where the alpine world meets the Karst: Slovenia's dividing line. Vienna lies in one direction, Venice in the other. The deepest cave in the former Yugoslavia is in this park; in another cave are the headwaters of its longest river, the Sava, named for Serbia's patron saint, an old shepherd who cared for wolves, Serbia's totem animal. We could follow the Sava south to the border between Croatia and Bosnia then east to Belgrade, where it flows into the Danube, creating what Vasko Popa called "the marriage/ Of the fourth river of Paradise/ With the thirty-sixth river of Earth."

The sun comes out, and soon we are sweating again. We hike past rock walls scored by runnels of rain and snowmelt, past lakes and springs that dried up this summer, then down through greening valleys. By early afternoon we are in thick forest, where we descend 1,500 meters in no time, gripping cables and spikes. At the top of a waterfall Prešeren celebrated in a poem, Aleš recalls a literary squabble, in the 1980s, sparked by a Serbian proposal to change the national curriculum. When writers and educators in Belgrade proposed amending the reading list to contain 50 percent Serbian writers and 10 percent apiece of writers from each of the other republics, the Slovenian literati erupted. How could Prešeren's most important epic poem be removed from Yugoslavia's canon? they cried. Strange to say, it may be argued that in some respects the writers' protest, which galvanized public opinion, culminated in war and, as Aleš wrote, "the birth of a Slovenian nation-state from the spirit of poetry." Now the Serbs think of themselves as England when she lost her colonies, says Aleš, gazing at the waterfall which, thanks to the drought, is flowing at only half its usual rate.

Down the last steep section of the trail we careen, grabbing cables, slipping and sliding on the gravel. At the base of the waterfall is an outdoor café, where old men are singing folk songs and drinking schnapps, preparing to hike up the trail we have just come down. It turns out that Kučan climbed all the way to the "Summit on the Summit." After all, says one old man, offering me a glass, he has an election coming up! I sip the schnapps. If you drink it fast, he says, you won't get drunk. I take him at his word, draining three glasses before I realize he is lying.

A short hike to Lake Bohinj, the largest body of water in Slovenia and the centerpiece of the region Peter Handke calls a "separate European country" in *Repetition.* "Difficult of access," he writes, "the Bohinj basin has been remote from the world down through the ages," a good description of the land into which my friend and I drag ourselves, a narrow stretch of pastures and dairy farms tucked under the mountains. Handke's recent statements about Slovenia trouble me. "I'm so disappointed Triglav is now on the Slovenian flag," he told an interviewer. "A beautiful part of nature has been misused for nationalistic purposes." Handke has forgotten that dreams, including the dream of national identity, are constructed of such stuff as bald eagles, wolves, and mountain peaks. Seeing only the nightmare of nationalism, he thinks the Slovenians are stylizing themselves as an alpine people. "Suddenly they want to be as dumb as Austrians with their 'Alpine Republic,'" he said. Yet in *Repetition*, published only five years before the Slovenian Spring, the narrator, Filip Kobal, is unreserved in his praise for the alpine republic: "How could I help wanting to count myself among this unknown people that has none but borrowed words for war, authority, and triumphal processions, but devises names for the humblest things—indoors for the space under the windowsill, out of doors for the shiny trace of a braked wagon wheel on a stone flag—and is at its most creative when it comes to naming hiding places, places for refuge and survival such as only children think up—nests in the underbrush, the cave behind the cave, the fertile field deep in the woods—yet never feels obliged to call itself 'the chosen people' and distance itself from 'the nations' (for, as their every word shows, this people inhabits and cultivates its land)?"

Handke's change of heart, Aleš believes, was born of the thinking he himself encountered earlier this week, in an interview with the cultural affairs editor of a German newspaper. Does a small country like Slovenia have the right to exist, she asked him, with 3,000 would-be nations around the world clamoring for the same status? Aleš quoted from Milan Kundera's essay, "The Tragedy of Central Europe," declaring that in this century the great works of art created in the small nations of Central Europe "can be understood as long meditations on the possible end of humanity." Small countries know they can disappear at any time; hence the importance they attach to their cultures, sometimes the only space in which to preserve themselves. While in Western Europe culture has given way to politics and the media, a measure of humanity resides in the small Central European states. No wonder Kundera sided with Slovenia during the war. "I am personally very close to the patriotism of Slovenians because it has never been based on militarism or political parties, but rather on their culture, particularly literature," he wrote in a letter published in *Le Monde*. Praising Prešeren and Slovenia's desire to join the West, he reminded his readers that at the Munich Conference British Prime Minister Neville Chamberlain had asked why the Western powers should care about Czechoslovakia, "a little-known and far-away land." Appeasement had not stopped Hitler. Why should it be any different with Milošević? In a small country we need to guard our identities, which is not true in large countries, Aleš told the editor. And I say that not as your typical parochial nationalist who hasn't been abroad.

By the edge of the lake two drunken old men weave past us, wearing identical black-and-red Guns N' Roses T-shirts, which read "Appetite for Destruction." Do you think they know what they're advertising? I say. Aleš shakes his head. They just know it's from America.

It is too late to catch a bus to Ljubljana, so we walk to the tourist office to book a room in a private house. All at once the weather changes. The sky turns black, rain falls in sheets, lightning flashes. There is a mysterious saying—*Bohinj has rain like little ones*—I now but dimly understand. We are huddling outside the tourist office, waiting for the rain to stop, when a British psychologist strikes up a conversation. He has taken leave to live in Budapest, where he is organizing a conference on nationalities and the war in the Balkans, the failure of a love affair having convinced him to return to the home of his Hungarian ancestors.

"It was a relationship that had to end," he begins.

"How do you mean?" I say. "You went your different ways?"

"The war," he corrects me. "It's all situational ethics."

"That's moral relativism," Aleš mutters.

"You can't cover all your bases," the psychologist replies.

"That's why your profession does so well," says Aleš.

"Doesn't Slovenia have something to answer for?" says the psychologist.

"For Serbs bombing us?" says Aleš.

"There wouldn't be a war in Bosnia if you hadn't seceded."

"Did we invent Milošević? Or Karadžić?"

"People get the leaders they deserve."

"Did the Russians deserve Stalin? After all the cultural glories they gave us in the nineteenth century—after the novels of Dostoyevsky and Tolstoy, they deserved seventy years of terror? Is that what you're saying?"

The psychologist takes a step back. "That's not what I mean."

"Then what *do* you mean?"

"Your president climbed Triglav today. Why would he do that?"

"What are you talking about?" Aleš demands.

The psychologist shakes his head. "People get the leaders they deserve, unless it's imposed by force."

Aleš storms off to the market to buy film for his camera. The psychologist and I stand there in silence until he returns. Out into the rain we walk, Aleš and I, and down the road to a private home in which to spend the night. The old woman who owns the house invites us into her kitchen for a snack—black bread and a bowl of *kislo mleko,* sour milk that has fermented into a kind of delicious yogurt. You're the right guys, she says, meaning we have come from Triglav. In the mountains, snow is falling. This year the climbing season will end earlier than usual.

5

Ljubljana II

The foreign minister, Dimitrij Rupel, was a novelist and sociology professor. Under the Communists the fervent secessionist had published nearly a score of books by the age of forty. But his output had dwindled lately because, as Aleš said before our meeting with him, "A writer's allowed to be ambiguous, not a politician." And his participation in the London Conference marked the summit of Rupel's political career. Seeking to distance Slovenia from the war in Bosnia, Kučan had sent the foreign minister in his stead, a diplomatic maneuver which impressed Slovenians.

What a relief it was to enter an air-conditioned building! I had forgotten about such amenities.

"We have to conduct business like the rest of the world," Aleš said outside Rupel's office.

While Aleš and the minister conferred about a joint research project at the university, I sat back in a comfortable chair. Two walls were covered with abstract paintings by contemporary Slovenian artists, a third with photographs of Rupel's family. The minister, a tall, husky man with a mustache and a deep voice, was pressed for time: he had to hurry home to supervise the installation of a new heating system. What is it like, I asked, to be a writer among professional politicians?

"I cannot dwell on moral dilemmas," he said, "if I want to do my work properly. For instance, just now the president and I were discussing police and military matters. I cannot keep my ambiguous, schizophrenic personality and get things done. I hope to go back to my previous life, but three years of constant excitement compensates for the lack of dimensions political life offers."

He lit a cigarette. "But I am writing a memoir on my political life, ending with the London Conference, in which I explain how I have overcome these moral dilemmas. For years I contemplated the idea of Slovenian nationhood. Now I'm in a position to test what I was preaching, and this is such a serious challenge that I cannot dwell on aesthetics. Yes, our defense minister has written a book, but mine will be more literary. I want to describe what it feels like to drive at night to a meeting on the future of our country—to give the texture of these times."

I asked him what room Slovenia would inhabit in the common house of Europe?

"We want to stay as detached as possible from Yugoslavia," he said, "though for the last several months we've had to cope with that link."

A chilling view. *Better last in the city than first in the village*, Slovenians said. The city was Europe, Yugoslavia was the village. And Slovenia? The country girl with bright lights in her eyes.

"We want to be one of the normal players in the EC," he said. "We haven't abandoned hope of making it in. Perhaps by the end of the century we will be a useful member of the community."

A small country has an important role to play in Europe, he insisted. "In London I suggested to Larry"—he gave me a meaningful look—"to Eagleburger—that we make a free-trade agreement. It might suit the U.S. to have a small partner in the heart of Europe. We're in a position to balance foreign influences; we cannot afford to make enemies. The Serbs made that mistake already."

I wondered if our new secretary of state would even remember this foreign minister's name. Not that it would matter for long: in the November elections Rupel's party would suffer heavy losses (like Eagleburger's) and he would have to return to teaching. The critics would pan his memoir.

"We do not want to be an island in Europe," Rupel was saying. "Americans may accuse us of being disintegrationalists with respect to Yugoslavia. But Yugoslavia will never be part of Europe. We want to integrate with Europe."

"I always *hear* the paintings I like," said Metka Krašovec.

She had returned to the capital for the fall term, and one afternoon she invited me over to look at her work. I had spent a disappointing morning at the Modern Gallery. Here was another short course in the schools of symbolism, expressionism, surrealism, abstract expressionism, new figurative art, and conceptualism. What struck me was modernism's unity, the speed with which artistic movements and vocabularies migrate around the West. In Central Europe, Milan Kundera suggests, the salient feature of the modern age—that culture replaced God as Europe's unifying principle—still holds, a fact borne out by this museum. In one room was a bad version of Braque, in another a fair copy of Motherwell; even the work from the last decade revealed the same confusion afflicting artists in New York and Paris, London and Cologne. I might have been in the modern museum of any Western capital, except that the clerks did not know how to sell the catalogues and books.

A single painting by Metka stood out, a fiery red church whose haunting shadows evoked the early work of Giorgio de Chirico. It was as if she had taken the ideas of the Italian master to a new conclusion, a nightmarish version of History titled "Silence," in which solitude was the protagonist. Here was a church no one would enter, I said, or leave. Metka smiled.

"I'm an outsider everywhere," she said in a living room crowded with books and paintings.

Tomaž Šalamun emerged from the kitchen with orange juice and a bottle of wine.

"In a way, you're a loner in your language," he said.

"The mainstream tolerates me," she said, "especially because I'm a woman. But they don't take me seriously, which is liberating. Perhaps it would be easier if I were an exhibitionist: Frida Kahlo, for example. That's expected of women."

Tomaž said, "Maybe we should wait on the wine?"

"No," Metka giggled.

"Never," I said.

"Okay, okay," he said, and poured.

In Metka's studio, a large, airy room, she showed me her playful sketches, surreal drawings, and the remarkable portraits of women that had occupied her for the last several years, a series of large canvases titled "Presences." In a typical painting she placed against a sky-blue background two women with short curly hair, one white, one black, naked from the waist up. These were other-worldly landscapes of desire in which one woman's eyes were usually open, the other's closed. Sometimes shards from a broken mirror were embedded in the sky, like stars, and often in the distance there were small snow-bound mountains.

"My drawings are spontaneous," Metka said, "but a painting, like a fresco, must be planned."

"Do the drawings inspire the paintings?" I said.

"Not directly. There was a crisis, and I started drawing. For two years I didn't show them to anyone. They frightened me. As far back as 1985 I was drawing scenes from the war. The paintings *come out* of the drawings. I always do something precise and something loose, alternating between constriction and expansion."

"The old division between Vienna and Venice," I said.

"I suppose," she said. "My landscape is Mediterranean, but I live in Ljubljana and Bled."

In the afterword to a book of her drawings, Metka wrote: "What I am looking for is the right form to express the inner sound I can hear and the light I can sense. Sound, light and colour."

"What do you mean by that?" I asked.

"What I want in my paintings is a certain sound I hear in my imagination," she said. "And it changes. You have to find people akin to your own notion of madness. In one period of your life you look at certain painters, then drop them. Some stay, like Mondrian, Piero della Francesca, Sienese painting, Matisse. It would have been easier for me if I had found women artists as role models."

The natural light streaming in through the windows was fading. Metka had to catch a bus to Bled, where she would spend the weekend working, and Tomaž was taking me to dinner, where we vowed not to drink too much. "I don't believe you," Metka smiled. She kissed Tomaž goodbye, then said to me, "I didn't like Bled until I discovered the lake and the sky above it, which gives the water a silvery, pinkish, salmon light. I like that."

The archbishop's return to Ljubljana, after more than thirty years of exile in Switzerland, was a signal political event in the Slovenian Spring. There was a special effects photograph taken of him the night independence was proclaimed, a double exposure in which two images of a priest reading a prayer, eyes fixed on Heaven, stand next to two images of the archbishop staring down at the photographer, a heavy-handed comment on the power the Church had regained. Old and frail, Monsignor Alojzij Šuštar was a skillful diplomat whose savvy impressed the faithful—and complicated the political situation. The separation of church and state, a Slovenian told me, doesn't go down our throats very well.

I visited the archbishop on a cold, rainy morning. His hands shook, but his mind was sharp. Equal rights for even the smallest nation was his point of departure.

"The Church worked for independence," he said in a quaking voice, "and when we were attacked by the JNA the Vatican was the first to recognize Slovenia."

He reminded me of the Church's long ties to the people. "But we distinguish between the country's politics and spiritual mission. We drew attention to the dangers of chauvinism and nationalism, because we didn't want to be independent at the expense of others. Our priests issued warnings against nationalism, but we cannot be sure that excesses did not sometimes take place."

"Some think the Church has too much influence on Slovenian politics," I said.

"Unfortunately, the waters have not cleared yet. But I want to emphasize: we are not connected to any political parties. We support the freedom of Catholics to choose their parties."

"And other religions?"

"We lobby for all religions," he began. "The Church wants to stress the principle of equality and the right to be different." All at once his voice rose in anger. "But due to the ignorance of our people, there is much spiritual confusion. We keep telling our people that, regardless of how much money the Mormons throw around, they must stay faithful. The Mormons and the Jehovah's Witnesses never sought out any discussions with the Church."

I wish I had known then about the infamous post-World War Two trial in Zagreb at which a number of Jehovah's Witnesses, charged with the crime of pacifism, had been sentenced to death. It was not until much later that I learned of the collusion between the Communist authorities and the Church, which had set the trial in motion; both had compelling reasons to rid Yugoslavia of people who, in the words of the official report of the trial, "called themselves faithful servants of Christ, to whom earthly life was of no concern." But because I did not know any of this I made a joke instead about having lived among the Mormons in Salt Lake City: the worst four years of my life, I said.

The archbishop smiled. "The Church is not a beacon for those in the void left by socialism," he said, recovering his composure. "They're looking for answers elsewhere. But we're painting a better picture of spiritual community."

"Dimitrij Rupel was a dissident, yes," the drama critic snorted, "but he was also a professor. The Communists were smart: they let the artists change the system, and they stayed in power."

Uršula Cetinski was large and square-jawed, with short red hair and a blunt speaking style. In the bar of the Hotel Slon, she did not hide her distaste for the new Slovenia, reserving the greatest scorn for the writers. "They never had it so good as they did under socialism," she said.

I raised an eyebrow.

"The state was very clever," she explained. "It knew the people needed an outlet, so it let dissident voices be heard, especially in the theater. In our political theater the ethical function was most important; now it's the artistic function that dominates. But who cares? Our theater's in trouble, because it's connected to the economy. The state doesn't need theater anymore."

The bar was filled with young businessmen in expensive suits. The street outside, Čopova, a pedestrian zone running from the post office to the river, was packed with shoppers. Originally known as Elephant Street—Archduke Maximilian once spent a night in Ljubljana, and the elephant he was bringing back from Spain was kept where the Hotel Slon now stands—Čopova was for these businessmen the most hopeful sign of Slovenia's future.

"We thought we were making something dangerous in the theater, but we were so naive," said Uršula. "The politicians stayed in their flats and laughed. They gave artists the feeling of freedom, but only within limits.

"This new society can't break with the Communists because it needs them. They didn't go to church before, now they do. Once we had a culture, now we have militarism. Our defense minister has written his own *Mein Kampf*. But war for Slovenia was like a children's game."

She gazed at the businessmen. "Sometimes I think every Slovenian should have his own political party," she sneered. "This is such a small place, yet there are so many different ideas. But without the Communists, nothing would have happened."

At the Opera Bar, a favorite haunt for writers, I had lunch with Boris Novak, the president of the Slovenian Writers' Association. Embarrassed by my drunken encounter with him at KUD, I was reluctant to order a drink, but I overcame my scruples as soon as Boris called for beers. He was a tall, bearded man with glasses and brown hair, a former dissident whose formal poems were viewed as innovations. That "radical" style suited his hunger for sonic values and a deepening perspective.

I had come to feel that my question to him at the reading in protest of the war in Bosnia—"What ethical system will Slovenians find to replace socialism?"—was not entirely frivolous. I recalled Václav Havel's remark that a billion people in the Eastern Bloc had subscribed to Marxism, a narrative that for better or worse gave meaning and purpose to their lives. What story would replace it? What forms and rituals—spiritual, aesthetic, political—would serve them now?

Central Europeans were looking to the West for answers, and what they discovered was a brief for free market economics, which ignored the great mysteries of existence. Octavio Paz writes that the market "knows all about prices but nothing about values." Yet the stories we live by depend on values. Perhaps the absence of a suitable narrative after the Cold War allowed tribalism to flourish. "Mirror of the fraternity of the cosmos, the poem is a model of what human society might be," the Nobel laureate suggests. A very different sort of poem was being written in the Balkans.

"Writers still have a moral obligation," said Boris. "They're expected to speak out on social and political issues. Slovenians couldn't fight with arms, so writers led the way to democratization."

In the independence drive Boris had made his mark editing *Nova revija* (*New Review*), a literary journal founded with the tacit support of the Slovene Communist Party. It was the first salvo fired in the name of democracy in post-Tito Yugoslavia; when the inaugural issue appeared in 1982, it was regarded (especially among younger Communists) as a forum for issues hitherto impossible to raise in public, including respect for human rights and the idea of a civic society. The federal structure needed to be reevaluated, argued Boris and his colleagues, among them Dimitrij Rupel. So in 1987 they produced a special issue, "Contributions to the Slovenian National Program," laying the groundwork for democracy, parliamentary elections, and independence; in 1988 they created "the writer's constitution," which provided the conceptual framework for Slovenia's first constitution;

in 1989 they published a debate among Slovenian and Serbian writers about the strengths and failings of the Yugoslav constitution; in 1990 they came out with "Independent Slovenia"; and during the Ten-Day War they turned their editorial offices into a clearinghouse for information, rallying support in the international community for their battle against "JNA aggression."

"Yes, sometimes the writer is invisible," said Boris. "The newborn politicians who were nowhere to be seen during the dissident period are now speaking the loudest. Thus literature is becoming marginalized. But there are still problems for writers to confront. Even before the war I knew we were only at the beginning of a long and difficult journey toward democracy."

And the writer's place in this new country? I asked.

"Writers should stay outside politics. Only an independent position allows you to maintain your critical distance and moral authority. I never attached myself to a particular ideology. Once the regime changed, I felt I was in the opposition again."

He remembered a golden period under socialism, when everyone with good intentions was against the government. Unfortunately, now that they needed to present something positive, their energies were dispersed. "Of course it was easier when we were united against Communism. Yet even then people criticized us for not being radical enough, people who did nothing themselves. Many groups and politicians pretend to have fought against Communism. Where were they during the war?"

"Were the writers tools of the government?" I said.

Boris glared at me. "Those who say that are cowards," he said, controlling his anger. "Career hunters for the Communists. For example, in 1987, after we published an article suggesting the West back a military coup, the political police called me in. I put on my suit for the interview. My father, a Communist who was nevertheless imprisoned twenty-two times, told me to take it off. 'You don't wear it every day,' he said. 'You'll feel awkward, you'll give them a psychological advantage.' And he was right. They didn't beat me, but they threatened me with three years in prison—six months before the fall of the one-party system.

"Then I saw one of the policemen right here in the Opera Bar. He had the icy look of a hunter stalking his prey. I realized it was not us who won: those people are eternal."

One afternoon Aleš and I visited a publishing house founded on the demise of a Marxist theoretical center. A slender blonde woman dressed in black leather was editing a manuscript by a presidential advisor who advocated shock therapy for the economy, tempered by state support—a system to replace workers' self-management, the Yugoslav variation on traditional socialist economic practice, invented by Edvard Kardelj. A Slovenian by birth, a schoolmaster by training, a true believer by temperament, Kardelj was by turns a Partisan, Tito's Marxist theoretician, and Yugoslavia's foreign minister. After World War Two, Kardelj and the Montenegrin writer Milovan Djilas were the intellectual linchpins of the new regime. But unlike Djilas, who grew disillusioned with Yugoslavia's political structure and in 1954 was expelled from the Communist Party, Kardelj remained Tito's heir apparent until he died in 1979. Ironically, his two greatest achievements came to nothing. The 1974 Constitution satisfied no one, least of all the Serbs, and by the 1980s his "self-managed, non-aligned" policy had spawned a new slogan to describe Yugoslavia: "self-aligned, non-managed."

Nor was this editor impressed by the new economic model proposed by Kučan's advisor. A boring book, she said. What about the defense minister's memoirs? Aleš said. She made a face.

"But he's our first real soldier," said Aleš.

"I do not like the promotion of these values," she said, lighting a cigarette. "This nationalism."

But it was Hegel himself who first examined the ways in which war creates nations. That his most famous disciple did not grasp the power of nationalism was one reason why this publishing house was now devoted to Slovenian topics, a change rooted in military action. "Could a Nation, in any real sense of the word, really be born without a war?" writes historian Michael Howard. The answer in Slovenia was no. But independence had its costs. This new house published its books in editions of fewer than a thousand. I asked the editor if the venture was state-supported.

"No," she said emphatically, shaking her head. "Private."

"Then how do you raise capital," I said, "if your print runs are so small?"

"From the ministries of information or culture," she replied. "They pay for the books they think will have national importance."

"Is it fair to say that most of your money comes from them?"

"Oh, yes," she said.

She had to edit the last two pages of the economist's manuscript, so she gave Aleš a goodbye kiss. She started to kiss me, then thought better of it, extending her hand instead.

"Why no kiss?" I asked.

"Because you're American," she said.

I looked at her, perplexed.

"AIDS," she explained.

"But I've been married for ten years," I protested.

"It doesn't matter," she said, and wandered off with Aleš.

Harris Burina, the comic actor from Sarajevo, was the proverbial man without a country. He had made his reputation in Sarajevo, performing in more than a dozen films, earning awards, acclaim, and enough money to live in style. But the crowning moment of his career in Yugoslavia—he and ten other Bosnian actors had spent the year before the war working in Belgrade, where his last role had been in a Serbo-Croatian production of Lanford Wilson's *Burn This*—seemed to have cost him his future in Bosnia: he believed his countrymen would always view him with suspicion.

When I met Harris and his wife, a quiet, blonde beauty, they were waiting for French visas; with luck they would fly to Paris, where he hoped to rejoin his theater troupe. They had no time to lose: they had only a thousand marks and one-week Slovenian visas.

"I played the golden years in Sarajevo," he told me one afternoon. "Now it is finished."

He was suffering from a severe hangover—he and his wife had drunk seven bottles of wine since lunch the day before—so he was alternating glasses of water with mugs of beer, chain-smoking at Drama, the popular bar. He was a slim man with black hair, thick black eyebrows, and an earring; his colorful shirt was the brightest thing in the dimly-lit room. His wife was talking with a Slovenian screenwriter. Stojan was their guide in Ljubljana, and it pained him that he could help Harris but not his own cousin who was still trapped in Sarajevo.

A hard rain was falling outside. Harris ordered another round of drinks.

"For that year in Belgrade I was a superstar," he said. "My job was just theater, film, and giving interviews. The best ever. Now they're killing us. Serb terrorists came to my home in Belgrade and told me to leave the country. It was a *theater* director," he cried, "waving his gun at me. He belonged to the White Eagles, the worst fanatics.

"I can't work for my people either, because I played in Belgrade. So I'm living every day in pictures. It's just memories, especially since my parents and friends are still there."

Despite his hangover and despair, he had not lost his sense of irony. Fleeing to Macedonia, he had made it across the border only because the guard, recognizing him from his films, had looked the other way. (Much the same thing had happened in Slovenia, according to Stojan. "Let Burina through," the border guard was told over the phone.) In Skopje, Harris took a job as a bartender in a pub, where the patrons kept looking for cameras: this must be a movie set, they thought!

That story reminded Harris of a friend who went to America with 20,000 worthless Yugoslav banknotes, each bearing a picture of Nikola Tesla, the Croatian electrician whose inventions included systems of arc lighting and wireless communication. Tesla emigrated to the United States in 1884, and in America he is remembered for constructing the first power station at Niagara Falls—a fact Harris's friend turned to his advantage. He set up a stand by Tesla's monument in the state park at Niagara Falls and sold his suitcase full of dinars to honeymooning couples for five dollars apiece. In three days he made $80,000. His only regret was that he had not brought a million dinars.

Harris told these stories with glee, gesturing wildly, now smoothing back his hair, now lighting another cigarette. Then his mood changed. "Maybe I can never return to Bosnia," he sighed. "I was everywhere, and now it's finished. People in New York and Paris and London gave me chances before, but I always said I must stay in my country. For ten years I was one of Bosnia's best cultural ambassadors to the world. 'I'm a Sarajevo actor,' I used to say. Now I just want to leave. Belgrade, Skopje, Ljubljana, they're no longer mine. And I played for Yugoslavia.

"Everything comes in phases. Now I'm in the second half of my life. What was before is over. I lived for five years in a hotel in Sarajevo, every day as if it were my last. I never knew if I would wake up in the morning. I was rich, and I didn't buy a house. I know ten millionaires with big houses, and they left with shopping bags. Now we're the same. I'm twenty-nine years old. I'm lucky I played half my life as an actor. Maybe it's all over. Or maybe I'll be good in Paris. Anything's possible. I have to learn French, then English. I'm living so fast." He took a drink. "In Belgrade they said I'm better than John Malkovich. It's true. I have sparks, hair, and energy, like the space shuttle!"

Perhaps only in the Balkans could a comparison be made between this comic actor and the demonic star of *Dangerous Liaisons*. But I said nothing, unlike his wife, who kept interrupting him to ask questions. His replies grew abrupt. All he had left, he said, were his memories and his wife.

"If you don't watch your step," I said, "you may end up with only memories."

Harris stared off into the distance. He took a sip of water, drained his beer, and ordered more drinks. "What can I do? I must help my parents," he said suddenly. "They have no food. You can't imagine. I called my father and said, 'I'm so upset I can't see you.' He said, 'Don't be upset. This is my second war. Everything is destiny.' And me? What I did in Sarajevo is history."

He lit yet another cigarette. "Do you know there are so many refugees here that every night at the disco is Sarajevo night? And you have the same divisions here you have in Bosnia."

It was Nina Zagorničnik on the telephone.

"I need you," she said breathlessly. "Hurry, please."

I rushed to the bus stop and promptly missed two buses in a row, because the man selling bus tokens in the kiosk was talking with a young woman. Finally she paid for her cigarettes and left. I bought a token, and waited twenty more minutes for a bus. Leaves had fallen overnight; autumn was in the crisp, clear air; the streets were crowded. It took a long time to get to the center of town, and then I jogged through the park by the law school. Nina had spread her notes across a table at an outdoor bar. She was enraged. What took you so long? When I tried to explain, she shook her head.

"Look," she said. "Read this translation. It is for CNN *World Report*."

I made some corrections in her text on education and values then followed her over to the law school, where we met Boštjan Zupančič. The law professor stood under the balcony from which writers, activists, and politicians had delivered speeches during the independence drive, and rehearsed his comments. He had forty seconds in which to answer the question, "The moral vacuum left by socialism: will the Church fill it?" He was unable to do that in less than four minutes. The camera crew Nina had lined up was nowhere to be seen. She was in despair.

"No matter what," she muttered, "I am a professional."

Like other parts of society, the media was changing. Private concerns were applying for radio and television frequencies, foreign investors were starting magazines, and the Italian newspaper *La Repubblica* was opening an office in Ljubljana. Journalists were at once excited by the opportunities and afraid of losing their security. Boštjan could have been talking about the media when he said the competition inspired by the opening of new private and Catholic schools worried many Slovenians.

"They don't know how to compete," he added when Nina looked at her watch. "They don't know how to perform up to the top of their abilities anymore."

Nina turned to me. "Do something, please," she cried.

"Why don't you try, 'We don't know how to compete anymore,'" I said.

Boštjan began to take notes.

I recalled a conversation with a Western diplomat, who thought Slovenians were living in a dream world. "If Slovenia wants to make it in the international community," he said, "it will have to act like a capitalist country, with an independent press that knows its responsibilities." Twenty-five hundred people were employed by the state radio and television station; the diplomat believed that to compete with private stations all but five hundred would have to be laid off.

Eventually the camera crew arrived in a Hertz Rent-a-Truck. Two tipsy middle-aged men ordered a young man to unload the equipment; the driver, a white-haired man, slept in the truck. The director stroked his grey goatee, his partner laughed and laughed, Nina shook her head.

Miraculously, Boštjan got his part right after only two takes, and then we moved on to the next setting, the university library designed by Plečnik, an impressive brick building from which jutted islands of black stones and windows resembling open books. Nina leaned her bicycle against a tree, and while we waited for the camera crew—they had stopped at another bar—she practiced her pronunciation of phrases like "gene pool" and "This is Nina Zagoričnik for CNN *World Report*." She brushed her hair. She looked at her watch. Unaccountably, she said, "I have lucky."

The camera crew finally arrived, set up a shot, then left for yet another bar. When we caught up with them, Nina argued with the director over atmospheric shots. She wanted footage from the castle overlooking the city, from the top of the city's tallest building, and from a park. Refusing to leave this part of town, he ordered another beer and promised to meet us in the Jewish Quarter in half an hour. By my calculation, the crew drank away two of the three hours they were supposed to be filming, then quit early. On their way back to the station they disappeared again.

In the station there was none of the frantic energy common to American newsrooms. No doors were locked, though microphones and tape recorders routinely vanished. Young men and women sprawled in chairs, smoking and chattering.

"No one knows how to work here," Nina muttered when the director, bleary-eyed and smiling, delivered the tape. Nina played it through twice, pleasantly surprised to find it in focus. She decided to celebrate by taking me to Drama, where we met her friend, Marko, the editor of a cultural affairs journal. A jealous man, he was not happy to see me. When she showed him my corrections of her text, he took issue with my decision to change "genetic pool" to "gene pool."

"It is 'genetic pool,'" he said sourly.

I shook my head. "Trust me," I said. "We say 'gene pool.'"

"You're wrong," he said.

"I don't believe this," I said.

"Then explain it to me," he persisted.

"What?" I said.

"Why you say 'gene pool'?" he said.

"We just do," I shrugged.

"No, you don't," he said.

"All right," I said. "'Genetic pool' is an adjectival phrase, 'gene pool' is a compound noun." I had no idea what I was talking about, but that did not keep me from explaining, at length, why we avoid "adjectival phrases" in English—an intricate and mysterious language, I assured Marko. "Does that satisfy you?" I said.

"You mean you don't have adjectives in English," he sneered.

I looked him in the eye. "No," I said with a straight face.

Nina tugged me away. "Don't mind him," she said, and clutched the tape. "This," she said, "will be for my grandchildren. Some day I can tell them I did something for CNN."

In the 1980s rock-and-roll was a powerful force for change in Yugoslavia, particularly in Slovenia. To learn what direction it might now be taking, I called on one of the most interesting new artists, Branko Mirt, the founder of Magirus, a group that made music with racing car engines. He and his girlfriend had just moved into a walk-up apartment, and they had not finished unpacking when I arrived one evening. Branko shoved some boxes off the couch and popped a tape of their last concert into the VCR. What appeared on the television screen was amazing.

The setting was a car engine factory in Maribor. Two men in red and yellow racing suits strode past a line of engines revved at different pitches, climbed up on a makeshift stage, waved large green flags, as if they were signaling the start of a Formula 1 race, and the concert began, with Branko on electric guitar, two men playing basses that looked like hub caps, and a drummer. The throbbing music was surprisingly melodic; the engines added a haunting dimension. The lyrics:

> We fill our hearts with gasoline
> Burning lava erupts from our throats
> Our blood's inflammable
> Our arms are a rotating turbine
>
> We turn the wheel
> On which you turn
> We hold the power
> We are the wild knights of maximum speed

"At our next concert," said Branko, a playful, wiry, dark-haired man, "we'll have tires burning and a stunt man going up in flames. We want the hall to smell like a petrol station!"

He punctuated his sentences with wild gestures. He was a classical guitarist by training, and the pedigree of Magirus, named after the engines the group "played," was equally eccentric: one flag-waver was an artist; the bassist had gone to law school. "Powered by Magirus" was the motto of the Tekton Motor Corporation—for technology, tectonics, and Teutonic people. Branko hoped to develop an engine fast enough to break through the horizon and "formalize its consciousness." Magirus's album-in-progress was called *Human Race Ignition*. And Tektonic music was the music of "explosive lungs, fired hearts, steeled minds, and motional spirits."

The Teutonic emphasis recalled Slovenia's best-known band from the 1980s, Laibach (the German name for Ljubljana). The band cultivated a neo-Nazi image, wearing brown shirts in concert and singing in German, partly to link Communism and Nazism, partly to bait German-fearing Serbs. How seriously Laibach took its Nazi trappings was hard to know. Less difficult to gauge was its influence, which was tectonic, increasing pressure on the cultural and political fault lines destined to reshape life in the Balkans. Impossible to imagine Magirus—or any new band, for that matter—acquiring that kind of power.

Branko switched off the tape. "We work only with computer samples," he said, "and video footage of Formula 1 races. At first we stole the footage from TV, but now Philip Morris has given us three tapes, which will be cleaner to reproduce. Racing cars purify our sound."

His influences ranged from Shostakovich to the sound track for *Star Wars*. "We listen to classical music to learn how composers work with orchestras and sounds, then have a pianist play a Liszt Variation on pipe organ. He wears makeup to look like Liszt, and I play along on engines. Then we do 'The Art of the Fugue' with a Ferrari on first violin: we want to translate some works directly into engines. Bach changed the direction of music when he wrote 'The Well-Tempered Clavier.' We do the same thing when we put an engine into his piece, paying homage to those breakthrough moments. People think engines are just noise. Not at all: we're teaching speed. Everything's in movement. We want to reach absolute speed. Ours is a battle with speed."

He started pacing rapidly around the small room. "We're taking information age ideas to their artistic conclusions. In our concert I'll have an electronic magnetic field attached to my helmet so the crowd can see my brain waves as we play. We want to put all our ideas on earth. And they must succeed on the philosophical level, so we're testing them through the engines, because they make music. Every idea has a motor, a body."

And gasoline engines were medieval. "Solar energy's next, and after that? Consciousness. But for now we use gasoline engines, because the world's based on them. And we're realists."

Indeed Magirus had a contract with a record company in England. Because the cost of putting on a concert with forty Formula 1 engines was prohibitive, Branko knew they had to sell enough records to catch the attention of MTV. Only then could they think about touring. Meanwhile, he was raising money; unwilling to take out a bank loan and unable to secure a grant from the ministry of culture—"We won't support their nationalist program," he said—he had gone to Philip Morris, which sponsored the Formula 1 racing circuit. Perhaps it would underwrite musicians, too.

At the same time, Branko saw his future in Japan. "Only the Japanese have the money and engines we need," he said. "If they like our ideas, maybe we can make a deal. They build ten to twenty Formula 1 engines to get one that works; the rest they throw away. Maybe we could use them. I have on tape all the Japanese factories: one fanatic did the whole thing, and that's what I'm trying to do. One fanatic can bring along a lot of people to realize his ideas.

"But it's hard to do that in Slovenia. We have to look outside. On the record our names will be in Japanese. We don't want anyone to know where we come from. We're working *against* the idea of nationalism. Ideas know no borders. Our idea is to build, with speed, for the human race."

Jana Schneider, the photojournalist, was describing the mortar attack in Sarajevo that maimed her and killed her Slovenian lover. "Twenty-two holes in my legs. Forty pieces of shrapnel in my body. Some brain damage. What can you expect after a stroke and two heart attacks in nine days?"

She was a passionate, outspoken woman. "Fuck this!" and "Fuck that!" were her favorite phrases; the businessmen near us in the bar of the Hotel Slon did not mask their indignation at her behavior. Not that she cared. "Slovenians don't give a damn about Sarajevo," she said loudly.

Schneider had appeared on TV, acted on Broadway, sung with Leonard Cohen. At the height of her fame she gave up acting to become a combat photojournalist, covering eighteen wars in the last four years. She said she did not miss the limelight.

"What I didn't realize was that the more money and fame you get," she said, "the more you think only about yourself. Now I'm focused on the world instead of having the world focused on me."

"But don't you ever get scared?" I said.

"If I have a fear," she said, "I go and find out what it's about. That's why I do war, which is not about death. It's about life. I made love to an Armenian commander on a sacrificial altar, and what I remember were the wolves howling outside."

The word among journalists was that she was too dangerous to travel with. And she had no faith in her colleagues. "Let's get a little hot at the press conferences in Sarajevo," she cried. "People are dying! But everybody has to be a nice guy. The journalists just swallow everything. They're afraid. And why not? The Serbs have been shooting journalists for a year, and the West has gone into denial, like a family hiding a murderer, just as it did before World War Two."

Everything I learned about her unsettled me. During the Gulf War, she was bitten and molested in a Baghdad prison, and she kept taking pictures. Her lover was killed by a shell Slovenians said was aimed at her, and she continued to live in Ljubljana, though she could not speak Slovenian.

"There's no reason I should be alive," she said softly. "In Vukovar I was hit by shrapnel, and I lost the twins I was carrying. Last night I heard a spy's heartbeat on my phone. When I said, 'What do you want to hear?', the heartbeat quickened! Slovenia's an eruptive, private place. Everybody's cheating everybody else. They think that's capitalism. They don't know how to question anything."

Jana paused. "I've seen the white light three times," she said, "but this time in Sarajevo it wasn't bad. It was peaceful."

My last stop in Ljubljana was the American Center, a nondescript building with a pair of photographs outside that could be mistaken for examples of social realism: Mount Rushmore and six combines harvesting a wheatfield on the Great Plains. For weeks I had tried to line up an interview with the senior USIA official, but he kept brushing me off, claiming either sickness or overwork. He was said to be allergic to the city, and it was true he had to prepare for the imminent arrival of our first ambassador to Slovenia. Nevertheless the diplomat's reluctance to meet with me had become a joke: I could talk to anyone here, except an American.

My Slovenian friends appreciated the joke. They were still angry that our last ambassador to Yugoslavia, Warren Zimmerman, had not supported their independence bid. Unity and democracy: one could not exist without the other, Zimmerman argued, recognizing that Milošević's anti-democratic efforts to centralize power only reinforced Slovenian and Croatian secessionist impulses. In his memoirs, Zimmerman reserved most of his scorn for Milošević and Tuđman, but his verdict on the Slovenians was also just: "Their virtue was democracy and their vice was selfishness. In their drive to separate from Yugoslavia they simply ignored the twenty-two million Yugoslavs who were not Slovenes. They bear considerable responsibility for the bloodbath that followed their secession."

In May, in protest against Serbian aggression in Bosnia, Zimmerman had been recalled from his post in Belgrade. Now I wanted to know what sort of diplomatic presence we had in the region. So I pestered the USIA official in Ljubljana until he finally agreed to a short meeting. I waited for him in the library, leafing through old copies of the *Herald Tribune*. The room was crowded with young Slovenians reading guides to American colleges, and I was trying to imagine what picture they had formed of America when the diplomat called me into his office. He was a bearded, dour man who said at the outset that his remarks were not for attribution. Nor would he provide any "deep background."

"Really," I said.

"You won't get anything out of me," he said.

"Not even impressions?" I said.

"No," he said.

"How about some stories?" I said.

"You're wasting your time," he said.

I closed my notebook. "Tell me," I said. "Why is it I can talk to anyone in Slovenia except you? I've interviewed the archbishop, the minister of foreign affairs—"

"You mean you talked to Dimitrij Rupel," he interrupted.

"A very productive talk," I said.

"Rupel won't talk to me," the diplomat whined.

"Oh," I said.

"Now you tell me something," he said. "Why do you think journalists don't visit me?"

"Go figure," I said.

Hoping to change the tenor of our conversation, I told him some of my impressions of Slovenia, and gradually he relaxed, even allowing me to quote him—once. "The Slovenians are on the verge of anarchy," he said. "It makes you think the Communist regime wasn't all that bad, next to this." He said he was trying to give a jolt to Slovenian society, but he would not explain how. He took a special interest in my friendship with Aleš.

"He seems to know everyone," he said, looking at his watch. "Would you introduce me to him?"

"Certainly," I said, rising to my feet. "May I use the phone?"

"Of course," he said. "A local call?"

I nodded.

"Any time," he said, then led me out to a free desk, where he dialed the number for me. When he heard me say hello to Jana Schneider, his face turned white with anger. He signaled for me to hang up; when I kept talking to her, he marched across the room and cut me off. "You can't use this phone," he said, hustling me out the door. And I never saw him again.

6

Venice

My idea of Venice came from a book. In college I read *Death in Venice* as a cautionary tale about the dangers that fame and single-mindedness pose to a writer. Gustav von Aschenbach's dissolution among "the feasts of the sun," the city's languorous ways, the determination with which Venetians—and Aschenbach himself—ignore or cover up the evidence of a cholera epidemic, the unspoken threat of world war in the distance: these were the notes of a chord that thrilled me.

"Solitude," Thomas Mann writes, "gives birth to the original in us, to beauty unfamiliar and perilous—to poetry. But also, it gives birth to the opposite: to the perverse, the illicit, the absurd." If once I thought I understood the hazards of solitude, I was no longer certain I could recognize either its risks or its glories. Venice represented that two-edged sword I had honed as a young writer. When Nina Zagoričnik invited me to accompany her to the Venice Film Festival, I felt as if I was making a pilgrimage to the birthplace of something vital in my own personal history.

Writers can justify almost anything in the name of their art, a central point of Mann's novel. Explaining away this excursion to "the land of satin dominoes" (Anna Akhmatova's phrase) was easy: I needed a change. My notebook was filled with impressions, my perspective had shrunk to the size of the country for which a single telephone directory sufficed. It would do me good to travel to the city some Slovenians called their spiritual home. And it was no stretch to imagine that Venice bore the same relationship to Slovenia as a novel's scenes and descriptions do to its plot: the pilings on which the city is slowly sinking into the sea come from trees cut down in the Karst. Besides, as Nina liked to say, "Slovenia is possible only as long as one thinks constantly of Italy."

On the first Saturday morning in September Nina arrived at my flat, four hours late. She had overslept, having stayed up until three to prepare for an interview with Gabriel García Márquez, a guest of the film festival. As we sped along the highway, through patches of thick fog, she did not hide her bad mood. The snow-capped Julian Alps vanished behind us, and from time to time Nina slammed on the brakes for no apparent reason. I expected to be killed in an accident. She could not wait to get to Italy. Our first stop would be for an espresso. She slapped on a pair of headphones, turning the music of the Scorpions up so loud I could hear the lyrics: "The wind of change."

The poet Joseph Brodsky called Venice "a work of art, the greatest masterpiece our species has produced." It was this rich source of invention and imagination I wanted to hold in counterpoint to the footage of Serbian concentration camps televised around the world for the last month. Summary executions of civilians; "ethnic cleansing" and forcible displacement; hostage taking; disappearances; mistreatment of prisoners; indiscriminate use of force; attacks against medical and relief workers, hospitals and ambulances; attacks against journalists; looting, burning, and pillaging: these were the Serbian war crimes Helsinki Watch was documenting in its first report. Here was a stark description of Jana Schneider's wounding and Ivo Štandeker's death in Sarajevo. Serbian paramilitaries manning a roadblock kidnapped the wounded journalists on their way to a downtown hospital; only when the UN intervened did the Serbs agree to take the couple to Pale, the mountain capital of the self-styled Serbian Republic, thirty-five kilometers away. Eight hours after the shelling, Štandeker died en route to the ski resort. And he was just one of more than fifty journalists killed or wounded in the first year of fighting. (The International Federation of Journalists deemed Yugoslavia "the most perilous site for journalists" in 1991.) In a document devoted to the cynical actions of men and women caught up in the inertia of war one attack against the media stood out: on 20 June Serbian forces near the Sarajevo airport launched incendiary bombs at the office building of *Oslobođenje*, the daily newspaper (which had already lost five reporters to the fighting, one shot in his office by Serbian paramilitaries). Fire crews arrived to put out the flames—only to be shot at by the Serbs. One fireman was killed; several were wounded. The news—the truth—was literally burning in Bosnia.

Responding to growing public uneasiness, in early August Democratic presidential nominee Bill Clinton said, "We cannot ignore what appears to be a deliberate and systematic extermination of human beings based on their ethnic origin." If elected, he promised to "begin with air power against the Serbs to try to restore the basic conditions of humanity." His strong words underscored the Bush administration's failure to act decisively in Bosnia. Indeed at the conclusion of the annual G-7 summit in July George Bush betrayed his feelings about the war: despite his promise to consider using military force to deliver humanitarian supplies, he said, "I don't think anybody suggests that if there is a hiccup here or there or a conflict here or there that the United States is going to send troops."

Not for the last time were observers reminded of Neville Chamberlain excusing the ceding of the Czech Sudetenland to Nazi Germany as "a quarrel in a faraway country between people of whom we know nothing." Chamberlain said Britain had secured "peace in our time" by yielding to Hitler. But the Munich Pact emboldened the Nazi leader to press his claims to other lands. Chamberlain is remembered as this century's chief architect of appeasement, a lesson lost on George Bush, who ascribed the Serbian death camps to "a blood feud arising from ancient animosities." Revising history in order to justify inaction was fast becoming a favorite diplomatic tactic.

Nor was the president alone in distorting the record. At the London Conference Lawrence Eagleburger said the war had "ancient and complicated roots." While ethnic and religious animosities had been stoked throughout Balkan history, only with the rise of nationalists like Slobodan Milošević and Radovan Karadžić, Franjo Tuđman and Mate Boban (leader of the Bosnian Croats), had talk of "ancient" differences impossible to settle become common; for theirs was a systematic campaign to illustrate Hannah Arendt's notion that in this century facts can be created—in this case, the "fact" that Bosnian Serbs, Croats, and Muslims could not live together.

"One sure way of judging the historical claims of the main perpetrators of violence in Bosnia is to look at what they have done to the physical evidence of history itself," Noel Malcolm writes in *Bosnia: A Short History*. "They are not only ruining the future of that country," he says of the Serbs and their war machine; "they are also making systematic efforts to eliminate its past.

> The state and university library was destroyed with incendiary shells. The Oriental Institute, with its irreplaceable collection of manuscripts and other materials illustrating the Ottoman history of Bosnia, was also destroyed by concentrated shelling. All over the country, mosques and minarets have been demolished, including some of the finest examples of sixteenth-century Ottoman architecture in the western Balkans. These buildings were not just caught in the cross-fire of military engagements; in towns such as Bijeljina and Banja Luka, the demolition had nothing to do with fighting at all—the mosques were blown up with explosives in the night, and bulldozed on the following day. The people who have planned and ordered these actions like to say that history is on their side. What they show by their deeds is that they are waging a war against the history of their own country.

And they did this in plain view of the international community.

This was becoming a story of diplomatic malfeasance writ large. By the winter of 1991 the U.S. State Department and CIA had ample evidence of impending strife in the Balkans. But Baker and Eagleburger (who had served as ambassador to Yugoslavia in the 1970s before steering Serbian state contracts to his next employer, Henry Kissinger Associates) ignored or underestimated the warnings issuing from the intelligence community. Then, in a secret meeting in March 1991, Milošević and Tuđman agreed to carve up Bosnia, and in May their representatives, armed with maps, met in Graz, Austria, to settle on their shares of the republic. In June, a week before Slovenia and Croatia declared independence, Baker made a fateful trip to Belgrade. Milošević came away from his meeting with the secretary of state believing he had a free hand to keep Yugoslavia together. What diplomatic leverage the great powers had with Serbia was squandered at the outset. "Without threatening the use of force," John Newhouse wrote in *The New Yorker*, "Baker could have said that unless Milošević changed course—ceased exploiting the hatreds and passions of old—the United States might feel it had to recognize the independence of Croatia and Slovenia and do its best to seal Serbia off from the rest of the world." Instead, the Balkans were plunged into bloodshed.

Timing is everything in war. Milošević and Tuđman chose a propitious moment to tear their country apart. Despite his success in the Gulf War, Bush would not let the United States become "the world's policeman." Yugoslavia's geographical, historical, and trading ties to the Continent allowed him to call the fighting in Slovenia and Croatia a European problem. Because the EC was negotiating amendments to the 1957 Treaty of Rome and the 1987 Single European Act, which in December 1991 would culminate in the Maastricht Treaty and the creation of an economic and monetary union, the administration argued that Europe should take the lead in Yugoslavia. One provision of Maastricht required the EC to unite on foreign policy issues, but Germany's decision to recognize Croatia and Slovenia on 15 December (one week after Maastricht, one month before the EC took that step) destroyed any hope of the Europeans working in concert on foreign affairs.

At the same time, the Western alliance was preoccupied with the Soviet Union's dissolution. The disposition of thousands of nuclear weapons in the Soviet republics agitating for independence was a priority for politicians in every Western capital. Milošević had used real and imagined threats to the Serbian population living outside the republic's borders to wreak havoc in Yugoslavia; far more ominous was the prospect of Russia fighting over the twenty-five million Russians in its near abroad. A Balkan war might be contained. The same could not be said of fighting in Georgia or Ukraine. "Yugoslavia got caught between Maastricht and the Soviet Union," said one diplomat, "between the processes of integration and disintegration."

Moreover, by the winter of 1992 Bush was campaigning for reelection, a distraction which may explain yet another diplomatic blunder. At the Lisbon Conference in February Bosnian President Alija Izetbegović agreed to an EC-negotiated partition plan, but with American backing he changed his mind upon his return to Sarajevo and urged his countrymen to vote for independence in a referendum concluded on 1 March. (99.4 percent of the electorate endorsed his view. Most Bosnian Serbs, under instructions from their leadership, boycotted the referendum.) The administration had once warned the Europeans against recognizing Croatia and Slovenia; now it "internationalized the problem," in Warren Zimmerman's words, partly to convince Serbia to leave Bosnia alone, partly to avoid setting the precedent of partition for the successor republics of the Soviet Union. But EC and then U.S. recognition of Bosnia and Herzegovina on 6–7 April prompted the Bosnian Serbs (supported by the JNA) to wage war on the new country.

Unlike Slovenia, Bosnia was not prepared to defend itself. The Security Council resolution in September 1991 to impose an arms embargo on all the republics of the former Yugoslavia preserved the superiority of the Serbian war machine; the outgunned Bosnians were no match for the firepower of one of Europe's largest military forces, as anyone could see on television. Yet on 19 May (the day before the UN recognized Bosnia), even as TV cameras broadcast pictures of people maimed and killed by artillery fire in Sarajevo, State Department spokeswoman Margaret Tutwiler said she was "not aware of" any American security interests at stake in Yugoslavia—a position Baker had to counter, less than a week later, when he called for stronger action at a NATO meeting in Lisbon. On 30 May the Security Council imposed economic sanctions on Yugoslavia, which by then consisted only of Serbia and Montenegro. The Serbian prosecution of the war was hardly slowed by Milošević's "order" to withdraw the federal army from Bosnia: he had transferred a sizable portion of JNA assets to the Bosnian Serbs before the fighting began.

French President François Mitterrand made a daring flight into Sarajevo on 28 June, seventy-eight years to the day after Gavrilo Princip assassinated Archduke Franz Ferdinand, setting in motion the diplomatic events that led to World War One. (Princip was hailed as a hero in Serbia; in Bosnia and Croatia, Serbs were robbed, beaten, and lynched. One month later, Austria-Hungary declared war on Serbia.) Mitterrand's attempt to open an air corridor into the besieged capital was the first of many dramatic efforts to alleviate the suffering; by the start of the London Conference, where Cyrus Vance and EC envoy Lord David Owen were instructed to negotiate Bosnia's partition (the amount of land allotted to the Bosnians was far less than what the Lisbon plan offered), the largest humanitarian operation in history was underway. The West refused to counter the Serbian assault. "As long as I'm alive, France will never make war on Serbia," said Mitterrand.

Chess was as popular as soccer in the former Yugoslavia, and while the world awaited the Serbs' next move on the political chessboard a reclusive grandmaster came out of retirement to add his name to their war effort. This week, Bobby Fischer announced his intention to flout the sanctions and play Boris Spassky in a five million dollar match on a Montenegrin island. His response to international pressure mirrored that of his host country. At a press conference he spat on a letter from the U.S. Treasury Department ordering him to cease and desist his activities in Yugoslavia.

Meanwhile, the fighting continued. Two days ago, for example, Serbian gunners shot down an Italian relief plane, killing all aboard. Perhaps that was why, once we crossed into Italy, we were stopped three times within a kilometer of the border. Nina's anger mounted steadily. One policeman rummaged through her suitcase, and when he examined a brassiere she cursed him in Italian. But she brightened as soon as we reached a gas station, where she drank two espressos. Before long we were in Venice, boarding a ferry—Nina called it "a fairy tale"—bound for Lido.

It was fine to chug along the Grand Canal. The Piazza San Marco was full of tourists, the sky was clear, salt spray rose around us. Across a stretch of open water we disembarked on Lido, the island separating the lagoon of Venice from the Adriatic Sea. I took a room in the Hotel Hungarica, and Nina, who would stay with a friend, went to the festival, promising to pick me up in the morning before her interview with García Márquez.

At sunset I strolled along the beach, past a long line of cabanas, teenaged boys playing soccer, families packing up their belongings. A couple was making love in the sand. Old men fished from the stone jetty leading out to the lighthouse. A half moon appeared above the Gulf of Venice, an arm of the Adriatic. The war in Bosnia was just across the water, less than 200 miles away, yet it seemed to belong to another world. I was thinking of questions for the Nobel laureate: how to depict our so-called "new world order"? "It is very difficult to write now," the Polish journalist Ryszard Kapuściński had said in a recent interview, "because the end of our century is marked by a tremendous acceleration of the historical process. Literature hates this." He went on to suggest that,

> The writer needs a certain quietness and evenness of perspective, a space of time for reflection. There is no distance now. There are some changes that we can watch on the television, but there are certain very profound, important transformations which we do not have the opportunity to see. Fiction writers avoid this by writing about marriage, love, things like that; they will not touch the volcano the world has become.
>
> If you try to touch it, to describe it, you find that it requires a new imagination. The problem is not so much with the writing itself as with the creative imagination. The acceleration of history proves that we have a very limited imaginary capacity. We never dreamed the world would become such a rich and various place.

How would García Márquez respond to Kapuściński's ideas? If anyone possessed the imaginary capacity to describe this new world, it was the author of *One Hundred Years of Solitude*, the magical realist who inspired writers everywhere to use their resources—imaginative, linguistic, political—in the cause of freedom. Nina wanted him to offer hope to her countrymen, who were staring, as she said, into the abyss of Bosnia. I had a personal interest in meeting him: his American editor was a friend, and the stories he told about "Gabo" led me to imagine I almost knew García Márquez.

I walked to the end of the jetty. A hard rain the previous night had cleared the air. In the stiff breeze windsurfers skimmed across the water and yachts glided along the horizon. By the tenth century Venice reigned over much of Dalmatia, the closest point of which lay just beyond the sailboats, and from these shores the Fourth Crusade was launched, first to seize the Croatian port of Zadar, then to storm and sack Constantinople. This was a place for journeys to begin, not end, as the Venetian ambassadors who created the modern diplomatic service knew.

Seven centuries ago, that quintessential travel writer, Marco Polo, set out from here to explore the Far East; the account of his journeys, composed in captivity, inspired adventurers and writers, none more brilliantly than the late Italo Calvino, whose novel, *Invisible Cities*, is a meditation on, among other matters, the nature of travel itself. Marco Polo describes for Kublai Khan the cities he has visited—Cities and Memory, Cities and Desire, Thin Cities, Trading Cities, Cities and the Dead, Hidden Cities. Are these journeys into the past or the future, Kublai Khan asks. "And Marco's answer was: 'Elsewhere is a negative mirror. The traveler recognizes the little that is his, discovering the much he has not had and will never have.'" In fact, each city Polo paints (or invents) for the Tartar emperor is another version of Venice: "Perhaps I am afraid of losing Venice all at once, if I speak of it," he admits. "Or perhaps, speaking of other cities, I have already lost it, little by little."

Walking back along the jetty, mourning Calvino's passing and the loss of his "imaginary capacity," I was hardly alone in thinking that despite the ravages of time Venice still deserved to be called "the Queen of the seas." In the gathering dusk I was in no mood to reflect on its decline as a maritime power, which Rebecca West reminds us "was really more spiritual than economic." Like any visitor, I was hoping someone would figure out how to keep the city from vanishing under the waves.

On to the film festival. Billboards advertising Ferre jeans lined both sides of the main street; only the tan lines revealed the sex of the topless models. I went up the steps of the white pavilion that was festival headquarters, past photographers surounding some celebrity, and checked in the press office for Nina. No luck. She planned to see as many films as possible; *Glengarry Ross*, *The Plague*, and an animated short by The Brothers Quay titled *Are We Still Married?* were three she could not miss. But her first report would begin with a joke about a film she did not need to cover, the one written by Peter Handke—another writer led astray by politics, she said. In this he joined Céline, Pound (whose mistress lived out her days in Venice), Éluard, Sartre. "Mistaken ideas always end in bloodshed, but in every case it is someone else's blood," wrote Camus. "This is why our thinkers feel free to say just about everything." Nina's audience would not forgive Handke for his mistakes.

I ate dinner at an outdoor café, where over a bottle of white wine I reread *Death in Venice*, struck by what Mann called "the voluptuousness of doom," a style of thinking that heralded World War One. In "Tonio Kröger," a story from 1903, Mann said a writer needs "a cool and fastidious attitude toward humanity" to break up the conventions of speech that trap people in outmoded ways of thinking. The Cold War was over, but who would dare to imagine they had broken free of the mindset that divided the world into two rival camps? García Márquez himself had been a victim of the McCarran-Walter Act; his virulent criticism of American foreign policy had at times prevented him from traveling to the United States, where translations of his books had made him a wealthy man. Would he now view the concentration camps in Bosnia as another study in "the voluptuousness of doom"? And was it true that great writers discover, word by word, the language with which to touch the volcano of this world—not in the manner of Aschenbach, whose crowning work, *The Abject*, draws from Mann this withering description:

> With rage the author here rejects the rejected, casts out the outcast—and the measure of his fury is the measure of his condemnation of all moral shilly-shallying. Explicitly he renounces sympathy with the abyss, explicitly he renounces the flabby humanitarianism of the phrase: '*Tout comprendre c'est tout pardonner.*' What was here unfolding, or rather was already in full bloom, was the 'miracle of regained detachment', which a little later became the theme of one of the author's dialogues, dwelt upon not without a certain oracular emphasis. Strange sequence of thought! Was it perhaps an intellectual consequence of this rebirth, this new austerity, that from now on his style showed an almost exaggerated sense of beauty, a lofty purity, symmetry, and simplicity, which gave his productions a stamp of the classic, of conscious and deliberate mastery?

"And yet," he concludes: "this moral fibre, surviving the hampering and disintegrating effect of knowledge, does it not result in its turn in a dangerous simplification, in a tendency to equate the world and the human soul, and thus to strengthen the hold of the evil, the forbidden, and the ethically impossible?" I was a little drunk, yet I saw "dangerous simplification" as the biggest threat in our new world order. Politicians and journalists were not the only specialists in reducing complicated situations and ideas to slogans and sound bites. I slept fitfully that night.

In the morning Nina and I drove to the Hotel Excelsior to pick up García Márquez. She parked in a no-parking zone, and when she went into the lobby to call him I stayed in the car to keep it from being towed. Soon I was shaking hands with the novelist, a handsome man dressed in a bold blue shirt and white sport jacket. In *The Paris Review* interview Nina used to prepare for her own interview he was likened to "a good middleweight fighter—broad-chested, but perhaps a bit thin in the legs." Ten years later, the description held, and I was too dazzled to say anything more than my name to introduce myself. He smiled, and then we set out for the Hotel des Bains, Aschenbach's residence in *Death in Venice*. This morning the ballroom was reserved for a film festival panel featuring, among other celebrities, García Márquez.

"*Nous parlons en francais*," Nina said as we drove by the sea, "*parce que c'est la langue nous pouvons parler ensemble*."

"*Bien,*" I said.

García Márquez answered her questions pleasantly enough until she turned around and asked me, in English, if I knew the French word for the billboards lining the road. "I can't remember," I shrugged. Nina went on with her questioning, but now she got no answers. Clearly upset, the novelist did not speak again. When we arrived at the hotel, he got out of the car, slammed the door, and marched off. Nina glared at me. "Did you say anything to him?" she said. "Not a word," I said.

She told me to park the car, and chased after him. I found her in the hotel lobby, shaking with rage. García Márquez had said to her, "I hate Americans. No interview." Then he had stomped away.

"What did you say to him?" she demanded.

"Nothing," I said. "You heard me."

In the ballroom, an elaborate affair of chandeliers, ornately carved wooden pillars, and thick red curtains hanging from the balcony, more than six hundred journalists were waiting for the panel to begin. The reporters adjusted their headphones to hear the translators, television cameras filmed the crowd, security officers with cellular phones scanned the list of speakers.

"They all have such beautiful women," Nina sighed as a group of men, including García Márquez, approached the tables arranged behind the lectern. The men kissed their consorts goodbye, paused to study themselves in the mirrors embedded in the pillars surrounding the room, and took their places at the tables. "I am so angry at you," Nina hissed at me.

García Márquez, Wim Wenders, Peter Handke, and Costa Gravas were among more than fifty luminaries invited to attend the panel. I pushed my ugly encounter with the Nobel laureate out of my mind and leaned forward to hear what the panelists had to say.

I assumed the subject of the panel would be Bosnia. We were too close to the fighting—the footage from the concentration camps was too fresh—for these famous *engagé* artists to avoid speaking out against the war. It was García Márquez, after all, who told *The Paris Review*, "the only advantage of fame is that I have been able to give it a political use." A panel including Italian President Paolo Portoghesi, France's flamboyant Minister of Culture Jack Lang, and filmmakers like Lina Wertmuller could not shirk the responsibility of addressing the slaughter of innocents.

I could not have been more wrong.

The celebrities had come to denounce the U.S. government for its failure to comply with the "moral rights" provision of the Berne Treaty, the world's copyright standard: American filmmakers had no protection against studios tampering with their work. George Lucas painted an "Orwellian future" for his industry, according to a press handout. "Releasing entities have gained experience through the video market and colorization, and it's not going to be too long in the future before an actor, who has become unacceptable for whatever reason of politics or marketability, might be in the future electronically replaced by another actor." Several speakers echoed Lucas's warning that "the theme of a motion picture might be altered to achieve whatever politically correct or economically driven objective its owners wish to impose."

I could not argue with the stars. Yet as they droned on through the morning I grew impatient: they had missed a chance to call attention to a genuine tragedy. I counted up the costs of bringing these people together: hundreds of thousands of dollars. In Mann's rendering of Venice Aschenbach loses his bearings, inflamed with his desire for the forbidden—a Polish boy named Tadzio, who will "likely not live to grow old"—and becomes infected with cholera. His journey, prompted by his recognition that "what he needed was a break, an interim existence," in order to recover his strength for more stylish writing, ends, foolishly, with his death; and all the while the menace of world war hangs over Europe. In this ballroom I decided I too had lost my bearings.

A sentence I had underlined the night before in my copy of *Death in Venice* haunted me: "For in almost every artist's nature is inborn a wanton and treacherous proneness to side with the beauty that breaks hearts, to single out aristocratic pretensions and pay them homage." I was done with the film festival, and I had not seen a single film. I walked out into the sunshine and down the beach until I came to a deserted stretch, where I stripped off my clothes and dove into the sea.

In the end Nina convinced García Márquez to do an interview, on condition that he not be asked to comment on the war, a situation he professed not to understand. In *The Paris Review*, on the eve of being awarded the Nobel Prize, he had made this prediction: "I'm absolutely convinced that I'm going to write the greatest book of my life, but I don't know which one it will be or when. When I feel something like this—which I have been feeling now for awhile—I stay very quiet so that if it passes by I can capture it." Nina rounded off her interview with him by asking if he still felt that way.

"No," he said sadly.

By then I was in Trieste. I had gone directly from the beach to check out of my hotel. On the ferry to Venice I decided to store my duffel bag at the train station and spend the afternoon in the city. I wound up at a retrospective of Robert Mapplethorpe's photographs. To find the exhibit meant following a dozen signs down narrow alleys and through piazzas crowded with German tourists. My journey resembled Aschenbach's feverish stalking of Tadzio in at least one respect: once I started searching for the gallery I could not stop. At last I came to the controversial show, and indeed I found it troubling, not so much for the sexual content of some images as for the self-portraits Mapplethorpe had completed just before he died of AIDS. The exhibit included a video in which one of his friends said the artist had wanted only to document his life and times, a world destroyed by another plague. In Mapplethorpe's self-portraits I saw the mask of Gustav von Aschenbach.

Outside, the light on the water was so brilliant I thought of returning to Lido to see Nina. But the ferrymen had gone on strike, and when a man offered to take me there for 10,000 *lire* I realized I would not find Nina for hours, if at all, so I boarded a train to Ljubljana.

In the early evening, during a layover in Trieste, I walked down streets long familiar to writers living in exile. Perhaps it is no coincidence that two of this century's literary masterpieces—Joyce's *Ulysses* and Rilke's *Duino Elegies*, the one a celebration of ordinary life, the other a meditation on the difficulty of faith—were partly composed in or near Trieste. This is a city with a rich, independent, and emblematic history. A free commune by the twelfth century, Trieste fought with Venice for two hundred years before placing itself under Habsburg control, where it held a favored position: the empire's only port was both a gateway to Central Europe and a magnet for artists and writers, a polyglot cosmopolitan center in which Joyce recreated his native Dublin and Rilke searched for eternity: "*Here* is the time for the *sayable, here* is its homeland./ Speak and bear witness."

One language spoken here was Slovenian. In 1918 there were only two cities, Ljubljana and Cleveland, Ohio, with more Slovenians than Trieste. Indeed the Slovenians originally planned to build their national university here, but in the 1920s they began to leave the city for the same reason they moved away from Carinthia: Fascist intimidation, in this case Italian. (Tomaž Šalamun's grandfather was one such émigré.) Slovenian could not be spoken in public, and the Slovenian National House, the main cultural institution, was burned down. The countryside remained Slovenian, and eventually Trieste become a place for Slovenians to shop, not to live and work. Now Italian authorities were threatening once again to outlaw Slovenian-language schools and periodicals in Trieste.

But the language I heard everywhere I walked was English: hundreds of American servicemen were in the streets, a common sight. In the final days of World War Two, when Yugoslavia claimed Trieste and its Slovenian province, the Western powers nearly squared off against the Partisans. A tense moment in history. The Allies stopped Tito from annexing Trieste (as well as Venezia-Giulia, Istria, and southern Carinthia) but not before the Partisans had massacred 2,000 Germans, Italian officials and police, and anti-Fascists in Trieste, in a campaign of terror that came to be known as "The Forty Days"; a mass grave in a nearby gorge was said to hold 500 cubic meters of corpses. All the more reason to maintain a military (and intelligence) presence in the new state created in 1947, the Free Territory of Trieste, which included the city and a coastal zone of Istria running from Duino to Cittanova. When the UN Security Council could not agree on who should govern this state, British and American forces occupied Trieste and its immediate surroundings, an area called Zone A. Yugoslavia took over the rest of the territory or Zone B.

It sounded familiar: in 1954 the Free Territory was formally partitioned and ceased to exist. Nearly forty years later, people on both sides of the border were still angry. Italians who had lost their country houses demanded compensation from the new Slovenian government—some, remembering "The Forty Days," called for Istria to be returned to Italy—while Slovenians in Trieste clamored for better treatment from the Italian authorities. The breakup of Yugoslavia reminded some that in addition to providing temporary lodgings to wandering writers Trieste was known in the Habsburg era as a center of irredentism. History was reappearing in some of its less appealing masks.

Harry's Bar & Grill was a good place to look for an American who could tell me what our military was doing in the region. And three pilots from the *USS Saratoga* sat down at the next table just as I was finishing dinner. I ordered another bottle of wine and struck up a conversation with the youngest officer. Are you going into Bosnia? I said.

"Who knows?" he said. "We just watch CNN and wait for our orders. Whatever happens it's just another day of business for us."

The oldest officer caught the young man's eye. He looked away from me, embarrassed. I poured myself another glass of wine. The pilots were drinking mineral water, and what I heard from their table—code words and numbers—was another language I did not understand. Rumors were circulating that George Bush had considered salvaging his disastrous Republican convention by announcing on the last night that he was sending troops to Bosnia. Perhaps the *Saratoga* was in port because the president had decided once and for all not to intervene?

I polished off the wine and stumbled to the train station. "And so near Venice" was how Nikola Koljević, the Shakespeare scholar from Sarajevo University and vice president of Republika Srpska, taunted the journalist who discovered the Serbian concentration camp, at Omarska, in Prijedor. "It took you a long time to find them, didn't it? Three months! And so near Venice. All you people could think about was poor sophisticated Sarajevo. Ha-ha!" Iago, for obvious reasons, was Koljević's favorite character, and the journalist heard a chill in the professor's voice when he added, "None of you ever had your holidays in Omarska, did you? No Olympic games in Prijedor!"

It was in late May or early June that the authorities in this Serbian village converted an open iron mine into "an interrogation center," where thousands of Muslims and a smaller number of Croats were beaten, tortured, and killed. "Dying was easy at Omarska," a survivor would write, "and living was hard." One man was forced to drink old motor oil and then bite off a fellow prisoner's testicles; three of the witnesses were killed with metal rods. Another man was burned alive, after two drunken, knife-wielding guards had sliced off his penis and half his buttocks. Yet another, an older man, was beaten to death for refusing to make love—in public—to a young woman. One woman was raped and beaten for four nights straight. Hundreds were shot or clubbed to death, hundreds more disappeared without a trace. And the guards made videotapes of their savagery to watch at home. "Wherever wolves feast," said one inmate, "they leave a bloody trail."

Needless to say, these pestilent scenes were far from my mind when I found a seat on the train and promptly fell asleep, only to be awakened minutes later by an angry Slovenian border guard, a woman who at first refused to let me into the country. She shook my passport at me, threw it into my lap, and walked away. I fell asleep again, and when we arrived in Ljubljana early in the morning, I did not recognize the station. I climbed down from the train.

"Is this Ljubljana?" I asked the conductor.

He stared at me, as if I had lost my mind.

"Ljubljana?" I repeated.

He did not answer.

7

Vilenica

Cool, dark, and damp. Water dripping from the roof of the cave. Whitish stalactites and stalagmites surrounded us, like a mouthful of misshapen teeth. The guide's voice echoed weirdly off the slick walls. I could not understand him; Aleš had sneaked me in at the native rate, over the objections of the guide, a suspicious man who shook his fist at me when my friend asked him when the tour in English would begin. He waved us toward a crowd of Slovenians; and as we walked deeper into the earth I occupied myself with a map of the Postojna caves. They had such vivid names: Whale's Gullet, Brilliant, Calvary, Russian Pillars, Canopy, Concert Hall, Collapsed Pillar. The guide was asking me a question. Aleš cursed him. The tourists edged farther down the path.

The Postojna caves lie forty kilometers southwest of Ljubljana, a short drive through a mountain range into the Postojna basin, deep in the Karst. My first Balkan journey was ending in what Auden might have called "a secret system of caves and conduits," a marvelous world of sinkholes and lakes which appear and disappear as if by magic. Of the 15,000 caves in the Slovenian Karst the Postojna are the most spectacular, extending more than twenty kilometers underground. Slovenians like to say, without any evidence, that these caves inspired Dante's *Inferno*.

And history records a local Partisan operation concluding with a fiery scene made for the poet's vision of Hell. Because control of the Postojna basin was critical to the Nazi war effort—the rail line to Trieste led through Postojna Gap—there was a premium on Partisan actions, none more dramatic than an ingenious attack mounted in April 1944. The Germans stored a supply of airplane fuel in one cave, sealed off its entrance, and posted guards outside. But the Partisans, slipping in from the other end of Postojna, followed the intricate network of caves until they came to the drums of fuel, which they punctured. One saboteur lagged behind to light the leaking fuel.

The fire burned for seven days.

And this is where history takes a mythological turn: the flames engulfed the last saboteur. He passed out, and when he came to, naked and scorched, his clothes burned up and skin blistering, his comrades were gone. For two days he wandered through the caves, shivering, searching for an exit, hiding from the German patrols. There was plenty of water to drink but little to eat besides insects, lizards, and the reclusive human fish, the prehistoric relict Slovenians call the symbol of the Karst. What must he have thought—lost in a labyrinth, with no thread to lead him out?

And yet he survived. How? The records do not say.

"I will set fire to everything," Tomaž Šalamun declared in his prophetic poem. "So I will remain./ So I will survive everything."

The mythology of the Karst: How to reconcile its colorful legends (dragons, witches) and brutal history (witch burnings, mass graves)? I walked up one damp corridor and down another, haunted by ghosts. The last European battle of World War Two was fought on Slovenian territory, when Nazis, Croatian and Muslim Ustaše, Serbian Četniks, and Slovenian Home Guards (who sided with the Germans against Tito) retreating to Austria were caught at a pass in the Karavanke Mountains, defeated by the Partisans, and massacred. Meanwhile, thousands of Yugoslavs, civilians and anti-Communists for the most part, were fleeing to Carinthia, where they surrendered to British forces. A bad decision. When Tito threatened to annex southern Carinthia, claiming to have Soviet support for Slovenia's historic homeland, the British reversed their policy on refugees and repatriated 30,000 Yugoslavs, by force or trickery, sentencing them to certain death. By May 1945, there were 24,000 Slovenians in British refugee camps; half were sent back to Slovenia the next month—and summarily executed by the Communists. Their corpses, hidden in caves in the Karst, have yet to be excavated. The other Slovenian refugees joined the half million opponents of Tito who emigrated to Argentina, Australia, New Zealand, and North America (including the most prominent Ustaša officials, war criminals all: the British inexplicably let them escape).

On my flight to Ljubljana, I had sat next to an old émigré widow, from Toronto, who was returning to Slovenia for the first time. She could not contain her excitement. I was wondering about her past. Had her late husband been a war criminal? Had she herself belonged to a resistance group—the Home Guards or the more conservative, church-supported White Guards? Or had they simply bet on the wrong side? The history of the Slovenian resistance is, of course, murky. Hitler's *Anschluss* in 1938 brought the Nazis to the Karavankes, and from there they went on to control the northern and western parts of Slovenia, where the Partisans inflicted the most damage—and Slovenians suffered the greatest reprisals. But in Ljubljana, which came under the jurisdiction of the Italian Fascists, the leaders of the middle-class Liberal and Catholic parties chose to appease the occupying army, a decision they regretted when Italy capitulated in the autumn of 1943 and Germany took over. These politicians and their followers were among the refugees in Carinthia who either lost their lives at the hands of Yugoslavia's new Communist leadership or went into exile.

Resist or accommodate?

It depended upon where you lived. "The way things played out, the model of partisans and collaborators captures only a tiny fraction of the truth," Czesław Miłosz writes. "It was a three-way game in which the vast majority of the population zigged and zagged, seeking insurance with both sides. In addition, there were also independent enclaves, separate statelets, as it were, cobbled together by little war lords." The Nobel laureate is referring to Nazi-occupied Poland, yet his words apply as well to the Balkans during World War Two, with this exception: in Yugoslavia the three-way game led to a civil war. "That ought to demoralize them," Edvard Kardelj is reported to have said after executing hundreds of White Guards. No wonder the Austrian border was a beacon to so many Yugoslavs, war criminals and civilians alike.

And the old woman on the plane? All she would say was that the Communists had killed most of her friends and dumped their bodies in a cave.

"This is my country!" she cried when we landed.

Now the Postojna guide was telling a story about the Russian Bridge, which connects Calvary and the Beautiful Cave—how it was built by Russian prisoners during World War One. When he was finished, he asked me another question in Slovenian. Aleš swore at him again. The tour was over.

A tale of betrayal. Contrary to local legend, Erasmus of Predjama was no Robin Hood. Historians say the knight was a robber baron who murdered a lord at the court in Vienna and fled to Predjama Castle, near the Postojna caves. Soon he was launching raids on his neighbors, stealing from rich and poor alike—a campaign of terror which led to an imperial order: capture Erasmus, dead or alive. No small task. Predjamski Castle juts out from a cliff 120 meters high, in front of a cave occupied on and off since the Stone Age. The only route to the castle, up a long, narrow staircase, was easily defended. If provisions ran low Erasmus had a secret route through the cave to get more.

Ravbar, the governor of Trieste, laid siege to the castle in the fall of 1483, hoping to starve out the knight, but by spring it was his own troops who were hungry. On Shrove Tuesday Erasmus sent them a roasted bull, on Easter they received a lamb, and when cherries and fresh fish arrived Ravbar's men began to believe the cave led to Paradise. The knight seemed to have endless supplies.

Then Erasmus died, betrayed by one of his own servants.

What happened was this: the servant hung a flag outside the castle, marking the place where the knight went to the bathroom. Ravbar trained four cannons on that part of the wall, and Erasmus, a man of regular habits, was in a squatting position when he was killed.

Some of our most homely gestures, according to Milan Kundera, are destined for immortality.

"Where, with my first look around, did that sense of freedom come from?" asks Filip Kobal, the narrator of Peter Handke's *Repetition*. "The only answer that occurs to me offhand is the Karst wind (and perhaps the sun as well)."

I felt that sense of freedom driving with Aleš from Predjama Castle to Istria and the medieval city of Piran. Ljubljana's overcast skies had followed us to the Karst, but once we reached the ancient saltworks along the coast the sun was shining. Vienna and Venice. In hours we had gone from one sphere of influence to the other. Here were signs in Slovenian and Italian. The Adriatic was a deep shade of blue, though closer to shore the water turned emerald. In Piran we climbed a hill to a stone church overlooking the sea. A fresco of the Dance of Death. A cross propped in an arbor vitae. A fishing trawler was returning to port. I made a list of those who had ruled over Istria:

The Greeks.
Rome.
Byzantium.
The Franks.
Venice.
Napoleon.
The Habsburgs.
Serbian King Alexander.
Mussolini.
Tito.

To put it another way: Religions. Empires. Monarchies. Ideologies.

And now another system was emerging: Democracy. Markets. Trade.

Piran had never really prospered. It depended on the sea for its staples, fish and salt, and if its standard of living had peaked under the Venetians, that was because Istria had a monopoly on salt, a commodity Vienna needed. Economic stagnation turned the coastline into a smuggler's haven and preserved the medieval character of the cities: main squares, narrow streets, sheltered ports. Piran was the unchanging backdrop for conferences on Slovenia's changing architecture. But how would it fare in the new economy? The government's decision to sell off the vacation houses belonging to the trade unions was the first test. Beautiful houses, some with as many as a dozen bedrooms, were going for $50,000 apiece. Italians were buying them up—proof that the invisible hand of the market likes symmetry, Slovenia having been enlarged at Italy's expense after World War Two. I myself caught a mild case of buyer's fever, which gave way as soon as we drove out of town.

Near Trieste, we veered inland toward Sežana, the site of the seventh annual Vilenica Literary Festival, a celebration of Central European literature. Vilenica had played a role in the breakup of the Communist order, and of Yugoslavia, by bringing together dissident poets and writers from the former Habsburg lands to explore the ways in which "the rigid structures are dissolving," as one writer put it. Accordingly, the theme of this year's discussions was the disintegration of universalism. Milan Kundera was to receive the Vilenica Prize, not so much for his writings, I was given to understand, as for his support of Slovenia during the Ten-Day War.

In *The Art of the Novel* Kundera spelled out the theme of this year's discussions:

> In the Middle Ages, European unity rested on the common religion. In the Modern Era, religion yielded its position to culture (to cultural creation), which came to embody the supreme values by which Europeans recognized themselves. Now, in our own time, culture is in turn yielding its position. But to what and to whom? Technology? The marketplace? Politics involving the democratic ideal?

Presumably, Vilenica would provide some answers.

In the early evening the countryside was a wash of green. Indian summer. We drove through the woods, down empty curving roads, with the windows open.

"Vilenica has to broaden its outlook," Aleš said. "During the Cold War, when Central Europe was asserting itself, Vilenica was the third way, a lightning rod for us."

For others, too. Was it a coincidence that excerpts from the Memorandum of the Serbian Academy of Sciences and Arts appeared two weeks after the first Vilenica, in September 1986? Just as the literary journal *Nova revija* paved the road to Slovenian independence, so the Memorandum was a rallying call for Serbian nationalists. The work of prominent Serbian intellectuals, this seventy-four-page tract declared that Slovenian and Croatian business interests were conspiring against the Serbs who, readers were reminded, had made the greatest sacrifices on behalf of Yugoslavia. "Weak Serbia—strong Yugoslavia," a common expression, was acquiring new meaning. Not since World War Two had Serbs in Croatia faced such danger, the writers warned. In Kosovo they were threatened with genocide. Yugoslavia was disintegrating. Who led this conspiracy?

None other than Tito and Kardelj, dead these many years.

Though he disavowed authorship, the Memorandum bore the fingerprints of novelist Dobrica Ćosić, the so-called spiritual father of the modern Serbian nation. His fame as a dissident derived both from his decision not to write in the social realist style and from his belief in the Serbian cause. He became a Serbian hero in 1968, when he accused Kosovar Albanians of anti-Serbian activities and of harboring separatist ambitions. Expelled from the Central Committee of the Communist Party, he began to hold monthly meetings for his fellow writers and intellectuals, clandestine events at which he called for democratic reform. By the 1980s Ćosić had assumed mythical status in Serbia. "If a man gives up poetry for power," Mark Strand writes, "he shall have lots of power." Indeed Ćosić was hand-picked by Milošević to become president of rump Yugoslavia in June 1992.

Yugoslavia's breakup began to look like a literary squabble with tragic consequences. They fed each other, Slovenian and Serbian writers—*Nova revija* and the Memorandum, Vilenica and Serbian mutterings about a German plot to restore Mitteleuropa, "the writer's constitution" and arguments over the curriculum. In the freedom born of Tito's death, Slovenian and Croatian writers revived the idea of Central Europe while Serbian intellectuals mourned their diminished standing in Yugoslavia. Centrifugal or secessionist impulses clashed with centripetal forces. Slovenian writers had nothing but invective for the Serbs. No doubt writers in Belgrade would describe Slovenians in the same way. And the Bosnians? You could say their concerns had been forgotten in this dispute.

Aleš would hear none of it. He pressed the accelerator. "We're late," he said.

"There are no majorities, only minorities," said Veno Taufer.

The small, frail, white-haired poet was recovering from open-heart surgery, and he spoke softly, as if to conserve his strength to greet the writers gathering in the lobby of the Club Hotel. Next door was the Lipica stud farm, which had bred Lippizaners for the Habsburgs, and in the evening you could see the white horses galloping around a pasture lined with lindens. What better place to discuss the disintegration of universalism? Vilenica was a notoriously decadent affair.

"Yugoslavia meant Serbia," said Veno. "When I was abroad it was frustrating to have to explain who I was. Once I was asked why I didn't take up a world language, like Serbo-Croatian. Impossible. We've had our own language, culture, and manners since the sixteenth century."

The bar next to the lobby was doing a brisk business with the writers. Cigarette smoke filled the air; the conversations grew more spirited. Veno, clutching his chest, fended off glasses of wine and brandy. Some said his heart had broken when his son, a talented artist, had committed suicide.

"Czechs, Slovaks, Austrians, Hungarians, Croats, all Central Europeans share this frustration," he said. "And we all have our so-called minorities: Jews, Hungarians in Romania, Italians in Croatia. Everyone belongs to a minority, not in an ethnic sense, but in a broader cultural way. We Europeans are to the world what Slovenians were to Yugoslavia, 8 percent. Nothing. And nothing is self-sufficient. Those who think otherwise are the aggressors in this world."

He greeted an Austrian poet, a Croatian journalist, a Polish critic. From one end of the bar Ludwig, the wild-eyed editor from Salzburg, hailed him as Vilenica's "brilliant" director. Famous for his Slovenian debaucheries, Ludwig had gotten an early start. The hotel's only other guests, wealthy Italian and Austrian equestrians, hurried their children past him into the dining room.

Central Europe as a governing idea—one writer described it as a "nostalgia for Europe"—was officially discouraged in Yugoslavia, but it gained momentum when the Slovenian Ministry of Culture agreed to support Vilenica as a meeting place for writers from small countries and smaller cultures.

"It was a lucky time for us," Veno said with a mischievous grin. "This was where we first met Czech, Hungarian, and Polish writers, dissidents from everywhere. Not a single regime writer. From the beginning Vilenica was an underground function"—he called the cave where the literary prize was presented a *tectonic cathedral*—"because literature comes from the subconscious."

A Slovenian novelist pulled me aside. "Do you know what you see when you get delirium tremens at Vilenica?"

I shrugged.

"White horses," he said.

Ludwig was holding forth on Slovenia's headlong rush to the West. He was waving his arms, and with each spill of his drink the circle widened around him. "Europhoria," he said contemptuously. "I despise this ideology of freedom."

"Ludwig, you're a drunk," said a Slovenian novelist.

Ludwig shook his head. "I am a smuggler of words," he said.

"A drunk," the novelist repeated.

The editor drained his glass. "The center cannot hold, because it only knows power," he said. "Life must come from the margins."

I pointed at Ludwig's empty glass. "How do you do it?" I said.

"It's like when your father teaches you to sharpen a knife," he explained. "If you go too far, it won't cut anything. One must find that fine edge."

Tomaž Šalamun handed me a list of invited guests. "Nature, horses, wine," he said. "I do not even feel the need to read anymore."

Peter Handke, one of the first recipients of the Vilenica prize, was among those invited this year. The names of the two Serbian writers on the list had been crossed off.

"Will Handke show up?" I asked.

The novelist muttered a curse.

"Handke's the number one pen in Europe," said Tomaž. "Kundera's his only competition; when he came to Slovenia's defense, Handke had to do the opposite."

The novelist rolled his eyes—and quickly changed the subject. He wanted to know if Tomaž was making money on the stock market.

Tomaž began to explain the intricacies of futures trading and the commodities exchange. The novelist stared at him in wonder.

"You were always so avant-garde," he said.

An older Slovenian woman approached us. "Once I fell in love with a Parisian ballet dancer, but she wouldn't have me. So I threw myself on her floor and begged like a dog for her."

The novelist asked her if the dancer had changed her mind. The woman laughed.

"A fine edge," said Ludwig. "Poetry is a balancing."

Just past the hotel parking lot, hidden by brush and brambles, was a steep cliff. The next morning I took a walk along it and then ambled up the road to the pasture where the Lippizaners were going through their paces. Too hungover for breakfast, I leaned on the fence and gazed at the horses, trees, and sky, recalling a conversation with a professor of English from Ljubljana. I had suggested a reason for the sheer number of Slovenian poets: the landscape was a constant inspiration.

"The *genius loci* is very strong here," he agreed. "In fact, there's such a strong attachment to the land it's almost sick."

A young man from Carinthia, who came over to look at the Lippizaners, was pleased to tell me they were the subject of Edvard Kocbek's most famous poem. Kocbek (1904–1981) was a tutelary spirit for Slovenian literati. He was by turns a pre-war Christian Socialist poet and essayist, a war hero who went on to write about the ambiguities of the Resistance (eschewing the strictures of social realism), and then a disenchanted political official of the new regime. The first writer to denounce the post-war massacre of anti-Communists, Kocbek was subsequently stripped of his political position and sentenced to internal exile. Unable to publish his work, still he became Slovenia's intellectual conscience; his place in Slovenian letters was not unlike Dobrica Ćosić's in Serbian literary history, though Kocbek was blessed with a richer sense of irony. It was in "Lippizaners," after all, that he called Slovenian a language "for peasants and horses." Nor was he content to describe war in the Balkans in anything less than tragic and ambivalent terms. I was remembering the last lines of his "In a Torched Village"—"now I see over/ the horror's shoulders"—when the Carinthian boasted of having already published two collections of poetry, one in Slovenian, one in German.

I was in an ornery mood, so I said, "How do you decide which language to use?"

"German expresses my intellectual life," he said with feeling. "But Slovenian is an innocent, archaic language, the language in which a flower is still a flower."

"And this?" I said, waving at the landscape. "German or Slovenian?"

He stared at me for some time before he shook his head and walked away.

The cultural center in Sežana was the setting for Vilenica's formal discussions. The writers were concerned with what would replace the socialist system, and more than one speaker thought the quest for identity was the key to understanding the end of this millennium. If my notes are any record, the word *democracy* was absent from almost every speech:

"Small nations must be the keepers of diversity."

"Why did Austria-Hungary disintegrate? Because Vienna didn't recognize the Slavs."

"The idea of Central Europe was the dream of culture replacing politics. What collapsed was not only the world of Yalta but also that of Versailles."

"Central Europe is like a heart that gives out when an athlete overtrains."

"We are witnessing not only the demise of ideologies but of the individual. To hold on to their identities small nations must walk along the verge of history."

"Writers from small nations must be suspicious of universalism."

"The small nations started their independence movements without consulting one another. We had all these meetings in Belgrade, Zagreb, Skopje, Sarajevo, and Ljubljana, trying to figure out how to live together in Yugoslavia. And look at what happened."

"Many mirrors were broken in Sarajevo. The war has put an end to what I remember. I must remember what no longer exists."

"If one universalism is disappearing now, another is creating itself."

"War is a symposium where Death has its lecturers. Fiction is the only space that can escape universalism."

"No metaphors before noon!"

"The most universal thing in literature is silence, and poetry acts on that silence in the most individualistic way imaginable."

"Literature begins with the individual and ends in the universal, which is akin to the process by which new universalisms are created."

"We all know what happens when poets get involved in politics, and vice versa: repression."

"I want to tell you something romantic: I used to live in a city where I had to understand five villages all at once. And I must tell you that to live as a refugee from Sarajevo in one of your so-called democratic national states is horrible."

The human fish, *Proteus anguinus*, is found only in the Slovenian and Dinaric Karst. But this symbol of Slovenia's subterranean life is an endangered creature: the only specimen in the pool Aleš and I came upon in Postojna had withdrawn from view.

Though it is said the human fish descends from the dragon, photographs of it suggest a lizard lineage. Thirty centimeters long, with skin the color of Caucasian skin, it "tans" in the light; the farther you go into the recesses of a cave the paler the skin of the human fish. With pairs of hands, legs, and eyes, a worm-shaped body, and a flat tail, it is a figure straight out of the dark imaginings of Henri Michaux. What it most resembles is a human fetus. It has a set of atrophied lungs. It breathes through its gills. Once it was fashionable to steal human fish and put them in caves around the world. Alas, such transplants never worked. The human fish survives only in its native place. What sort of country has as its symbol such a secretive creature? I did not see it so I could not say. Perhaps I had seen nothing at all on this journey. Or is *Proteus anguinus* related to the god who changes appearances at will? In which case I saw it everywhere.

"The human fish is retreating deeper into the caves closed off to tourists," Aleš said. "It can't stand human beings."

My last night in Slovenia passed in an alcoholic blur. There was a reading at the Franciscan monastery in Piran, after which Aleš and I hitched a ride with a journalist from Ljubljana and his girlfriend. On our way to the car the journalist stopped for a brandy, and on the drive to a dinner party for the writers he opened a bottle of red (or "black") wine from the Karst. Twice in the last year he had lost his license: once for drunk driving, once for driving without a license.

"But that's nothing," he insisted. "When my friend got caught in a police ambush, they asked him if he would take a breathalyser. 'No,' he told them, 'I'll throw up.' And he did."

We headed south along a twisting road above the sea, toward the Croatian border. The journalist's girlfriend pointed at the water.

"Not that I'm nationalistic," she said, "but I would even pay Croatia for the rest of this gulf."

"This wine has no bones in it," said the journalist, passing me the bottle.

"What do you mean?" I said.

"Drink," he commanded, weaving across the road.

The sun had set over the Adriatic, and the Slovenians were telling Bosnian jokes. I had heard them before: how the public-address announcer at the Vienna airport must instruct the Bosnians not to feed the airplanes, and so on. While they were no crueler than other ethnic jokes it seemed a strange time to tell them. I gazed out the window, hoping we would not drive off the road.

There was no reason to stay for the next night's closing ceremonies. Kundera had skipped the festival, and the rumors of his wife accepting the award on his behalf had proved unfounded. None of the invited luminaries, including Peter Handke, Ryszard Kapuściński, and Adam Michnik (whose parody of Lenin—

nationalism is the last stage of Communism—should have been emblazoned across the top of Vilenica's program), had shown up. And who could blame them? Vilenica was one more conference in a small country desperate to enter the common house of Europe.

No one could remember where we were supposed to go, so we drove to the border and then back toward Piran, stopping at every restaurant. We found the writers in a shabby inn off the main road, on a hill overlooking the sea. Our waiter, an insolent old man, refused to serve us, and it was some time before another waiter brought our food. By then we had polished off two more bottles of wine. The journalist added a splash of mineral water to his glass—"I have to drive back to Ljubljana tonight," he joked—and drank it down. After dinner Aleš and I declined his offer of a ride and took the bus to the hotel, arriving well after midnight. In the bar, Ludwig was on another roll, toasting the new world order. And no one stopped him when he stumbled out to his car at four in the morning. Ludwig will do anything! A crowd gathered on the balcony to see him off, making bets about how many kilometers he could drive before the police pulled him over.

He did not get very far. Having made the mistake of parking his car by the cliff, when he fished his keys from his pocket—no small matter—and started the engine, he drove straight ahead, crashing into a rock. It took him ten minutes to climb out of the front seat, and then he fell into the brush just below the cliff. The writers hurried to the parking lot. Aleš waded into the bushes, helped Ludwig to his feet, and steered him toward the hotel. But Ludwig pushed him away. He punched at the air, he swung around, he swayed back and forth until he fell again, tumbling over the cliff and onto the jagged rocks some fifty meters below.

"Leave him there," said a poet who had just come from America. "He can sleep it off."

Wiser heads prevailed. A rescue team was summoned, and Aleš climbed down to the fallen editor, who lay unconscious for some minutes. Ludwig had broken several bones and ruptured his spleen. He did not know where he was when he came to.

"I can hear the stones talking," he said over and over.

A fire truck, an ambulance, and two police cars arrived. The rescue workers laid out ropes and a stretcher then belayed down the cliff. The crowd looked on in silence. Ludwig would have to be hoisted up the rock face—it was perhaps a blessing to fall in a mountaineering country—and at the hospital surgeons could begin to piece him back together. But for now he was too drunk to feel any pain. He drifted in and out of consciousness. Aleš ordered him to stay awake.

"Where am I?" Ludwig kept saying. "This is my home."

Part II

December 1992–February 1993

8

Croatia

The lector was asleep. The crowd began to titter.

The Hofmusikapelle's year-long celebration of the three hundredth anniversary of Mozart's death was concluding with his unfinished *Requiem*. The Vienna Boys Choir had joined the orchestra for this performance, a liturgical text had been added to the score, and between the first and second readings the lector, seated on stage, nodded off. I was watching the concert on a closed-circuit television in a room by the balcony, happy to be inside after spending a rainy day visiting the tombs of the Habsburgs. Vienna itself had the moribund feel of a museum, with a gaping hole: the Spanish Riding School, which once trained Lippizaners for the court, had just burned down. The newspapers said it was a miracle the horses had escaped.

Kakania, Robert Musil's fictional name for Vienna, was the jumping-off place for my next Balkan journey. The former home of the Holy Roman emperors, this melting pot of Central European peoples and cultures was the lodestar for many South Slavs, and I wanted to gauge what impression the Third Balkan War was making on it. Precious little, it seemed. Despite the rain, this afternoon the streets were crowded with Christmas shoppers. The conversations I overhead brought to mind Ulrich, the hero of *The Man Without Qualities,* who in the opening pages of Musil's unfinished novel cannot decide how to renovate the little château he has just purchased. "He was in that familiar state," Musil writes, "of incoherent ideas spreading outward without a center, so characteristic of the present, and whose strange arithmetic adds up to a random proliferation of numbers without forming a unit." In short, Ulrich cannot put his house in order. Soon he realizes that politics and, in fact, all his relations in Viennese high society, have been drained of meaning: this on the eve of the Great War. A world is on the verge of disappearing. No one in Vienna sees the storm gathering on the southern horizon.

It was not hard to find parallels between Ulrich's predicament and the West's response to the war in Bosnia. The assassination of Franz Ferdinand in Sarajevo, that dark cloud (never mentioned) hanging over *The Man Without Qualities*, buried the Concert of Europe, which had emerged from the Congress of Vienna in 1815 to help preserve peace on the Continent. No such concert was working in the Balkans, not with the EC in disarray on foreign policy matters, the UN mired in a failed peacekeeping mission, and the United States adrift in the post-election interregnum between the Bush and Clinton administrations. It was perhaps too much to hope that in his last weeks in office George Bush would use military force to stop the fighting (though outgoing Secretary of State Lawrence Eagleburger had branded Milošević, Karadžić, and Mladić as war criminals). And the incoming Commander in Chief? Bill Clinton was devoting his attention to domestic concerns.

Thus the siege of Sarajevo tightened. And the killing went on. And prisoners released from the death camps were telling their stories. They were not alone in bearing witness to crimes against humanity. A Bosnian Serb soldier, arrested in Sarajevo after making a wrong turn in his car, had just confessed to raping and murdering scores of Muslim women. He said he was only following orders. There was indeed nothing memorable about him, unlike Moosbrugger, the sex-murderer whose case, at the heart of *The Man Without Qualities,* fascinates Viennese society. (Is he responsible for killing a prostitute? Or is he insane? "His presence," one critic suggests, "keeps recalling the catastrophe that awaits the elites who can no longer give sense or direction to their experience.") The soldier languishing in a Sarajevo jail cell was all too ordinary. Bosnians called him a monster.

Laughter filled the concert hall. The lector, waking, gazed groggily at the audience. He shook his head, pulled himself upright, and walked to the lectern, where he began to read.

"Tito," said the former model, "could point the tip of his little finger at you"—here she pointed her little finger at me—"and snuff out your life."

The train from Vienna was climbing through snow-covered mountains. Ana, a tall redhead who looked too young to be the mother of a thirteen-year-old son and too thin to have delivered a baby girl only months earlier, was telling me about modeling in Skopje, the capital of Macedonia. Before the show, a government official had appeared in the dressing room. Marshal Tito, he told the models, would be pleased to speak with one of them. They should be prepared for his signal, which could come at any time. The show was almost over when the dictator, who did not have long to live, chose Ana. She could not remember anything he said to her.

"He lived in another world," she said. "Our social fabric meant nothing to him."

A revealing choice of words. Tito's interest in fashion was well known. Born in Kumrovec, a village in the Zagorje district on the Croatian-Slovenian border (his father was Croatian, his mother Slovenian), late in life he recalled that as a boy he hoped "to be a tailor, a natural result of the wish of every little peasant in Zagorje to have nice clothes." He left home at fifteen because he wanted a better wardrobe; and his first major purchase in Zagreb, where he found work as a mechanic, was a new suit, which was soon stolen. During his dictatorship, Tito's love of finery, coupled with his fear of being robbed, translated into an inordinate fondness for ceremonial pomp and secret police.

First the Great War. Tito fought on the Austro-Hungarian side. Captured by the Russians, he went from convalescing prisoner to overseer of a POW camp in the Urals and then soldier in the Red Army; what little action he saw in Russia's Civil War he made up for in Yugoslavia's. Upon his return to Croatia he worked as a labor organizer until he was charged with political agitation and imprisoned for five years, during Alexander's dictatorship. In 1937 he reorganized the Yugoslav Communist Party at the behest of the Comintern, and when World War Two broke out he emerged as the leader of the Partisans. From his stronghold in the Bosnian hills he built a resistance movement, supported first by the USSR and then by the Allies. A pattern was taking shape. Tito knew how to play one side off another, in the international arena and at home, a strategy enhanced by Yugoslavia's crucial geographical position between the East and West. Whatever misgivings the Allies had about Tito's Communism they sacrificed to the exigencies of war: the Partisans inflicted more damage on the Nazis than the forces supported by King Peter's government-in-exile, the Četniks under the command of Draža Mihailović. Fitzroy Maclean, the British officer who parachuted into Yugoslavia and, after making contact with Tito, recommended that the Allies back the Partisans, records a marvelous exchange with Churchill over Tito's politics: "'Do you intend,' [Churchill] asked, 'to make Yugoslavia your home after the war?' 'No, Sir,' I replied. 'Neither do I,' he said. 'And that being so, the less you or I worry about the form of government they set up, the better.'"

The Yugoslavs had little to choose from in 1945. The murderous Ustaše were in flight, King Peter was in London, and Mihailović had a date with the firing squad. Tito moved quickly to set up a Communist state—executing his enemies, taking over the media, confiscating private property and church lands. Western observers thought he was acting on instructions from Moscow until in 1948 Stalin expelled the Yugoslavs from the Cominform for refusing to enter into a Balkan alliance with Albania and Bulgaria. Tito's courageous resistance left a chink in the Soviet armor (like the Partisan actions that contributed to Hitler's downfall) and opened the door for Yugoslavia to Western financial institutions. Tito then joined in the founding of the non-aligned movement, the "third way" between the capitalist and communist blocs, and acquired the veneer of a world-class power broker; a cult of personality developed, which persisted in some quarters. I had seen his photograph in Slovenian mountain huts; from those who stood most to lose from a rekindling of nationalist feelings (Bosnians, Macedonians, children of mixed marriages) I frequently heard his praises sung. He was "an improviser," according to the scholar Fouad Ajami, "and his skill went with him to his grave"—a tailor, that is, who used the threads of his personality, an anti-nationalist policy known as "Brotherhood and Unity," and the secret police to stitch together a country torn apart by civil war. His followers in the collective leadership lacked his abilities: one reason why they invented the slogan, "After Tito—Tito!" But after Tito came nationalists and politicians with no interest in the good of the whole.

The Serbs ripped up the fabric of Yugoslavia, Ana declared near the Slovenian border. That was why she and her husband—a clever but unimaginative man was how she described him—had moved to Vienna just before the war. She was traveling to Rijeka to visit her parents.

"First they went mad," she said, "and now they're ready to die for this meaningless war. Yet they're a generous people, much warmer than Croats. In fact, Serbs are better men than Croats. They have more soul, more capacity for deep thoughts and poetry than we do, but they brought this all on."

She recited complaints I was to hear repeatedly in the coming weeks: that Serbs had dominated Yugoslavia's police, military leadership, diplomatic corps, and foreign press (this was true). In Zagreb, Ana claimed, 95 percent of the teachers were Serbian (false).

"Do you know what Aristotle said?" she asked. "If you have one bad cobbler in your village, then ten people will go without shoes. But if you have a bad teacher, he can wreck the whole society. That's what happened to us."

"I had the evening program, but at the station a soldier was barring the way. I didn't know what was going on, and when I went on the air I just tried to be calm. All during the show I got calls from people threatening to kill me. You're a Četnik, they'd say. Or: You're Ustaša. They were mad at an institution associated with Yugoslavia. And when I signed off I said, good night. For that I am suspended for two years. 'How can you say good night when your Croatian brothers and sisters are suffering?' the director cried. Do you know, for the next year no one said good night at the end of the evening program! Can you imagine? That's why I say I was the first victim of the war."

Jadranka Pintarić grinned. She was a young woman dressed in black, with a weary voice and a weary kind of beauty. Her dark hair was cut short; her sole ornament was a necklace made of twisted pieces of tape, the censored outtakes of a colleague killed in Vukovar. Jadranka despaired of ever falling in love. The pool of eligible men was shrinking daily: thousands had gone into exile to avoid military service and thousands more were planning to leave. An elderly émigré architect was courting her—she reminded him of a woman he had known before fleeing abroad thirty years earlier—but he bored her. Yet for her sake he had moved back to Zagreb, leaving his wife and family in Toronto.

During her suspension from the evening program, Jadranka hosted a weekly arts program. With hyperinflation cutting into her salary, which had fallen from 900 deutsche marks a month to 100, she made meals out of the hors d'oeuvres served at the exhibition openings she covered, dispensed with new clothes, and sewed her own Christmas presents. Yet in the Palace Hotel café she insisted on paying for our drinks. At the next table were two young men from Herzegovina wearing Italian suits and designer sunglasses, a wad of marks piled high between them.

"I don't like the way Zagreb is changing," said Jadranka. "The trams are always full, and bars like this have been taken away from my friends and me."

The young men tried, without success, to draw her into a conversation.

"What business are they in?" I asked.

"Oh, import-export," she sighed.

"What do they export?"

"Almost nothing!"

"And import?"

"Everything!"

The assassination of King Alexander in 1934, which prompted Rebecca West to travel to Yugoslavia ("We had passed into another phase of the mystery we are enacting here on earth, and I knew that it might be agonizing," she wrote), was the brainchild of the ultra-nationalist Ante Pavelić. A French court sentenced him to death in absentia (the regicide had occurred in Marseilles), and his sponsor, Mussolini, imprisoned him in Italy. But church bells rang in Zagreb when Pavelić returned from exile in 1941 to lead the Ustaša regime. The majority of Croats enthusiastically embraced him. For pious Catholics he was a bulwark against Orthodox Serbs; for artists, intellectuals, and politicians, a hedge against Soviet Bolshevism. Within a week of his arrival the roundup of Jews and deportation of Serbs had begun. Then came the forced conversions and death camps. The Irish writer Hubert Butler said Pavelić was "a desk murderer like Himmler . . . the epitome, the personification, of the extraordinary alliance of religion and crime which for four years made Croatia the model for all satellite states in German Europe." Indeed he attended Mass every morning, and in the name of Christianity he had thousands of Christians, Jews, and Gypsies executed. But Pavelić was never prosecuted for his crimes against humanity. He fled to Rome, disguised as a Spanish priest, and then to Argentina. He died in his bed in Spain, in 1957, presumably with a clear conscience. For in the NDH he had "applied the simple creed of One Faith, One Fatherland," Butler wrote, "with a literalness that makes the heart stand still." Many Croats considered him a hero. There was even a movement to have his remains returned to the soil he had helped to "liberate." *Contaminate* might be the better word.

A cold, overcast afternoon. The smell of burning coal in the air. I had moved into the spare bedroom of Jadranka Pintarić's flat in Dubrava, a suburb carved out of the farmland east of Zagreb, where rural habits endured. Fenced gardens pressed up against the stuccoed walls of the houses; chickens picked in the wet earth, among the cabbages; scrawny German shepherds growled at us when we strolled to the tram station. At one end of the street were piles of garbage and debris, about which Jadranka remarked, "We have a saying, *Wild will overtake culture*"—and across the meadow was the former JNA hospital. In the first days of the war, Jadranka had waved from her balcony at a man she thought was a Croatian soldier shooting at the hospital. He had promptly fired at her, and missed. Her cousin had not been so fortunate: felled by shrapnel when a missile struck his house, his liver and kidneys scattered across the floor, he was still recovering at a popular spa given over to war invalids.

Policemen patrolled every corner of the route the tram took to the center of Zagreb. Croatia's first elected president, Franjo Tuđman, had inherited from Tito respect for a vast security network as well as a taste for pomp, with national colors—Croatian heraldry and hussars. Tuđman came by his nationalism in a circuitous fashion. A Partisan general (from Tito's own district), when he left the JNA to become a historian he argued that Serbian scholars wildly inflated the numbers of Serbs killed by the Ustaše. This did not go over well with Tito who by 1968 was again enmeshed in the Croatian national question: the Croatian Spring, which began with a literary declaration, became a political movement to win more autonomy for the republic, and concluded with Tito purging hundreds of politicians and jailing intellectuals, including Tuđman. The historian's writings landed him in prison again in the early 1980s, which made him a national hero. The émigré community backed him in the founding of the Croatian Democratic Union (HDZ) in 1989. At the party congress in February 1990 Tuđman vowed to fight for self-determination and sovereignty:

> Our opponents see nothing in our program but the claim for the restoration of the independent Croatian Ustaše state. These people fail to see that the state was not the creation of fascist criminals; it also stood for the historic aspirations of the Croatian people for an independent state. They knew that Hitler planned to build a new European order.

What an order Tuđman was creating. After his election (with only 40 percent of the vote) he purged Serbs from the government, police, media, and schools; invited exiled Croats, including war criminals, to return; renamed streets after Ustaša figures; and restored the flag flown in the NDH. He was correct in asserting that the flag's checkerboard design (ubiquitous in the new country) dated back to medieval Croatia, and Croatia's minorities were also right to fear him. "Thank God my wife is not a Jew or a Serb," he said in a speech delivered in Jadranka's neighborhood.

Sovereignty remained his primary objective. But his plans to declare independence on 10 April 1991, the fiftieth anniversary of the NDH's founding, went awry in early March, when huge protest rallies were staged against Milošević in Belgrade. Although they disliked each other, Tuđman and Milošević stuck together for their own political ends, producing a combustible binary system in which Tuđman's anti-Serbian policies gave Milošević more license to preach Greater Serbianism, which in turn fueled Croatian nationalism, and so on. Mirror images, with this difference: Tuđman was an ardent nationalist, Milošević an opportunist who exploited his people's nationalism. Within days of the Belgrade protests, Tuđman and Milošević took a decision to divide Bosnia along the same lines that Serbian and Croatian leaders had drawn on the eve of World War Two. The Sabor (or parliament) in Zagreb anticipated by hours the Slovenian declaration of independence, the only advantage Croatia had over its northern, homogeneous neighbor. For Tuđman not only lacked Slovenia's battle readiness but also its mandate to secede. One in five of his people was non-Croatian, the majority of these Serbs, and he could neither fend off their rebellion nor withstand the JNA offensive launched in their support; hence a third of his country lay in Serbian hands (separated by a UN military force (UNPROFOR) of 14,000 peacekeepers), and in the territory under his control one in five people was now a refugee. The father of Croatia had begotten a disaster.

The train took us to the central square, which was teeming with refugees—old men in threadbare overcoats smoking, and stamping their feet to keep warm, and trading stories by the equestrian statue of Baron Joseph Jelačić von Buzim, for whom the square had lately been renamed. The Communists had removed the statue from its plinth in 1947; its restoration six months after Tuđman's election was a decisive event in the independence drive. Thousands of people turned out to see men and women on horseback, dressed like hussars, escort the bronze statue through the city, accompanied by a choir in folk costumes.

Who was Jelačić? "Like many nations that try to invent or improve their history, the Croats had got it all slightly wrong," writes Tito's biographer, Richard West. Jelačić belongs to the pantheon of Croatian heroes whose relationship to the idea of Croatia can best be described as ambivalent. Three centuries after Slavs migrated into the area, Tomislav (d. 928) forged the short-lived Triune Kingdom of Croatia, Slavonia, and Dalmatia, which curved around Bosnia like an inverted question mark, and which for a thousand years posed an unanswerable question to Croats: how to unite three distinct regions in one independent nation? In 1102 the Croatian tribes agreed to recognize a Magyar, Kalman, as their king, retaining some autonomy, including the right to elect their own *ban* or viceroy, and for eight centuries they belonged to the kingdom of Hungary, a Habsburg possession after 1526. The inevitable deterioration of Croatian-Hungarian relations was as nothing compared to the vassals' long history of war, principally with the Ottomans, who razed their villages and towns, advancing as far as the outskirts of Zagreb. But when the Croats did not fall to the Turks they became for the Vatican the ramparts of Christendom. Jelačić (1801–1859), a German-speaking frontier officer, won a name for himself with a reprisal raid on the Turks in Bosnia, though he lost almost as many men as he killed. Appointed *ban* in 1848, in the spirit of the age he freed the serfs. Then he marched on Hungarian revolutionaries, earning Marx's wrath and a chestful of medals from the Emperor. Jelačić had supported the Habsburgs in the hope of creating a confederal Slavic Austrian Empire, but it all came to naught. Not only did Croats lose most of their liberties, but to Jelačić's chagrin relations worsened with the Serbs who had fought alongside him. He was at heart a Yugoslav.

Now he had been restyled as a Croatian hero to suit a nation led by a historian of dubious merits. Tuđman had begun to fulfill his people's millennial dream of independence, along the same geographical lines as Tomislav's kingdom, but at what a cost. The question now was where, not when, he would resume fighting, his specious argument that Bosnia was historically Croatian territory having laid the groundwork for another front. Rumors were flying of a general mobilization, though few thought Tuđman would act before March, when the peacekeeping mandate in Croatia ended and the weather cleared. Every historian knew the danger of initiating a winter offensive in the Balkans.

Andrija was moving to Prague. The sociologist had given up his position at Zagreb University before he was fired for being Serbian. Though his research funds had been cut off and anonymous threats had convinced him to leave town sooner rather than later, he wanted no part of any discussion about discrimination against Serbs. Tuđman's Croatia-only campaign had cost him his livelihood and his homeland—he was born and raised in Zagreb—yet his anger was directed at Slovenia: in his mind it bore more responsibility for the fighting than the international community would acknowledge. Andrija was obsessed with the differences between Slovenia and the rest of Yugoslavia.

"It's impossible to imagine a Slovenian sense of humor, isn't it?" he said.

He told a story about Václav Klaus, the Czech prime minister, addressing the Slovenian parliament. The translator was hard of hearing, and Kinkel had to repeat everything. Finally, he said, Why don't we do this in English? But only Dimitrij Rupel, Slovenia's outgoing foreign minister, could speak English. Kinkel threw up his hands in despair.

"Slovenians are doing everything they can to distance themselves from us and join Europe," Andrija said. "One day they'll realize they need our markets, because Italy and Germany aren't interested in their products. It's not good for them to act like they were never part of Yugoslavia."

The one thing Andrija regretted about his visiting professorship at Charles University was that he would have to stop studying the ways in which refugees were shaping life in Zagreb. But the Croatian government would no longer fund such research, even if it was carried out by a Croat.

"There's no money for anything, is there?" I said.

"Only arms," he said.

Jadranka found the money-changers under a colonnade, at the edge of Jelačić Square. These were good times for them, with inflation spiraling upward and the government strapped for hard currency. Jadranka's best defenses were the deutsche marks she sometimes earned for free-lance work and her American Express card, which she used in order to take advantage of the devaluation of the original price by the time the bill came. Unfortunately for her, many merchants, wise to this strategy, accepted only cash; likewise restaurateurs, who penciled in prices that seemed to rise every few days. Jadranka pulled a young man aside and began to haggle with him; fifteen minutes later, she had changed a hundred dollars for me at 20 percent more than the official exchange rate.

The economic role in the Yugoslav wars of succession cannot be overestimated. Tito's "socialism with a human face," propped up by generous international loans, enabled the country to go on a spending spree in the 1970s. While the government imported expensive technology and invested in factories, Yugoslavs bought cars and washing machines and television sets, acquired a taste for foreign travel and shopping in Italy, built new and bigger houses. But public debt began to mount before Tito died, and in the early 1980s, when Yugoslavia could no longer service the interest on its loans, the IMF forced the government to devalue the dinar and make large spending cuts. Inflation continued to rise, however, and austerity measures (including a new tax on foreign currency that hit Croatia particularly hard because of its extensive tourist trade on the Dalmatian coast) fueled discontent. It was easy to blame other ethnic groups for shortages of goods, power outages, labor strikes, unemployment. With the Communist party losing its authority, the market-oriented reforms instituted by the last federal premier, Ante Marković, could not withstand the nationalist charge. Marković, a Croat and perhaps the last true Yugoslav, had U.S. support and an EC promise of $4 billion for Yugoslavia if the country would stay together. But in the end even Croats turned on him, saying he would have no chair to occupy when he returned to Zagreb. For by then Milošević had imposed trade boycotts on Slovenia and Croatia and looted $1.8 billion from the national bank; when the war began, he seized control of Yugoslavia's hard currency reserves. And Tuđman understood that one antidote to the economic misery accentuated by his failed policies and his encouragement to friends and family to buy up Croatian businesses and factories at bargain basement prices was war.

"Our only hope," Jadranka sighed, "is if someone will assassinate our president."

Famous among her friends for her ability to foresee the future, she sadly admitted to having no special knowledge about Tuđman's future. But when she opened her mail that night and learned her gas bill had gone up 250 percent Jadranka made a prediction.

"War is coming," she said.

The strange and emblematic career of Ivan Meštrović (1883–1962), the shepherd from Dalmatia who became a world famous sculptor. His biography approaches the fabulous: a descendant of *hajduks,* the bandits who waged guerrilla war against the Ottomans, he trained at the Viennese Academy. Then he joined the Secessionists, who had no use for historical subjects, and made his reputation with a series of works on the Battle of Kosovo. Among his statues of fallen Serbian heroes was a relief, "The Artist of My People," in which an old blind man with a *gusle* (the one-stringed instrument that accompanied medieval ballads) leans on a boy's shoulder, a figure of the Serbian artist passing on his people's foundational myth. But Meštrović was Croatian. Nevertheless one critic said he "entered into the spirit of the Kosovo epopee so deeply that even the easternmost Serbs could not have done better, making him a greater Serb than most Serbs." How was this possible? Meštrović considered the Yugoslav state, its groundwork laid by artists and writers, his people's greatest achievement, so he devoted himself to forging for them a unitary Yugoslav culture.

A hopeless task, even for the "Prophet of Yugoslavism." His most ambitious design was the Kosovo Temple, but his monument to the sufferings of the South Slavs was never built. A failure rich in symbolism: the temple, combining Catholic and Orthodox architectural elements, exists only on paper—or many papers, since Meštrović made drawings of it for the rest of his life. And with a dome bigger than Saint Peter's and a five-tiered tower of ages (one for each century of Ottoman rule), the temple borders on the pretentious. Indeed Meštrović's works on Yugoslav themes, scattered through the republics, are often bloated, as though size alone could compensate for poverty of inspiration or mask his waning faith in the Yugoslav idea.

In Alexander's dictatorship the sculptor found it difficult to champion a secular cure for the South Slavs' misery. Like Auden (and at about the same time), Meštrović returned to his childhood faith (in his case, to Catholicism); his subsequent religious works are among his most powerful. If once he described his spiritual yearnings as "mysticism without God," his disenchantment with politics led him back to the New Testament, more so after he was imprisoned by the Ustaše. The price of his release was his proposal to travel to Rome to execute a bust of the Pope. From there he escaped to Geneva, only to become seriously ill. Bedridden for months, when he recovered he set to work on a wooden relief of the Crucifixion. All the available oak, however, had been bought up by a local manufacturer of rifle butts, and at the factory he could find only a few boards and one withered trunk, probably rotten. But when he cut into it he discovered solid wood. The tree came from Croatia. And the carving Meštrović sent back to his homeland was of an emaciated, disfigured Christ, which he said represented "the crucifixion of His idea." It was a subject he knew well: his first wife, whom he had divorced twenty years earlier, and thirty members of her family, all Jews, perished in the NDH.

The gulf between Meštrović's youthful ideals and post-war sensibility may be measured in his refusal to accede to Tito's entreaties to return to Yugoslavia—a blow to the new leader, still consolidating his power, who promised the sculptor riches and liberty. Meštrović instead obeyed a calling familiar to South Slavs, moving to the United States in 1946 to teach at Syracuse and Notre Dame. More symbolism: Yugoslavia's greatest artist died an American citizen. Now Croats despised his Yugoslav works; Serbs dismissed his Catholicism. His Church of Saint Savior, an elegantly simple hilltop design near Split, which housed the Meštrović family vault, had not survived the war.

"The Serbs stole the sculpture, destroyed the mausoleum, and shit on the floor of the church," Jadranka told me one morning over a breakfast of cheese-and-spinach pie washed down with homemade blueberry *rakija*. "Can you imagine what kind of people can do that?"

So much for breakfast. I reached for the bottle of *rakija,* one of several Jadranka had placed on the table for me to sample. Her brandy, which she also made from pears, peaches, grapes, apples, and even walnuts, gave me as much pleasure as the way she answered her own question, inventorying the stereotypes underlying Yugoslavia's ethnic jokes. Serbs were nasty, Slovenians, stingy. Bosnians were stupid, and Albanians couldn't learn languages. But Montenegrins were lazy.

"Good morning, Montenegro," she called out. "It's twelve o'clock."

And Croats? They had an excessive concern with hygiene.

But Jadranka had forgotten the Macedonians. When I asked her to tell me a joke about them she could not think of any.

"Would you like to try *rakija* made from walnuts?" she said. "It's very good."

The HDZ controlled three of the four national newspapers as well as the state television and radio stations; the three independent newspapers were under pressure from the government, and Radio 101, the only independent station, was also in trouble. Ironically, in the 1980s Tuđman had used Radio 101 to attack the Communist authorities, and now he wanted to close it down to muzzle any opposition. What was true of the Communist era—state control of the media and the arts drove people to value independent culture much more than in an open society—held for nationalist Croatia, too. So what are the artists doing? I asked Jadranka. In the new establishment the social realists had given way to a school of "naive" artists whose sentimental rural scenes kindled national pride. Jadranka, who had seen enough primitive art to be able to write her reviews in minutes, welcomed the chance to cover the Labin Art Express. These ten artists wanted to convert an abandoned mine in Istria into a cultural center. And they dreamed of autonomy for Istria, the mountainous peninsula, smaller than Rhode Island, divided among Slovenia, Italy, and Croatia. Art is an underground activity, they said, governed by ideas more than by technique. Their work: photographs of a mining operation; four videos of a young man hooked up to an EKG, beating his chest with a metal prong, his arms bleeding profusely; drawings of the proposed cultural center. Jadranka thought the food at the reception poor: figs, deviled eggs, potato chips, grappa. She had no sympathy for the group's political aims.

Will the nationalists worry about this exhibit? I asked.

"They think artists are just children," she replied. "Harmless."

One artist, however, had made some interesting drawings, with coal dust, of inchoate figures and labyrinths on mirrors painted the color of fire.

Jadranka was unimpressed. "How far can you divide something," she said, "before it ceases to have meaning?"

Maps. In the kiosks updated maps of Croatia were for sale, and at the Museum of Ethnography, an imposing yellow building in Secessionist style, an exhibit of Croatian cartographical history had just opened. The colorful maps, some dating back to the sixteenth century, along with globes, sextants, and compasses, were arranged according to aesthetic principles, not chronology. I had the exhibit to myself, and what struck me were the changes history had wrought upon Croatia's borders. "More delicate than the historians' are the mapmakers' colors," Elizabeth Bishop wrote. But what historians these were! The short history of Croatia included in a book of maps distributed by the Croatian Information Center made no mention of Ustaša atrocities. The main World War Two entry:

> Germany attacks Yugoslavia, which capitulates after only a few days. Proclamation of the Independent State of Croatia (NDH). Dr. Ante Pavelić returns from Italy and takes over the government. One part of Croatia under Italian, the other under German and Hungarian rule. Serbian terrorist Četniks, under the command of Draža Mihailović, carry out horrible massacres in eastern and southern parts of the Croatian state. Throughout that year, they executed tens of thousands of Croats and Muslims.

"Omissions are not accidents," warned Marianne Moore.

"He's not a terribly attractive figure to most Croats," the American scholar was saying.

The bar was crowded, and to the historian's mounting anxiety he could not catch the waiter's attention, even though he spoke good Croatian. I was just happy to be inside on a rainy night, hearing about the subject of his Fulbright research, Josip Strossmayer (1815–1905), the Catholic bishop and founder of the first Yugoslav Academy. Once revered in Zagreb, the clergyman had fallen out of favor. The historian was hoping to restore his credibility. No small task. Strossmayer, after all, had argued for spiritual reconciliation between the Eastern Orthodox and Roman Catholic faiths; at the First Vatican Council he spoke vigorously against adopting the dogma of papal infallibility. Strossmayer knew how the Orthodox would react to what they viewed as heresy.

"He was guilty of everything Pope Pius IX wrote in the *Syllabus of Errors*," said the historian. "But he's the bridge between Croatian culture and Western Europe."

For contemporary Croats he was guilty of much more because, as one scholar wrote, he "was incapable of bigotry." Strossmayer wanted not only to Westernize the South Slavs but also to unite them, through language and religion, in a Yugoslav federation, within the Habsburg Empire, which could liberate lands under Ottoman control for Christianity. What better way to protect the small nationalities? he argued. But his Yugoslav idea no longer had any currency in Croatia, reconciliation with the Orthodox Church was out of the question, and among ecclesiastical figures Strossmayer ranked much lower in the public imagination than Alojzije Stepinac, Zagreb's archbishop in the NDH, the man responsible for the Church's immediate acquiescence to the dictates of the murderous Ustaša regime: a terrible sinner who, in the Irish writer Hubert Butler's words, allowed "a whole vocabulary of Christian goodness [to be] blown inside out like an umbrella in a thunderstorm."

And Strossmayer? A short walk away was a sculpture of him by Meštrović. "He was the last romantic," said the historian, rising to his feet to stand in the path of a waiter heading for the bar. "He wanted to bring the Slavs into the greater Western sphere, but unfortunately it never happened."

The weather continued cold and overcast. Some afternoons I liked to walk up the cobblestone street of Radićeva to the Upper Town, around which Zagreb was settled in 1094. Legend has it the city got its name from a military commander who, plunging his sword into the ground near Jelačić Square and seeing water gush from a spring, cried to his thirsty troops, "*zagrabite*"—"scoop it up!" Through a stone gate, which housed a shrine to the Virgin Mary, I headed for Strossmayer promenade for a view of the city the bishop thought should become Europe's cultural capital. But here was a provincial capital with none of the ferment and diversity common to artistic centers. Destroyed by an earthquake in 1880, rebuilt in Austro-Hungarian style, and presenting a sleepy array of red-tiled roofs, Zagreb would never rival Paris or Vienna for cultural riches (though Croatian mercenaries provided the necktie or cravat—from *hrvat* for Croat—which caught the attention of seventeenth-century French fashion mavens, like the modern vogue for combat fatigues inspired by "Che" Guevara.) It was said the truest Yugoslavs were Croats (Strossmayer, Meštrović, Tito); their mistake was to attempt to transcend cultural differences among the South Slavs. Better to honor diversity than to unite the unwilling: one of the century's lessons lost on the Balkans. Croatia's greatest modern writer, Miroslav Krleža, called Serbs and Croats one people afflicted with two national consciences, each defining itself against the other, sometimes with disastrous consequences. "God save me," Krleža wrote, "from Serbian heroism and Croatian culture": the martial deities of the Third Balkan War. But these were purifying times. The Croats were greedily drinking from a spring tapped long ago.

The wall of bricks outside UNPROFOR headquarters in central Zagreb commemorated the war dead. Candles and plastic flowers were scattered among the bricks; on each one a name was chalked, more than ten thousand in all. Even more than an indictment of the peacekeepers, the wall was a constant reminder of Croatia's greatest loss, Vukovar. The siege of this beautiful city on the Danube, the door to Slavonia, had already assumed mythic proportions in the Croatian imagination. Meter for meter, more shells landed on Vukovar than on Stalingrad; thousands of civilians and soldiers were killed; when the city fell to the Serbs in November 1991, after eighty days of bombardment, only rubble remained. Then came what international war crimes investigators called "an orgy of beatings": from the Vukovar hospital Serbian forces removed hundreds of wounded men and boys—civilians, hospital workers, political activists, militia—and mercilessly beat them. A uniformed soldier blew a whistle to summon fresh troops whenever his men grew tired, then the patients were shot and buried in a mass grave (one of the war's first) in the nearby village of Ovcara. The wall of bricks, then, was a monument to "ethnic cleansing" (80,000 Croats were expelled from eastern Slavonia) and the death of Yugoslavia, according to Branko Borković, Vukovar's last commander. Our conversation, in the lobby of the hotel next to Croatian military headquarters, began with his long-winded history of Yugoslavia, the JNA, and the war, confirming what one of his colleagues told me: it was impossible to imagine him having a childhood—he must have sprung full-grown from his mother's womb. A short powerful man with large ears and a crew cut, Borković was called *Mali Jastreb,* Little Falcon, but he would always be known as the man who lost Vukovar.

"A Pyrrhic victory," he said derisively. "Vukovar was Serbia's Stalingrad."

He had statistics to prove it: inflated numbers (extrapolated from Serbian media accounts) of Serbian casualties, destroyed tanks, and downed aircraft. He could even put a dollar amount on the total losses to the city—$3 billion—which a Serbian novelist proposed to rebuild in Byzantine style. But his most impressive figures came from the JNA itself: during six waves of assault on Vukovar, the army used up 800,000 shells, or about 110 per house, attacking with tanks, artillery, rocket launchers, MIG fighter jets, and helicopter gunships, which rose from the woods on the Serbian side of the Danube, fired, and dropped back to earth. The invasion took ten times longer to complete than military planners had expected. What the JNA finally seized was a city from which even birds, rats, and mosquitoes had disappeared.

"The Serbs wanted to destroy everything," said Borković.

He knew his enemy's methods, having spent his career in the JNA, ironically, the lynchpin of Tito's "Brotherhood and Unity" policy. And it was true, as he said, that Serbs and Montenegrins made up the majority (60 percent) of the officers, though contrary to widespread assumptions the percentages of Croatian, Macedonian, Muslim, and even Albanian officers grew in the twenty years before Yugoslavia's breakup. (The number of Slovenian officers declined slightly.) Borković rose steadily through the ranks until 1989, when he left the Communist party; stripped of his command in Zagreb, he blamed his demotion on his ethnicity. Then the daily questioning began (Slovenian independence? Democracy? Your grandmother lives in Munich?), until he was confined, during the Ten-Day War, in the psychiatric ward of the military hospital. Discharged before the end of the war (which was waged, he said, so that the JNA could test its officers' loyalties and discover the weak points in its command-and-control systems), he became one of fifty ex-JNA officers who built a Croatian army. They had little guidance from above. The defense minister, Gojko Šušak, a hardline émigré from Herzegovina who had made a fortune selling pizzas in Toronto, had a prankster's sense of military matters. In April 1991, for example, before Croatia seceded, he sneaked through the cornfields outside a Serbian village near Vukovar to fire a shoulder-launched missile at a peasant's house. Four months later, when Borković declined a posting to Dubrovnik in favor of Vukovar, from which tens of thousands of people had fled, the only way into the besieged city was to follow a tractor lane, through another set of fields, in the sights of Serbian gunners.

Borković's first task in organizing Vukovar's defenses was to arm the remaining civilians. His only equipment was a set of American combat fatigues, but from JNA weapons depots he rounded up Kalashnikovs, grenades, rocket launchers, and anti-tank missiles. Everyone was mobilized, some to fight, some to transport goods around the city; with no outside help or supplies, Borković had to appropriate civilian sources, rationing everything from milk to cigarettes to ammunition.

"Every shell had to count," he said. "I was always bluffing, feeding propaganda to the Serbs. *We are Ustaše, and we have so many soldiers.* It bought us some time."

Whether the propaganda worked or not, it was in Vukovar that the JNA perfected the black art of besieging cities from a safe distance, incurring the wrath of the international community when the medieval city of Dubrovnik and then Sarajevo fell victims to this curse. Paradoxically, the sacking of Vukovar marked the end of Greater Serbia (though this would not become apparent for several years), for with only 1,500 soldiers Borković wreaked havoc on the vaunted JNA, luring its tanks down side streets from which they could not escape, stalling its Slavonian campaign in the same way that the Partisans had once slowed Hitler's march across the Balkans. The damage his ragtag army inflicted on one of Europe's largest armies, which was also responsible for laying waste to one of Yugoslavia's prettiest cities, sparked resistance in Serbia to the war even as it gave rise in Croatia to a useful myth: that a plucky band of fighters could hold out against far stronger forces—until they ran out of ammunition. Needless to say, I understood none of this at the time. The fall of Vukovar weighed heavily on Croatian thinking; and I was more impressed with Borković's fate upon his return to Zagreb—he was placed under house arrest before being promoted—than with his prophetic announcement that he and his colleagues were building a small professional fighting force, modeled on the American Army, which would one day retake almost all of Croatia's occupied lands. But even this paled next to the verdict he delivered upon himself: "Vukovar is dead."

Alenka Mirković was happy to spend the evening away from her room in the Hotel Laguna. It was her idea to go to a pizza parlor in the Upper Town, where she refused to order anything. You will never understand the people from Vukovar, said the people in Zagreb, because they will always be at war. And Alenka was besieged by her memory. She managed the local office of an Austrian firm which produced telephone books, but in her imagination she was trapped in a city which had no use for directories. She was a schoolteacher by training, and though she had worked during the siege as a journalist for Vukovar Radio she remained a pedant at heart. Through a cloud of cigarette smoke she delivered a rambling lecture, of which I remember three things: how at the beginning of the siege "my boys," as she called the Croatian defenders, launched missiles with noisy cloud-seeding guns, which made the Serbs think they had the latest American weapons; how school lessons were broadcast over the radio to keep the children inside; how one night in the shelter a friend found a cookbook and read aloud the instructions for roasting a chicken. Alenka told her to shut up. There were costs to living for three months in a cellar, with up to sixty people at a time.

"You can see how each of us has been marked," she said. "One has a twitch, another can't keep his leg from shaking, a third smokes too much. I'm one of those people who doesn't have to eat, but I have to smoke. In Vukovar a cigarette had the same importance as a bullet."

And "her boys" received extra cigarettes for destroying tanks. By the last weeks of the siege, however, when their ammunition was almost gone, they were reduced to smoking corn leaves.

"The day before we left Vukovar," Alenka said, "a palm reader told me I'll have lots of children. Ha, I thought. We realized everything was over, but at the meeting to decide whether to leave the city I went back and forth. I wasn't afraid. I was in shock. It was pretty much the same: you could get killed going or staying. Finally I didn't want to wait for them to come and shoot me. We knew what they did to journalists, and as a woman I would be in bad shape. So the boys gave me a grenade to blow myself up if I was caught. I keep it in a drawer in my room."

"You're lucky to be alive," I said.

"I have plenty of chances to make up for that," she replied. "My uncle, my aunt, and my cousin are in Sarajevo."

Oil wells poked through the heavy fog. Flooded fields of corn, rye, and tobacco stretched from either side of the Brotherhood and Unity highway, which once connected Zagreb and Belgrade. A drive through flat and fertile country, past vineyards and brick houses all in a line. Pumpkins, said the schoolteacher from Vukovar, are considered hills in Slavonia. And her friend knew a man growing marijuana here in plain sight of the police; confiscating the plants was out of the question—the Serbs had laid mines around the field. Edo Popović lowered the volume on the tape deck, cutting off the refrain to Nick Cave's "One More Man Gone," and rolled down his window to listen for gunfire. Not since the first air raids had there been so many police checkpoints. Something's up, he decided. The cold air tasted of chemicals. A wall of smokestacks loomed in the smog. For some reason the Serbs had never shelled the factories in Kutina, ten kilometers from the front.

"Maybe they wanted to capture them intact," I offered.

"Serbs don't think like that," he replied.

The war had changed him, Edo admitted. A tall young man with a scraggly beard and long black hair, in his oversized U.S. Army jacket it was easier to picture him organizing a street patrol in Zagreb or filing dispatches from the front than writing the self-consciously literary fiction that brought him public attention. The first casualty of war is the truth, he liked to say. He took the line from Philip Knightly's study of war and propaganda, *First Victim*, a book he admired enough to found a publishing house in order to print it in Croatian translation. Edo burned to know the real story of the Yugoslav wars of succession, and his spirits quickened as he drove Alenka and me toward the UN-protected town of Pakrac, one of the war's starting points.

Balkan observers were not surprised when sporadic fighting broke out in the ethnically mixed regions of Krajina and Slavonia, months before the Ten-Day War. Croatia's underbelly had been a source of friction since the sixteenth century, when the Habsburgs turned these lands into a military frontier to stop the faltering Ottoman advance on Europe. The ramparts of Christendom being little more than a series of castles manned by German mercenaries and Croatian refugees from Bosnia, the Habsburgs encouraged Vlachs from the Serbian Orthodox Church to move into the area, a minor policy decision that was to have major political repercussions. For in 1690, when Leopold I allowed 30,000 Serbian families fleeing from Kosovo to settle in Vojvodina, expanding the military frontier far to the east, he let the proverbial wolf in at the door. A semiautonomous region now stretched from Dalmatia to Transylvania; to the Croats' consternation Serbs began to migrate into Slavonia, followed by Germans, Czechs, Jews, Hungarians, Ruthenians, Slovaks, and Albanians, often with Vienna's encouragement. The monarchy thus created one of its most ethnically diverse regions—and a symbol of Croatian dispossession no less potent in the Serbian imagination than Kosovo.

Time to invoke the Law of Unintended Consequences: what for three centuries served as the dividing line between Islam and Christianity became a Christian battleground as soon as the Muslim threat receded. Catholicism and Orthodoxy. Rome and Constantinople. Croats demanded control of another area carved out of their kingdom, Serbs wanted religious freedom—differences exacerbated by the rise of nationalism in the nineteenth century. In World War Two fighting between the Partisans and Ustaše started here. Nor did the Germans' expulsion after the war improve relations between the Croatian and Serbian communities, for Tito moved Serbs from Bosnia and Kosovo into the empty houses, another "Brotherhood and Unity" gesture with a short life span. For the rich ethnic history of the region now lay in ruins. The old women carrying baskets of food along the side of the road, the men on bicycles, the boy leading a pig—all were Croats. Drive twenty kilometers in any direction, and this peasant tableau might be Serbian.

The next checkpoint, manned by peacekeepers, marked the entrance to UN Sector North, one of Croatia's four UN-protected areas. And every house and barn for the next several kilometers was gutted, many with the word **Ustaša** scrawled on the side. Cemeteries and cornfields separated the villages we drove through, some of which were razed, others not so damaged. Outside a dynamited church, Edo professed to see no method to the destruction, but it was clear that only ethnically pure villages were burned to the ground; where Croats and Serbs had lived side by side the cleansers chose their targets with more care. A little further along one house was blown up, the next untouched, a third dynamited in the middle. A mixed marriage, Edo joked. At the heart of the Third Balkan War was the bitter enmity between Croats and Serbs, Catholics and Orthodox: everyone else, Bosnians in particular, got caught in the crossfire.

Horses, too, it seemed.

Our first stop, a Lippizaner farm in Lipik, bore no resemblance to Club Lipica in Slovenia, the writers' watering hole where I had first seen the white horses. The charred ruins of this barn housed only unexploded ordnance. We could not venture beyond a narrow roped-off area where a contingent of Canadian peacekeepers had set up camp, because the paddock and fields were mined. The horses were dead, their throats slashed one night by Serbian paramilitaries. Habsburgs, be gone!

"It's silly to say, but they even found a mass grave of horses," said Alenka.

Edo was eyeing the peacekeepers. "They all have M-16s," he said wistfully. "That gun is the dream of every Croatian soldier." The Canadians, however, found nothing enviable in their situation: they were spending Christmas in the mud, surrounded by land mines.

On to Pakrac. Near a gutted building named Future was Gold's Bar, where we drank whiskey with several unemployed men, who took turns telling the story of the Serbs' attempt to seize control of this market town. By August 1990, the Serbian revolt, centered in Knin, the capital of Krajina, had taken a curious form: playing on overblown fears of a Croatian invasion, the mayor of Knin dispensed weapons to the largely Serbian population and sealed off the town, ordering boulders to be dropped onto the railroad tracks and barricades of logs rolled across the roads to the Croatian towns of Sinj and Drniš. Thus was born the Serbian Republic of Krajina. And the success of the "log revolution," as it was dubbed by the Croatian media, emboldened the rebel Serbs to occupy the police station in Pakrac, in March 1991. Croatian paramilitary units retook the town within hours, the Serbs fleeing into the hills north of Pakrac. No one was killed in the operation, but the Belgrade media painted a picture of Serbs massacred by Ustaše. Soon there were tanks in the center of town, their guns aimed at the police station. For the first time the JNA had moved against its own people. Then we knew there was no turning back from war, said the men in Gold's Bar.

Fortified with whiskey, we strolled through the town.

"If I can't take you to Florida, I'll take you to Pakrac," Edo joked with Alenka.

"I'd rather be here," she said, morose.

Here the Orthodox church lay in ruins. Likewise the priest's house: according to Edo, it had doubled as an arms cache. An old man eyeing the devastation said relations with the five hundred Serbs who had returned since the cease-fire were not good. Though Pakrac had been predominately Serbian before the war, Edo and the old man insisted otherwise. The Serbs wanted to make this place uninhabitable, said the old man. Total war, said Edo. Like Sherman's march to the sea.

"You could predict what would happen," said Alenka. "Serb TV would report that Ustaše had burned down a Serb village, and two days later a Croatian village would burn to the ground."

"Our problem was nobody countered the Serbian propaganda with the truth," said Edo.

From Pakrac we drove north through a gently rolling landscape of fields and woods, corn cribs and hayricks and vineyards. Dusk was falling by early afternoon; cigarette smoke filled the car; Nick Cave's "Who Will Be the Witness?" blared from the tape deck. This area was cleansed of Četniks, Edo said in a flat voice. At one end of a village was a bare tree covered with rags, a Czech custom to announce the birth of a child; at the other, a long line of mourners, first men then women, marching toward the Catholic church. And in Darvar, where the only building destroyed on the main street was an insurance firm, we drank more whiskey in a bar styled after an American train station. *The Captain's word is law here*, read the brass nameplate. A different law held at the Serbian-owned café around the corner. Two weeks earlier, a drunk policeman had bombed it, wounding fifteen, one critically. The soldiers who told us this story had nothing good to say about UNPROFOR. They liberated every centimeter of the holy land, Edo explained. They don't understand why the peacekeepers won't let them wear their uniforms. Alenka called UNPROFOR an occupying army.

Then to the village of Velika Peratovica, emptied—save for one old couple—for the second time since World War Two. The Serbs who had taken over the houses and fields of the expelled Germans in this and scores of nearby villages had fled last fall. The suspicious old man and his silent wife were resigned to living out their days alone. They had overslept the morning of the exodus.

"I have no reason to leave my poverty," the old man declared, casting furtive glances at the window.

And they were poor. Lime-green paint peeled from the walls of their single tiny room; their furnishings consisted of a table, two chairs, and the bed on which we sat; an ancient stove provided heat. Outside, wood was stacked up against a tree to dry, and the yard was given over to cats and chickens. There was one cow. The old man refused handouts from the Red Cross.

"If I have a piece of bread, I can cut it into three pieces, and it will last for three days," he said defiantly.

Yet he was very generous with his *šljivovica*, the strong (up to 60 percent alcohol) homemade plum brandy cherished by Serbs and drunk throughout the former Yugoslavia. But this year he had not harvested any plums. After three glasses of *šljivovica* he pulled up his shirt to show us why: pus from a kidney infection oozed out of a hideous hole in his side. His monthly social security check did not cover the cost of his medicine. At night he was wracked with fever. That was when he really missed his neighbors. He and his wife were visited now only by Croatian soldiers and police.

"I tell the soldiers, 'If I'm guilty of something, come and talk to me during the day. Don't bother me at night,'" he said. "You see, I tell them, 'We're touching, and nothing happens to you.'"

What he did not say was that the soldiers were looking for his son. They have his name on their list of Četniks, Edo said when we drove off. He would be dead if he ever returned. Rain was falling, and on the tape deck Lou Reed sang "Take a Walk on the Wild Side." Twice within minutes we were pulled over, once by the militia, once by the police. The next UN checkpoint was a tourist agency specializing in taking Serbs to see their houses, said a peacekeeper. Through the gathering darkness I could make out more ruins interspersed with yards where butchered pigs hung from hooks. Alenka was still fuming about the old Serbian couple when we stopped at a roadside bar.

"I know their son wouldn't sit with my parents and drink *šljivovica,*" she spat.

Among the bottles lined up on the shelf above the bar were nine bullets and a grenade. Edo ordered coffee and Cokes for us. Three soldiers commemorating the anniversary of a comrade's death grew quiet when Alenka raised her voice to denounce her aunt for trading Christmas cards with her former Serbian neighbors from Vukovar. She took no notice of the soldiers staring at her.

"My aunt thinks we must try to forget. I can't," Alenka cried. "She's old. She just wants to go back to Vukovar so she can die there. Not me. I can never go back."

I confess: the depth of her bitterness had begun to bore me. It was easier to sympathize with her aunt and the old Serbian couple than to listen to her tirades.

Even Edo seemed wary of her now. "This is a war between two ideologies," he said, changing the subject. "Communism and the embryo of democracy, the Ottomans and Austro-Hungaria."

But here was another truth: the previous fall, in the town of Poljana Pakračka, a stop on our tour, a Croatian paramilitary unit called Autumn Rains had set up a death camp for Serbs and unsympathetic Croats. Their interrogation methods were particularly gruesome. One member of the Autumn Rains, who eventually confessed to torturing and killing more than seventy people, described how they burned their victims, poured salt and vinegar over their wounds, and plugged field telephones into them to deliver electric shocks. The Serbs were forced to sing the Croatian anthem before they were executed and buried in mass graves. And in the neighboring villages, some of which we visited, the Autumn Rains killed Serbs in their kitchens, looted their belongings, then blew up their houses to destroy the evidence. Not once did Edo or Alenka mention the Autumn Rains. It was some time before I learned that Croatian authorities had imprisoned some of the paramilitaries earlier this year only to release them without filing any charges. No wonder the Serbs had fled the area.

—∞—

"I hate them," said Marina, a student at the university. "I hate the refugees. They don't want to work, and they smell. I hate the smell of the buses and trams."

Hers was a common sentiment in Zagreb, which applied in unequal measures to Bosnian Muslim and Croatian refugees alike. And it is worth analyzing the differences in the ways the two groups were despised, if only to insure that each entry in the History of Hatred is correct. The Muslims were reviled, of course, for religious reasons, but also because of the spurious belief that they were Croats who had converted under the Ottomans. Then, too, they were from elsewhere (read: the South), which united them in the public mind with the Croats arriving from Slavonia, Krajina, Herzegovina, and Bosnia—with this difference: the Croats were poor relations who had appeared unexpectedly, demanding to be taken in. Eventually they would be sent packing. The Muslims were another matter: war with them was inevitable. But all the refugees, Muslim and Croat, were considered peasants even if they hailed from metropolitan areas. Marina said they were ruining Zagreb.

"There are advertisements in their hotels for cooks and cleaners, but they refuse to work. Young women could go take care of children in the camps, but they won't. Instead, *our* women are cooking for *them*. Awful!"

The young woman also hated Americans: they were fat, they were stupid, they imposed their will on Croatia. "Why do we have to watch your movies every night on our TV station?" she cried.

"I can't imagine Hollywood producers losing sleep over the Croatian market," I said, "not when their films are being pirated."

She shook her head in anger. "And why should George Bush decide our fate?"

"Ask your government," I replied, but she would not consider any other point of view.

"You see, I'm twenty," she explained. "I don't have to change my mind."

Jadranka kept every copy of the *Feral Tribune*, an independent weekly published in Split. Dedicated to Protestants, heretics, and anarchists and, according to its masthead, loved alike by God and the Devil, the paper had an iron-handed enemy in Franjo Tuđman. He did not appreciate its humor. On one cover, for example, beneath a headline reading **Croatia's Economic Program**, was a photograph of two boys peering into their swimming trunks. Another showed primary-school class pictures of Hitler, Stalin, and Tuđman; in each the future dictator stood in the middle of the top row. The headline, **Same School, Same Leader,** so infuriated Tuđman that he instructed his public prosecutor to charge three editors with the crime of insulting the president. Their case had yet to come to court. In the meantime, the government was levying on the *Feral Tribune* the same high tax that pornographic magazines had to pay. After all, Jadranka said with a smile, those boys *were* looking at their penises.

The entrance to the Cathedral of Saint Stephen's lay beyond a maze of scaffolding stretching toward a pair of neo-Gothic spires more than a hundred meters high. It was standing room only for evening Mass, and from the back of the cathedral I kept an eye out for Grozdana Cvitan, a leggy blonde known to her friends as *Walking Theater.* Sweeping into the café in the Hotel Dubrovnik earlier that afternoon, Grozdana had presented a study in black and white—white fur coat and muff, black blouse and jeans; the other patrons would look up in alarm whenever her deep husky voice flew off into cackling laughter. She viewed everything in her life—the Croatian Spring, her philosophy degree, her work as an artist, actress, and peace activist—as preparation for the war with the Serbs.

"I knew they wouldn't let us go without a fight," she said. "And if we didn't fight I would say Croats are not deserving of freedom."

Her sense of drama depended upon sharp contrasts: before Croatia declared independence, she organized a peace mission of nearly a thousand women to the Council of Europe in Brussels, even as she sent written appeals to world leaders for military equipment. She returned from an audience with the Pope to join the Artists' Brigade entertaining troops at the front; when that unit was disbanded she enlisted in the regular army—and became a theology student. She thought Croats had no business telling Bosnian Muslims they were Croats, though in her view Bosnia *was* historically Croatian. With a smile Grozdana said that war with the Muslims was inevitable.

"War is war," she explained. "I don't apologize for the bad things we've done."

No doubt many in the cathedral shared her feelings. I took up a position near Meštrović's relief of Archbishop Stepinac and scanned the pews for Grozdana. Dozens of parishioners crossed themselves before the shrine, praying for the soul of the cleric—a candidate for sainthood, according to the Vatican, despite his role in the NDH. True, Stepinac had not personally advocated violence (unlike Sarajevo's archbishop, who published an ode to Pavelić praising him for ridding Yugoslavia of its "miserable traitors"). Indeed sometimes he preached tolerance and once even said to Pavelić, "Thou shalt not kill!" (which Hubert Butler took to mean, "Thou shalt not kill too much!"). After the war, though he insisted his conscience was clear, a theological problem remained, which the Church defines as the difference between invincible and vincible ignorance. It is one thing to be ignorant of the truth—that your government is committing genocide—quite another to deliberately ignore evidence of that truth. This was the guilt of vast numbers of Croats during and after the NDH, best symbolized by Stepinac, who was far more knowledgeable about the extent of the Ustaša horror than he let on. "It is not possible," Butler writes, "that the Church leaders were ignorant of the wild and murderous excesses of hundreds and perhaps thousands of fanatical priests and monks in Yugoslavia."

Stepinac had a legal problem, too, which Tito dispensed with in a show trial—after failing to convince the cleric to go into exile. Sentenced to a long prison term, Stepinac became a martyr to Catholics the world over, and upon his release the Pope made him a cardinal. Absolution was also the theme of Meštrović's relief: Christ laying his hand on the head of the kneeling supplicant. Even Tito seemed in a forgiving mood when Stepinac died, allowing him to be buried in the cathedral. Another irony: Stepinac was one of Yugoslavia's fiercest opponents, yet he turned down every chance to leave his homeland, just as Meštrović refused to return to the country he had helped to create. In this marble tablet, however, the artist and cleric were united, presumably for eternity.

The cathedral began to empty. The service was over. No sign of Grozdana.

The minister of information was not happy to see me. This severe woman had expected only local journalists at the press conference, and as she examined my credentials, her lip curling into a sneer, she muttered, "This is not correct." But she could not come up with a reason to turn me away, so I took a seat alongside the other journalists in a conference room at the Hotel Intercontinental.

It was hard to understand her dismay. Two European Council representatives presented their findings on the destruction of cultural monuments, focusing on Serbian aggression in and around Dubrovnik and Mostar. An architect from Mostar had coined a term to describe the bombardment of his city, *urbicide*, which Colin Kaiser planned to use in his report. Kaiser was a tall precise man who pronounced the last letters of many words with a particular emphasis: "There is only on*e* wor*d* to descri*be* the situatio*n* i*n* Dubrovni*k* and Mostar: catastrophi*c*." Or: "Thi*s* i*s* cultura*l* cleans*ing* o*n* a lar*ge* scale." He meticulously described how thousands of houses, churches, historic monuments, and industrial facilities had been "visit*ed*" by the JNA; the deterioration of the damaged buildings; the international community's lack of interest in preserving places on the UNESCO Register of World Cultural Heritage. The Serbian destruction of Mostar's cultural legacy was deliberate, said Kaiser: how else explain the damage to almost every mosque, the felling of five minarets, the razing of the Franciscan church? Only the famous Old Bridge was intact. This was a systematic attempt to rid the world of a way of life, of everything important and symbolic to a people's identity.

Nor was this evil confined to the JNA. In one Serbian village, Croatian forces had blown up a sixteenth-century Orthodox monastery. And the new Orthodox church in Mostar was in ruins; a collective grave of Serbian soldiers had been dug up and looted.

The destruction of the educational system appalled Kaiser's partner, a French photographer. Every instrument in one music school had been destroyed; everything in every school—furniture, books, blackboards, pencils—had been burned. Classes had not been held in months. The schools were filled with refugees with nothing to do. Why would they not turn to prostitution, drugs, and crime?

"I can make an emotional appeal, but is the world ready to listen?" he concluded.

"If you're so moved," one Croatian journalist said with contempt, "why don't you give your money to the refugees?"

And thus began a series of speeches from the journalists, all denouncing the suggestion that Croatian soldiers had committed any sacrilege. They took particular exception to Kaiser's assertion that much of the damage to Catholic churches had amounted to symbolic vandalism.

"We've seen their reports on TV," said a TV reporter. "They spare the façade of a church so they can say they didn't touch it. But inside you see it's gutted."

Others chimed in with lengthy stories of Serbian atrocities, all for the benefit, it seemed, of the minister of information, who kept nodding her head in approval.

The vehemence of their denunciations took Colin Kaiser aback. "If I were to draw a ratio of destruction, it would be a hundred to one in favor of the Serbs," he said defensively.

But the Croatian journalists were still not satisfied. They continued their monologues—the few questions they asked were rhetorical—until the minister of information, satisfied that the truth was finally out, called the press conference to an end.

The park near Jadranka's flat was deserted when I ran there in the morning, occasionally startled by the sounds of gunfire from the nearby police firing range. At the top of a set of moss-covered stairs was an abstract monument to Partisans killed in the war, several of its planes defaced by angry nationalists—a cruder manifestation of the countrywide attempt to rewrite Croatian history. From politicians and journalists, artists and historians, I heard that there were many more Croatian Partisans than Ustaše, as if Pavelić's regime were an aberration imposed from without and not, as it was, eagerly welcomed by Croatia's elites. "Hopscotch," Charles Simic calls it. "Pierre leapt from Stalin to Mao to Pol Pot to Saddam. I hope after the experience of this century that no one in the future will still believe the myth of the critical independence of the intellectuals."

She had left town abruptly, without saying where she was going, and no one at the research institute could tell me when she might return. If you wait, said her assistant, perhaps she will call. I was more relieved than disappointed to lose out on an interview with the Bosnian lawyer: how to ask her about her internment in a rape camp? I knew only that in exile she was coordinating a study of Serbian propaganda, so I took a seat to listen to her colleagues describe Milošević's mastery of the Communist tactic known as *special war*: of inventing enemies, that is, internal and external, in order to unite and discipline the people. Thus he painted a picture, through the media, of Ustaše and fundamentalist Muslims preparing to overrun the Serbs, and so on.

Special war, it occurred to me, was what Serbian paramilitaries and soldiers waged when they began a systematic campaign to rape Muslim and Croatian women (as many as 20,000, the European Commission of Inquiry would conclude). What better way to ethnically cleanse a town or village than to subject its citizens to the most intimate, and heinous, crime short of murder? In *Rape Warfare*, Beverly Allen unearthed an official document, from the army's special services, which suggests the Serbian rationale for raping, in public view, three-year-old girls and eighty-two-year-old women:

> Our analysis of the behavior of the Muslim communities demonstrates that the morale, will, and bellicose nature of their groups can be undermined only if we aim our action at the point where the religious and social structure is most fragile. We refer to the women, especially adolescents, and to the children.

Genocidal rape was the term coined by Allen for what Serbian forces did to Bosnian and Croatian women and girls, a new crime against humanity which generally took three forms. First, paramilitaries would enter a village, drag women from their homes, rape them, and leave, so terrifying the other villagers that when the regular army soon arrived they readily agreed to abandon their homes. Then there were concentration camps, where women were raped and often killed. Finally, the rape camps: the restaurants, hotels, hospitals, factories, schools, and stables in which women were impregnated (the raping continued until the pregnancies could not be terminated, resulting in the births of hundreds of unwanted babies) or executed. This was indeed a special war.

As for the Bosnian lawyer I had arranged to interview? She never called.

Everyone who was anyone was at the opening of Ivan Lacković's show at Gradec, a state-owned gallery near Saint Catherine's Church. More than five hundred people turned out to see his work, and for those who could not attend Croatian TV filmed everything—the paintings, the string quartet, the wounded veterans delivering speeches in praise of Lacković, then the artist giving them money and paintings. On his sixtieth birthday local critics were calling Lacković "the greatest naive draftsman in the world." While Jadranka and her friends made a dinner out of the hors d'oeuvres, I went through the show, which took up three stories of the building, paying particular attention to fifty new Christmas scenes (proceeds from these would go to war orphans—the artist had grown quite rich during the fighting), all without feeling any quickening of emotion.

Snowbound villages encircled by peasants in traditional dress. Groves of bare trees filled with black birds or large swatches of blue blossoms. An Olympic torch surrounded by rings of stick figures pushing wheels and trees bearing men armed with torches. One of Lacković's champions described his work as "a catalyst of emotion." Milan Kundera's musings about kitsch seemed more appropriate:

> There is a kitsch attitude. Kitsch behavior. The kitsch-man's (*Kitschmensch*) need for kitsch: it is the need to gaze into the mirror of the beautifying lie and to be moved to tears of gratification at one's own reflection.

And:

> In the realm of kitsch, the dictatorship of the heart reigns supreme.

And:

> Kitsch is the absolute denial of shit, in both the literal and figurative senses of the word; kitsch excludes everything from its purview which is essentially unacceptable in human existence.

That night it took Jadranka less than five minutes to write her review.

"We couldn't save our architecture or land, but we could save our paintings," said Zvonko Maković, in answer to my question about when the National Museum would reopen. The poet and art critic praised the decision to hide the museum's holdings in a cave. I remarked at how much of Balkan history lies in its caves, in which Tito plotted, thousands of people lost their lives, writers gathered, paintings were stored. Zvonko shook his head impatiently. His mind roved from subject to subject with startling speed, never resting anywhere for long. Now in short order he spelled out why a certain woman journalist was "dirty and dishonest"; defended Tuđman against nostalgia for Yugoslavia; and described the mechanism by which avant-garde ideas are turned into kitsch. Then on to a new art form he was exploring: the last seconds of footage shot by a TV cameraman killed at the front, with his camera running, recording gunfire and smoke, then grass and silence, and then the sky.

We were in the Ban Café, which had just opened on Jelačić Square. An attractive space, I said when Zvonko fell silent. Momentarily bewildered, soon he waxed enthusiastic over the café's design. The central idea was marble—rose marble in the main room, black marble in the bathrooms. It did not faze him that the Ban was the favorite haunt of the security services.

"Cafés are fundamental to culture and civilization," he said, then caught sight of the refugees milling around the monument to Jelačić. "They have nowhere else to go," he said with a sudden force. "So they come here to gossip and exchange news. Each part of the square represents a different village or city—Vukovar here, Banja Luka there. Men and women who worked their whole lives, and now they have nothing to do. You see their faces? They're peasant faces. Sad."

The last thing Croatian TV broadcast one night was video footage of the occupied territories—Čelije, Palaca, Silas, and other villages overrun in Slavonia—overlaid with funeral music.

"I was not a nationalist before the war," said Jadranka, lighting a cigarette with considerable agitation. "I didn't think of myself as a Croat. I had Serb friends, even Serb boyfriends. They thought we were more sophisticated than their Belgrade girlfriends, and we thought they were wilder. But everything has changed. How could I be anything but a nationalist after this?"

The video, filmed from the passenger side window of a speeding car, showed gutted houses, burned-out cars and trucks, piles of debris. Jadranka stood up and paced the room, too upset to sit down again until after the video had ended, in the ruins of Vukovar.

"Before the war," she said when the national anthem began, "no one had ever heard of these villages. Now we know their names by heart."

"Exile is the nursery of nationality," Lord Acton wrote.

In 1990, when ads for guns outnumbered those for wedding dresses in the classified sections of the Zagreb newspapers, Maja Razović knew war was coming. Her friends thought she was paranoid. She was actually a keen observer, a journalist whose independent views had won her many enemies. Her long black hair was streaked with grey, and in her leather jacket and jeans she cut a bohemian figure. We met from time to time in the City Café, a Bauhaus gathering spot for what she called the ex-people—ex-politicians, ex-teachers, retirees. She herself was a former member of the mainstream press, having lost a succession of jobs on account of her "Yugo-nostalgia." In one respect Maja was a quintessential Yugoslav: her mother was Serbian, and her late father, a high government official, had been a Croatian Partisan. But it was her zeal for the truth, not any allegiance to the past, that caused her problems. That was why I liked to see her. She could not stop asking questions about Croatia's role in the war—even in the bomb shelter where, to the horror of her neighbors, passionate Serb-haters all, Maja had read Voltaire on the subject of tolerance.

"If an angel were to bring us real democracy, we'd be in deep shit," she said. "We forgot what real journalism is. Communism was dying a natural death, and freedom was growing. But the war changed that. Instead of a thousand blossoming flowers we replaced one single-party system with another. And now our journalists are war criminals."

Out of necessity she had turned to free-lancing for *La Repubblica*, the Italian newspaper, and earned enough now to support both herself and her mother, who was dying of breast cancer. I steered the conversation to the future, hers and her country's. Maja thought that Croats were infected with the same Oriental fatalism as Serbs, despite their so-called Western attitudes.

"It's not just a question of rebuilding cities," she said, looking around the café. "We have to rebuild people's minds. They don't understand that when they blow up a Serb's house it doesn't hurt him anymore. It's another house we could have put refugees in! Even the Orthodox churches, which were built in Croatian Baroque style—they're blowing up their own heritage."

Maja came from the island of Hvar, and before long she would return there to join an independent radio service broadcasting news reports into the former Yugoslavia from a ship in the Adriatic. For now, though, she was trying to make her mother's last days peaceful.

"The first time I was fired," she said wearily, "my mother started throwing up, because when she was my age that was the first step to the concentration camp. At least some things have changed."

It was the smallest book-publishing party in memory. The Yugoslav firm that issued Zvonko Maković's books used to host a gala holiday affair. But that house had gone out of business. Zvonko was at once relieved to be publishing with a new private firm and skeptical about its future.

"I'm so confused," he said when we met late one morning at a restaurant near the market. "Sometimes it worries me."

A single table was reserved for the six of us who had come to celebrate the firm's first list, two handsomely produced books of poems by Zvonko and Branko Čegec, and a novel by a writer who did not show up for the party. I was seated across from the publisher, a young man introduced to me as the son of a former high-ranking official. He himself was an economist, with no interest in literature. What better way to launder money than through the publication of poetry and fiction? He kept a close eye on the amount of *šljivovica*, grappa, and white wine brought to our table. No one was happier to learn that the mineral water came from Sarajevo.

"What does an economist do in times like these?" I asked the publisher.

"Smuggle," he answered without hesitation.

"And what do you smuggle?" I said.

"That's a business secret," he said.

"Then what, in general terms, is smuggled now?" I said.

The publisher smiled. "Foreign currency and arms."

He laughed at the idea of the government cracking down on smugglers or reining in the black market. "They need us," he said. "They need hard currency and arms, and this is the only real market operating in the country." All at once he glared at me. "And about this I will say no more."

I leafed through Zvonko's book—the production values rivaled those of American publishers. The poet seemed happy with how his poems appeared on the page, and if he had any qualms about where the money came from to produce his book he did not let on.

After lunch, I walked for several kilometers through the falling snow—the trams were not running—to the Croatian Homeland Foundation, an institute serving the émigré community from which it secured funds for the war effort. At its Christmas party a woman in a white fur coat urged me to admire the prints of the "naive" artists lining the walls. She had just returned from New York.

"I would not want us to copy the superfluous parts of America," she exclaimed. "Those supermarkets: too many goods! It made me sick to walk into them. Or in the movies, when you see someone fired from his job. I would never want it to be that way here."

She had faith in Croatia, great faith. Yes, the country was in a transitional—a *prenatal*—phase, but there was great potential, even if the people didn't realize what all these changes would mean. My attention was drifting. She introduced me to a young woman, Živana, who was on her way to a Scientology training course in Ljubljana. After the New Year Živana would have to find another job.

"On the black?" I said.

"Of course," she said. "No regular firm can hire anyone, because if they pay you a hundred marks a month they have to pay the government a hundred and fifty marks a month in war tax."

"How's that?" I said.

"That's normal," she said.

That seemed to be her favorite phrase. She used it to explain the problems with the trams and why the clerk at the train station had to rewrite her ticket to Ljubljana three times. When I told her that the founder of Scientology, L. Ron Hubbard, was considered a lunatic in my country, she replied, "Yes, I know he had a lot of strange ideas, but that's normal."

The poet Boris Maruna caught my eye. After twenty years of Croatian Spring-induced exile in Los Angeles, he had moved back to Zagreb only to discover that his colleagues at the Homeland Foundation and his countrymen drove him to distraction.

"You don't need to hear any more of their bullshit," he said, taking me to his office.

He closed the door and brought out a bottle of brandy.

"The problem is, we got rid of the pyramid at the top," he said, "but the bureaucracy's still there, and we don't know how to work. Some people put in long hours, but what do they produce? Nothing. The only economy we have is on the black. And the government is giving companies away, so the national treasury is empty. It's as if you sold Pennsylvania and had nothing to show for it."

There was a knock on the door. A former Yugoslav foreign aid officer joined us. He had just founded a publishing house specializing in books in translation. He planned to start with French poets, St.-John Perse first.

"That's amazing," I said. "I have a copy of *Seamarks* with me now."

His eyes widened. "I don't believe it," he said.

But when I pulled the book out of my satchel, he took it from me and ran out of the office.

Boris shrugged his shoulders. "Do you see all the bullshit surrounding me?" he said. "And I'm a Beat poet!"

His secretary came in with a message from another publisher. "He wants some of your new poems," she said. "What shall I tell him?"

"Fuck off," said Boris.

"Me?" she said in mock surprise.

"No," he smiled. "To you I would say something very different."

It was three days before my copy of *Seamarks* was returned.

A U2 concert, soon after Tito's death, was what opened Maja Razović's eyes. The journalist could not understand why the band's name was removed from the headline of her review. "U stands for Ustaša," her editor explained. And that was when Maya first felt the powerful tug of History, the rope entangling everyone and everything in the Balkans. Croatia, she came to believe, would not lay its ghosts and demons to rest until it had reckoned with its Ustaša legacy.

"People here don't have a healthy attitude toward their past," she told me the last time we met in the City Café, "so every twenty years we try to become what we were at our worst."

It was happening again. Communists, be gone! said her countrymen. Serbs and Muslims, too.

Easy to imagine, impossible to accomplish without bloodshed.

You could not miss the signs that war was about to begin again: hyperinflation; rising taxes; rumors of new Serbian missiles and Russian soldiers massing on the Montenegrin border; Tuđman's Christmas promise to the troops that 1993 would be the year they reclaimed their occupied lands.

"As long as the flags are waving, the people will stay quiet, even if they're starving," Maya declared. "They can always take more suffering. 'What are the poor

Croats going to do when the Hungarians are gone?' said a poet at the turn of the century. 'Who will they blame then?' Now I say the same thing about the Serbs. Who will we blame? Other Croats? The Bosnian refugees?

"We've always had someone to blame," she said with a wry smile. "But you have to own up to who you are, or else you'll repeat your history."

9

Dalmatia

The walls are what you notice first. Dating from the thirteenth century, they form an irregular polygon of grey stone, six meters thick in places and more than twenty meters high, around the historic center of Dubrovnik, the islet once known as Ragusa, on which the Ragusans, a proud seafaring people, established a powerful maritime republic. The western wall gives onto the Adriatic, and just beyond the eastern wall rises Mount Srd, which separates Croatia from Herzegovina. The vine-covered slopes below the ridge were once thick with oaks (*dub* is Serbo-Croatian for oak), but Dubrovnik remains an essay in grey stone. And the walls? Ragusans call them a metaphor for freedom.

One warm December afternoon I climbed the steep steps to the top of the walls and walked, counter-clockwise, from gate to tower, wall curtain to arsenal, circling the city which had never actually been independent. Yet for much of its history Dubrovnik enjoyed a level of autonomy unrivaled in the Balkans. Illyrians may have been the peninsula's first inhabitants, but Roman refugees fleeing from Slav and Avar invaders founded the city in the seventh century. The Slavs settled in the forest at the foot of Mount Srd, and as they moved into town they forged a link, strengthened by skillful diplomacy, between the Slavic and Latin worlds. Protected by its walls, the mountains, and good fortune, the Dubrovnik Republic thrived for hundreds of years.

"The only measure for liberty is liberty," a Ragusan wrote. Even so, Dubrovnik had to buy its measure of freedom. Salt, silver, and lead from mines in Bosnia and Serbia, shipbuilding, gold- and glasswork, cloth: these were the staples of the republic, which paid tribute to a succession of foreign masters—Byzantium, Venice, Hungary, the Ottomans—until Napoleon abolished it in 1808. From there it was all downhill: the Congress of Vienna assigned Dubrovnik to Austria in 1815; in 1918 it became part of the Kingdom of Serbs, Croats, and Slovenes; and in World War Two it was occupied first by the Italians then by the Germans. But the

Nazi occupation paled in comparison to what this "city of poets and a poetic city" (as one writer called it) had endured over the last fifteen months; only powerful earthquakes, in 1667 and 1979, caused more damage than the JNA. And what Dubrovnik suffered was nothing next to what happened in the surrounding villages, overrun and occupied by federal soldiers and irregular Montenegrin forces.

From the top of the walls I inspected some of the damage. The bombardment came from the sea, air, and mountains, striking one in five buildings. When electricity and water were cut off during the three-month siege, the sewage system established in the thirteenth century stopped working and the streets filled with wastes. Several walls suffered direct hits. *Catastrophic* was how the Council of Europe representatives described the city's losses. What I saw were sheets of nylon and tar paper covering holes in about half of the roofs, white and black islands in seas of red or the city's distinctive beige-ocher-and-brown tiles. Before the war UNESCO had placed Dubrovnik on its Register of World Cultural Heritage. Now it was on the Register of Endangered World Heritage.

It was disconcerting to have the walls to myself. The tourist industry had made Dubrovnik Yugoslavia's richest city, its 70,000 residents (90 percent Croatian, 6 percent Serbian) well-off even by European standards. Venice is the Queen of the Adriatic, goes a local saying, Dubrovnik, the Princess. Hard to see what strategic value these medieval churches, palaces, and fountains held. Why the best hotels were leveled and scores of yachts sunk was anybody's guess. Rebecca West, who found little to recommend here, believed that Dubrovnik, "lovely as it is, gives the effect of hunger and thirst." Prophetic words: during the siege, those who stayed to defend the city sometimes brewed their coffee with salt water.

An uneasy truce brought the restoration of water and electricity, but the hotels spared major damage were filled with refugees. Tourists would not soon return: more fighting was expected, train routes through Serb-occupied Krajina were closed, and the Serbs had stolen the landing equipment from the airport; whenever the authorities tried to reopen the airport the Serbs shelled the runway.

So I had flown to Split then taken a bus down the Dalmatian coast. On one side of the narrow, two-lane road were steep limestone cliffs, slopes of scrub oaks and cypresses, vineyards in which the leaves had turned bright red; on the other, the sea. Gardens surrounded the tidy houses, beyond them orchards of orange and olive trees. Salt marshes—and then the first of the razed villages. The sun was shining, the houses were roofless. Such beauty and such devastation. The passengers grew silent. Here was a bay in which even the floats for the mussel traps had been sunk by Serbian snipers. The bus driver thought nothing of passing cars and trucks, even when the road hugged the cliff. Just outside Dubrovnik, he sped by an entire UN convoy. The passenger next to me whispered, "What can we do? If we ask him to drive more carefully, he will only say, 'But there's a war on.'"

The Stradun, the main thoroughfare, "is not a street but the world," said a poet. A world, then, of stone façades and goldsmiths busy in their workshops; of shrapnel marks, craters from grenades, crumbling stone and stair decorations. The cafés were closed. The monuments to local poets were sheathed in protective planks of wood. Piles of stone, brick, and broken glass blocked the entrances to boarded-up businesses. At dusk, Croatian soldiers in U.S. Army issue combat fatigues swaggered by. The crowd out for a stroll would be off the street before the 9:30 curfew: no one wanted to give a drunken soldier an excuse to start shooting.

"This is not a town for soldiers," the historian Slobodan Novak was telling me, "The walls are pure poetry. They have no military significance. But we need the soldiers now, and in the future we'll need to remember why they were here if we want this to be a city again for poets."

Novak, who lived in Zagreb, had returned to his home town for a conference on the historical connections between Dubrovnik and America. A large, gregarious man, he was careful to distinguish between his own optimism and the feelings of the Ragusans who were still psychologically besieged. The historian did not think it was too soon to organize a cultural event.

"We must teach these children they don't live alone," he said. "Yes, Dubrovnik is liberty, but that means being prepared. We're only seven hundred meters from the Troglodytes."

The thin line between civilization and barbarism fascinated Novak. In his view it was Vatican efforts to evangelize the tribes surrounding Dubrovnik that gave the city its special character. The Church's influence had extended far beyond Dubrovnik's walls, with mixed results: Dalmatia was Catholic, but Herzegovina was a patchwork of Catholic, Muslim, and Orthodox communities; and Orthodox Montenegro was literally within shelling distance. The levelers of his family home Novak called barbarians. But he described the Bosnian crisis in much different terms, as the last colonial war.

"Bosnia was never its own state," said the historian, blithely ignoring the historical record: that the medieval Bosnian kingdom had enjoyed two centuries of independence; that under Ottoman rule Bosnia was a legally defined provincial entity; that it was accorded special status in Austro-Hungaria as well as in the former Yugoslavia. "It was always the colony of some other power—the Vatican, the Turks, the Serbs, and now we want it, too."

We came to the statue of Orlando. From the *Chansons de Roland* we know the French knight, betrayed by his stepfather, lost his life commanding Charlemagne's rear guard on its return from its invasion of Spain. From Novak I learned that for four centuries the white flag hoisted from Orlando's Column was the symbol of Dubrovnik's freedom. Only when Napoleon conquered the city did the flag come down for good, ending the golden period of Ragusan history. The Dubrovnik Republic was finished, destined to be revived only as an historical motif once the city gave itself over to tourism after World War Two. For forty years the Yugoslav flag was raised from Roland's effigy at the start of the summer festival, the centerpiece of the tourist industry. Now the statue was coffined in wood: the destruction of Vukovar had taught the town fathers not to take any chances.

At the Stradun's eastern end was Ploće Gate, near the Lazaretto, a complex of buildings dating from the seventeenth century, where foreign traders used to be quarantined and their goods were stored. If once the city had taken pains to confine visitors, in the age of the tourist the Lazaretto had become an amusement center. But it was closed for the war.

Novak was explaining why there were five gateways to the center of Dubrovnik. "To frighten strangers. By the time you reached the main square you knew you were in a mysterious place."

He hoped to create a different feeling at the international PEN conference he was organizing. For the sixtieth anniversary of the 1933 conference in Dubrovnik, at which Nazism was denounced, the Croatian PEN Club planned to target Serbian aggression. But they had encountered resistance to this idea—and to the conference as a whole—from those in the international literary community uneasy with Croatian nationalism. Novak dismissed their concerns with a wave of his hand. Dubrovnik had been rebuilt once with stones brought from afar. Perhaps the PEN delegates would bring books to rebuild the libraries.

I told him how sad I was to see the city only in a damaged state.

"That's perfect," he said. "Poets understand absences."

Opening night at an exhibit of Sephardic memorabilia in the Franciscan monastery. Dubrovnik's cognoscenti had turned out in full force. The speeches droned on and on.

"Yews," said a sailor.

"What?" I said.

He pointed at three old men. "Yews," he repeated.

"Jews?" I said.

The sailor nodded.

The synagogue, founded in 1408, in a narrow two-story house built in 1300, when the Jews first settled in Dubrovnik, was struck by shell fragments during the siege, another attack against one of Croatia's oldest Jewish communities. Nor did they find a warm welcome in the new Croatia. The Serbs were not the only ones to feel threatened by Tuđman's revival of Ustaša symbols. Which may explain why these three old men kept their speeches short. Surely they were not alone in remembering that in World War Two Dubrovnik was a haven for Jews (including five hundred or so who had fled from Sarajevo) only because the occupying Italian authorities refused to turn them and the local Serbian population over to the Ustaše. All that ended, of course, with the Italian capitulation.

"Yews," said the sailor. "There are only twenty left in Dubrovnik."

He himself had escaped one night during the siege by swimming, without a wet suit, from island to island, mindful of the JNA practice, in the early 1950s, of shooting anyone caught beyond Yugoslavia's territorial limits. Refusing to say whether he had come under fire or when he had returned to the city, the sailor spoke of losing three of his four charter boats on a single day of shelling, some of the five hundred boats destroyed in the harbor at Gruž. To finance repairs, he worked on other sunken hulls, earning up to $400 a day, all under the table. There was enough work, he said, to keep him going forever; to feed his family, he had started to fish again, like many in Dubrovnik.

The sailor repeated a proverb. "The sea will always provide."

An ethnologist pulled me aside. "Things are moving very fast," she whispered. "The Russian Parliament just voted to back the Serbs if the West intervenes. This goes back to 1914. You can't kill off a system just like that."

You could not miss the EC military monitors. They dressed in white (Croats called them ice cream vendors), drove white Land Rovers, and went everywhere from Trebinje, Montenegro, to Knin. The monitors who came from the intelligence community—and they were legion—must have found this public assignment unsettling. The affable Scotsman in charge of the Dubrovnik contingent was at once keen to talk to journalists about the progress of the war and wary of them.

"I have no wish to go down as anyone in history," he told me.

We were at the opening of an exhibition celebrating Croatia's role in Columbus's discovery of the New World. The cognoscenti had gathered in another unheated building, this time to look at maps, paintings, and shipping paraphernalia—a sextant, an hourglass, a small telescope—illustrating a local saying: "Stick your finger in the sea, and you are in touch with the world." Through the broken windows came a strong breeze off the Adriatic. In the next room *Batman Returns* was playing. Croats, it turned out, had navigated for Columbus.

"I know who started this war, yes," said the monitor, "but no one is innocent here. I told the Serbs this summer, 'It's not a question of you losing the propaganda war. You have already lost it.'"

"They could win back some ground if they stopped targeting journalists," I said.

He offered me a cigarette. "Look," he said, "these people—on both sides—think their war is the world. It's not. We can pack our bags tomorrow and leave. Some Croats know that. That's what we have over them. They want to join the EC, which means acting in a civilized manner. But the hatred here is familial. Generational. It's so ingrained no one can be rational. Our only hope is to get them to talk and maybe heal their wounds. Six months ago we had to fill the room with armed guards to bring together the police chiefs of Dubrovnik and Trebinje. Now they meet on their own."

Whatever faith he once had in the international community was gone.

"The diplomats have no balls," he said, his voice rising in anger. "What's needed is for honest, objective people to speak the truth whenever they can. Don't wait two days to tell us the Serbs have violated the cease-fire! And don't tell me it's your character that prevents you from responding."

He looked around the room. "I expected the city to be leveled when I got here, because of what the press had said. But it's not. In fact, I've recommended to the town fathers that they not try to restore everything. Leave some of it damaged or destroyed. To remind themselves not to get fat and arrogant again. This is now part of their history, part of the rich fabric of life."

After the earthquake of 1667, the nobility decreed that Ragusans must pull together to rebuild the city: "Anyone who leaves the territory of the Republic shall be punished for perjury." Perjury is used here in its older sense—to break an oath—and while no such law was passed after the 1979 earthquake pride and mercantile ambition hastened Dubrovnik's rebuilding. To find out how the damage from the siege would be repaired I visited the Institute for the Preservation of Cultural Monuments. The first floor was covered with debris, and Matko Vetma's office on the second was so cold I took him up on his offer to show me some of the buildings under renovation.

"The city and surrounding villages belong to the same organism," he said, "but it will be more difficult to restore the villages occupied by the Serbs. They burned down too many houses and vines."

Our first stop was a palace that could be repaired. (Nine other palaces had been demolished in the fighting.) Up a steep staircase we climbed to a well-lit room with very thick pine floors. A shell had ripped through a wall, revealing, layer by layer—paint, plaster, brick, stone—five hundred years of architectural changes.

"Why did the Serbs attack Dubrovnik?" Matko said. "Maybe they thought we would leave."

He had worked through the siege, and now, when he was not making drawings of the city, creating a visual record from every angle, he traveled to Rome to study traditional ways of building. After the 1979 earthquake, Dubrovnik was rebuilt with modern materials (red tiles replaced the city's characteristic tricolor-tiled roofs, and so on), but it turned out that traditional materials were just as strong.

Specialists in the old techniques had thus been hired to refurbish the palaces and monuments. Pine would be found for the floors. Tricolored tiles could be imported from France. Stone masons might turn their workshops into schools. Matko thought the bombardment even had its good side.

"Out of traditional ideas and materials," he said, "we're creating a new architectural language and values."

Then to the Franciscan monastery, which housed Europe's oldest pharmacy (a fourteenth-century institution lately transformed into a Red Cross dispensary) and Dubrovnik's largest library of original manuscripts. They were collected after the 1667 earthquake, when 4,000 Ragusans were buried in the rubble of collapsed buildings, tidal waves sucked the harbor dry then pounded ships to pieces, and fire burned up much of the city, including the Franciscans' library. An incalculable loss. The widow of the last Bosnian king had donated a sizable collection of illuminated choral texts in the Bogomil tradition. "If these books had survived," Rebecca West wrote, "they would have been a glimpse of a world about which we can now only guess: but the whole library perished."

A monk escorted us through the monastery, refusing to explain why he would not show us the library or the pharmacy. We could see only the belfry of the church, which had been struck by a rocket. He tapped me on the genitals and started up the stairs.

"You see," Matko chuckled, "things *are* changing!"

Up a steep, narrow road winding into the mountains, through fields strewn with white boulders, we came to the front line, a village of a hundred houses, all destroyed. And all the villagers were gone. Soldiers huddled around a fire in a garbage can. In the sheepfolds were charred tree trunks and the rusting chassis of burned-out cars. Someone had taken a hammer to a plaque commemorating Croatian Partisans killed at the end of World War Two. They were mistaken for Ustaše, said my translator, a pleasant woman who used to bring tour groups here for a traditional meal. Četniks murdered them, she added in the same cheerful voice she adopted to describe village life. The broken pieces of the plaque lay in a looted open grave; iron bars were driven through the vaults. "You asked for it," read the message scrawled in Cyrillic, in the register of the gutted church, "and you got it."

Illyria. Once upon a time the word conjured up a magical realm along the Adriatic coast, populated by warriors and pirates. The Illyrians were probably the ancestors of the Albanians (the capital of the Illyrian kingdom, established in the third century B.C. and then conquered by the Romans, was in Škodër)—which did not keep the Croatian scholar, Ljudevit Gaj (1809–1872), from suggesting that the South Slavs descended from the Illyrians, too. Taking his cue from Napoleon, who revived the name for the Illyrian Provinces of his empire, Gaj founded the Illyrian Movement to counter German and Hungarian claims in the Balkans. Like other Slavic intellectuals, Gaj sought a history and a language to unite the South Slavs; only his linguistic work survived.

Gaj chose a dialect spoken by most Serbs and a majority of Croats (except in Zagreb and districts farther north) to replace the literary languages (Latin, Hungarian, or German for Croats, Church- or Russo-Slavonic for Serbs) and serve as the official language of a new land which, by his reckoning, would include Slovenia, Croatia, Serbia, Montenegro, northern Albania, southern Hungary, and Bulgaria. But Slovenian intellectuals had their own linguistic aspirations; and Serbian scholars suspected Gaj, because the history he offered was Croatian, because he preferred the Latin alphabet to Cyrillic, and because his universalizing sentiments threatened Serbian nationalism.

Nevertheless Illyrianism was the prelude to the Yugoslav idea. A strange turn: when Viennese authorities, bowing to Hungarian pressure, outlawed the use of the word *Illyria* in 1843, Gaj and his followers called themselves nationalists (with a Yugoslav orientation), changing their orientation with the same alacrity with which Croatian Communists—150 years later—became nationalists seeking independence from Yugoslavia. Which is to say: the Croatian people abandoned their own idea when they seceded. Now the search was on for histories and languages with which to *separate* the South Slavs. (More irony: Ante Pavelić, who dabbled in philology, published a Croatian dictionary purged of words with Serbian roots, an exercise tantamount to excising from English the Latin-based words imported after the Norman invasion (*tantamount, excising, invasion*). Yet the Croatian Spring began with this fiction, when writers claimed that Croatian and Serbian were different languages. It took only another quarter century for Pavelić's dream to come true: millions of Serbo-Croatian speaking Bosnians, Croats, and Serbs now said they spoke Bosnian, Croatian, and Serbian, respectively.)

To put it another way: Ragusan poets were the first to write in the newly codified language of Serbo-Croatian. The first use of the word *Yugoslavia* appears in a poem, published in *The Dalmatian Dawn* in 1848, by a Serbian revolutionary from Dubrovnik, Matija Ban. Like so many Balkan writers, Ban attempted to translate his words into deeds, fomenting rebellions against the Ottomans in Bosnia, Kosovo, and Macedonia. He was unable to convince Montenegrins to join the Serbian rebels, but his example was not lost on his successors, the writers and politicians responsible for the JNA laying siege to Dubrovnik—and to the Yugoslav idea—all in the name of Yugoslavia.

A crowd had gathered in the schoolyard. Five boys were chasing and cornering another boy, hurling stones and rocks at him, some the size of bricks. No one stopped them.

"A game," someone explained. "Just a game."

Feel free to look in, read the hand-lettered sign nailed to a board covering the doorway of an artist's studio. Next to the inscription were two painted eyes, a play on Milošević's remark that Serbia and Montenegro were two eyes in the same head, which was also Hitler's line, after the *Anschluss,* about Austria and Germany. The studio windows were gone, and beyond the stone façade—nothing.

It was on the name day of Saint Nicholas (6 December), patron saint of Russians and Serbs, sailors and the young, that the JNA mounted a furious ten-hour attack on the port and old city center, damaging the Cathedral, Sponza Palace, and Onofrio's Fountain, destroying the cease-fire agreement that Cyrus Vance had just negotiated with the Serbs. Three shells (one containing phosphorous) struck the artist's studio, where he lived with his elderly mother. The roof caved in, all the family belongings caught fire, the artist's mother collapsed and died. From the rubble the artist retrieved a copper pot used for cooking beans, put it on his head, and waited to be killed.

But fortune was with him then and throughout the siege. Nor did he lose his sense of humor. To commemorate the anniversary of the Serbian assault, he hosted a party in his ruins. It was raining that day, so his guests drank brandy under umbrellas. The artist, who no longer owned an umbrella, kept his head dry with the same pot he had once worn to protect himself from Serbian shells.

The Dubrovnik Cathedral was the setting, in 1499, for a series of meetings on the nature of angels. The entire government, aristocracy, and literary community participated in these discussions, which were the inspiration of Juraj Dragišić, a Bosnian Franciscan who published the proceedings as *De Natura Angelica*. This same Dragišić was also in the relics business (surprise, surprise), and once, returning from a journey to Palestine, he presented the Ragusans with what he claimed to be the hand of Saint John the Baptist. Dragišić was guided in this transaction by the Greedy Angel (the one who "dreams of mines," Rafael Alberti writes), having promised the relic to the Pope. Even the Sultan could not convince the Ragusans to give it up, and so to this day it is in Dubrovnik.

Memento mori. "Shall we go see the reliques of this town?" Sebastian asks in *Twelfth Night*, a reference to the Cathedral treasury holdings, according to Croatian scholars. (Did Shakespeare have Dubrovnik in mind when he set his play in "A city in Illyria, and the sea-coast near it"? A book of travel writings, popular in his day, mentioned Dubrovnik and its reliquaries.) And in *Black Lamb and Grey Falcon* a memorable scene takes place within these dark precincts. A corpulent priest, leaning over the low spiked barrier separating the tourists from the cupboards filled with relics, does not realize he is pressing down on Rebecca West's hand until she shrieks in pain:

> I held out my hand, which was bleeding freely from a wound in the palm. 'Ah, pardon!' said the priest, coming forward bowing and smiling. He was taking it lightly, I thought, considering the importance which is ascribed to like injuries when suffered by priests.

This is the last straw for the writer. She leads her husband out to the square, where they begin discussing "what was perhaps the false finality of the town."

Alas, the damaged Cathedral, built, it is said, by King Richard the Lion-Hearted as a gift to Dubrovnik after his escape from a shipwreck along this coast, was closed during my stay. But a nun let me in long enough to see where a triptych of paintings had hung above the altar: three blank white spaces in the middle of a wall darkening with dust.

"Did you know your country has started another war?" the Scottish military monitor said. He lit a cigarette and stared at the map covering the wall above his desk marking the front lines in Bosnia and Croatia.

"Where?" I said.

"Somalia," he said, with a look of wonder.

Then he showed me the results of a poll taken in Serbia about the front-running presidential candidates, Slobodan Milošević and Milan Panić, the Serbian-American millionaire who had returned to his homeland to become prime minister. In five of the six categories—starting the economy, getting sanctions lifted, ending the war, promoting democracy, offering a better future—Panić outpolled Milošević by wide margins. But the Serbian strongman had twice Panić's support when it came to protecting the national interest. The race was too close to call.

"Very strange," said the monitor.

The man in the harbor was fishing at dusk for his family's dinner, and what he did was this: with his line tied to a float tucked in his pocket, he would cast a three-barbed hook out into the grimy water, let it sink a little, then tug hard, hooking fish in their gills or fins or eyes. These he yanked into shore and dropped on the quay, where they flipped about, twitching and bleeding until they died.

Three old men watched the fish pile up around him. One turned to me and said with contempt, "Who would eat such fish? They feed on oil and gasoline. He must come from a village."

There was a long line of refugees outside Pavo Handabaka's office. They had come to the Center for Social Work to pick up their monthly checks, but Pavo, a refugee himself, was in no hurry to attend to them. He was in such haste, though, to soothe his sore throat with brandy that, rising from his desk, he stumbled into the broken electric heater. He ran his hands over his loose-fitting leisure suit, and when he had filled our glasses he returned to his desk and began to fiddle with his pen.

It was hard to picture him as a captain in the merchant marines, sailing around the world—Houston was his favorite city—and even harder to imagine him rich. Yet both had been true. In the war he had lost his house and belongings, including a speedboat. Also gone was a lucrative business of renting out hang gliders and windsurfing equipment. Catering to tourists, however, did not prepare Pavo for administering to the needs of another transient population. Refugee work taxed him beyond his abilities, and he compensated for this by fussing over statistics. He wrote them down on a sheet of paper, which he studied as he spoke. The numbers added up to a disaster: 30,000 displaced in Dubrovnik, 50,000 from the surrounding villages. Thousands were living in hotels; thousands more were abroad, waiting to return to their homeland.

Pavo rose to his feet and stumbled again into the heater. Retrieving a stack of reports, he showed me how neatly produced they were. When I nodded in appreciation, he refilled our glasses.

"The old people say World War Two was a child's game next to this," he said.

The siege was not to be described. "Every day we waited for the bombs to come through our windows. We were expecting to see Četniks on the streets, cutting our throats, but they didn't want a second Vukovar. Why would they want to conquer a city of 6.7 percent Serbs?"

Pavo, who was proud of his three Serbian employees, claimed to harbor no ill feelings toward Serbs. But there was no room for more Muslim refugees; 350,000 of them (he showed me the figures) were already straining Croatia's resources. Hundreds of years ago, Dubrovnik had granted asylum to all refugees. Now only ethnic Croats were housed in the city; the Muslims, 1,692 at last count, were in a camp on a nearby island. War was coming with Bosnia, Pavo was certain of that.

Perhaps the city government made a mistake in pinning all its hopes on tourism, he began to say, and then abruptly ended the interview.

"Maybe I talk too much," he said anxiously, leading me to the door.

"Not at all," I said.

"Dubrovnik, like Vukovar, will have its place in the myth of Croatia," he said loud enough for the refugees to hear, "because now we have soldiers who want to fight for their country."

In the Rendezvous bar at the Hotel Argentina, with a Canadian military monitor, a tall, nervous man. "It's dangerous everywhere," he said, sipping brandy. "The other day, in one of the villages, a woman drove ten centimeters off the road and she was blown to smithereens."

He had just published an essay likening the breakup of the Soviet Bloc to the Reformation. "There you had a hundred sixty years of bloodshed, and after that the Church came out stronger than ever. The same thing may happen with Communism. All the signs point in that direction."

Two drunk Croatian soldiers took the seats on either side of us. One refilled our glasses from his own bottle of brandy, which he called "Croatia whiskey," and asked the monitor what his salary was in Dubrovnik. Four hundred deutsche marks a day, said the monitor. The soldier sank to his knees, pretending to take his bottle of brandy away from us.

"I make two hundred fifty marks a month," he said. "Can we trade jackets?"

Against his better judgment, the monitor let him try it on. The soldier went through a model's poses in the white jacket, dancing around the bar, preening in the mirror, waving his pistol, and then disappeared into the kitchen. We could hear the chefs and waiters laughing and toasting him.

"This could be the end of my career," said the monitor.

The other soldier spoke for the first time. A pregnant Serbian woman had shot him in the leg, and he had not fired back. It was very important for us to understand that he had not shot her.

"Everyone thinks something big will happen in the spring," the monitor said to me. "They're digging in now for winter, but the Croats won't let Krajina go without a fight. It could be a mess."

Yes, I agreed. A splendid view.

The ethnologist Marina Desin and her brother Andro were driving me down the coast, and we had stopped to take in the view from a cliff overlooking the sea. On this mild, sunny afternoon, the parking lot was empty except for a young man with a video camera. Andro and I were leaning against the rusted guardrail, the only protection against a steep drop-off, while Marina paced back and forth. She was a handsome woman with shoulder-length brown hair; her bronzed face was dotted with moles, and her voice was lined with bitterness, whether she was describing Serbian "butchery" or the way her family—fourteen generations of seafarers—had lived in Dubrovnik since 1490.

"We had three houses," she cried. "A library of two thousand books, tools, shipping equipment, all gone. Can you imagine? The Četniks went to the bathroom in every room. They shit on the paintings of our ancestors. This is something satanic."

"I understand them now," said her brother, a grey-haired young man, "because I am in the army. After two days in the mountains I come to Dubrovnik and I feel different. After wildness this is something else."

Marina slumped to the ground, crossing her legs on the tarmac, then stood up again, all in less than a minute. I could not tell if she did this to calm down or to launch another monologue.

"This may not look like much," she declared, "but it was a culture we sustained for centuries. There is an inborn beauty in Dubrovnik people. We don't want to destroy."

"We were educated to be equal, but how can that be?" said Andro. "Dubrovnik is something everybody envies. We don't envy their land."

"They admire us and hate us at the same time," said Marina. "They want to marry into Dubrovnik, and they want to sack the city."

"A woman asked a Četnik why he wanted to steal a TV remote control," said Andro. "He thought it was a computer he could give to his son."

"Primitives," Marina spat, recalling how the besieging forces would pump folk music into Dubrovnik from a loudspeaker on the hill, with Serbian war lyrics:

> You say Serbia's small
> Liars one and all!
> She's not small at all.
> Not small at all.

"The Serbs are not ashamed of starting this war," she said, "only of losing it."

But surely Tuđman has something to answer for, I suggested.

Marina glared at me. Andro stepped away from the guardrail. All at once they closed in on me. I had just enough time to straighten up before Marina shouted, "The world will not let us defend ourselves! How would you react if we attacked you right here? How would you like it if we pushed you off this cliff?"

She clenched her fist. Her brother, hesitating for only a moment, raised both hands.

"If you do push him over the cliff I'll have it here on film."

It was the young man with the video camera. He was dressed in black—black leather jacket, black pants, black shoes—except for white socks. His camera focused on Marina, he spoke in a soothing manner, explaining why, as a refugee living in Toronto, he had decided to visit his birthplace. He would never live here again, he said, edging into the space between the Desins and me.

And that was the end of it.

"The only thing I admire about Tuđman is his patience," the refugee said, and focused his camera on the sea. "Yugoslavia was a bad marriage that needed to end in divorce, not a bloodbath."

Marina, Andro, and I got back in the car and drove in silence, passing villages and vineyards laid to waste. The JNA had followed a scorched-earth policy along the coast, burning down houses, vines, and cypresses. Around a corner was an old man pushing a wheelbarrow, followed by two old woman clutching blankets. Then a sign—*Happy Journey Come Again*—where the road became rubble.

In the historic center of Čilipi, our destination, no house or building was intact, save for the façade of the church, which was spared, Marina insisted, for Serbian TV. The church interior was gutted. The priest's house, too. On the stone bearing an inscription about the church's dedication in 925 was a scribbled message: *God Keeps The Serbs.*

We walked through the village, among heaps of bricks, stones, and broken tiles. A grove of burned pines. A dead pheasant. A white camellia between two burned-out houses. A broken record—*Carmen*—in the road. Shoes in the ruins of a school. Ravaged vineyards. Someone had taken a knife to the seats of a bullet-ridden Yugo. The Serbian motto was scrawled on every façade:

Э Є
Э Є

Only Unity Saves the Serbs (*Samo Sloga Srbina Spašava*).

Marina, curator of the Institute of Cultural Preservation, stared at her gutted museum. Though most of its holdings, cultural artifacts of southeastern Croatia, were evacuated before the village was overrun, the traditional costumes with silk embroidery were destroyed.

"You could almost laugh at the destruction," she muttered.

A prosperous town, Čilipi. The airport was nearby. Young professionals who worked in Dubrovnik liked to live where mimosa, kiwi, and palms could be cultivated. Every Sunday was a fair. Thousands of tourists came to watch the villagers, Marina among them, tilling and pressing grapes or olives and dancing in traditional dress: white blouses and skirts, black vests and trousers, red caps. Wine was served free of charge, "as much as you wanted," said the curator.

"Everything we worshiped—our coastline, our costumes—must now be ugly, simply because we are Croatian," she said. "What are we to do? Advertise our ruins? Our money is in a Serbian bank. I don't know how we will start over."

But soon we came upon a woman hanging laundry next to a garden flattened by a telephone pole sawed off at its base. And further on a wrinkled man, in a grey sport jacket and beret, balanced himself in a fig tree, hacking at the branches with a hatchet. His tools had been stolen, he explained. His roofless house had doubled as a café, and in the debris stood a rusted espresso machine; the scorched frames of chairs lined the bar, surrounded by smashed china. A charred bedspring was draped over the wall, like a rug. The old man had salvaged a rack of empty wine bottles and one ice cube tray.

"I don't even have a hammer anymore," he said sheepishly.

Marina smiled. "If that man comes to cut his fig, he will be back," she said. "That's what we need—people starting to work on their places again. They can't stay in the hotels forever." She was looking for stone masons to rebuild her museum; a Christmas exhibition was in the works. Yet she could not resist telling me about the farmer who returned to Čilipi only to drown himself in his well.

When it was time for the Desins to leave—they were continuing down the coast to inspect the damage to Andro's house—Marina said, "Try to understand us when you go to the other side."

I crossed the road, where two soldiers kept watch in a dump truck overflowing with rubble, and walked to the other end of the village. The devastation was complete. Under the steep mountains, in the sights, perhaps, of a bored recruit, I found the silence more unnerving than the prospect of hitchhiking back to Dubrovnik. It was some time before a soldier in a camouflaged pickup stopped.

"The old people knew this war was coming," said Ivo, the soldier, a former airline ticket agent who had helped to defend the city, armed with a hunting rifle. Now he liked being in the army. "For years my father told me there would be war with Serbia. But I did not believe him."

What did he believe? God would punish the Serbs. Only America could stop them.

"They have attacked us three times in the last two hundred years," he said when we reached my hotel. "The next time we will be prepared for them."

The sign by the front door—*No Weapons Please*—did not apply to the soldiers drinking in the lobby. And the hotel manager was in no hurry to give me my key. He took pleasure in telling me I had no messages even when someone had called. Was he the one who listened in on my overseas calls, cutting me off whenever I criticized his government? I thanked him anyway and went up to my room.

The single light flickered. The sea wind was blowing through a gaping hole in the window. I had what the military monitors called an air-conditioned suite. They had them, too. The Argentina, the only hotel open to the public, served as their home and office, and they liked to tell stories about their colleague who had recorded the shelling of the city from the terrace. "Incoming, incoming," he would cry, then call to the waiter for a gin-and-tonic. Only rarely did he say, "Outgoing."

A refugee family had vacated my room the week before, leaving behind a toy car, a doll, a sock, and the rank smell of meat cooked in oil. I walked out onto the balcony to escape that smell. The temperature was dropping, but I decided to sit here for a while. To calm my nerves, I opened a book of poetry—St.-John Perse's elemental ode, *Seamarks*—and as the sun set over the Adriatic I meditated on the poet's emblematic fate. His birth in Guadeloupe, a diplomatic posting to the Far East, the publication of *Anabasis* (translated into English by T. S. Eliot), advancement in the Foreign Service—the first half of his life culminated in an unforgettable moment at the Munich Conference, when he stared Hitler down, believing appeasement would lead to war. When the Nazis invaded Paris, they ransacked his flat, and fifteen years worth of manuscripts disappeared.

Perse was in Washington by then, in exile, working as a consultant to the Library of Congress. After the war he stayed in America, where he composed his major poems. In 1957 he began to divide his time between the coast of Maine and the south of France. The double nature of his life intrigued me: how he shuttled between the Old and New Worlds, his refusal to mingle poetry and politics. Who could not admire his courage in the face of the Nazi threat, and his conviction, as he declared in his Nobel Address, that "it is enough for the poet to be the guilty conscience of his time"? The sweep of his vision, his musical phrasing, the way his work reads, in his words, as a "single and long sentence without caesura"—here was "Poetry to fire our watch in the delight of the sea."

Gunfire interrupted my revery on the poet's role in the modern world. Three loud bursts off to my right then an exchange of automatic weapons. I put my book down and headed for the bar.

The hotel was conserving electricity, so I had to pick my way along the darkening corridor. Over the sound system came the fitful strains of a popular song, which grew stronger as I neared the Rendezvous. Around a corner I stopped in my tracks, straining to hear a hauntingly beautiful guitar solo. A voice was singing, "Every man has to die." A wave of peacefulness washed over me. I looked through a broken window at the sea, grey in the gloaming, and wished the music would never end.

A waiter passed, eyeing me suspiciously. I stood there until the song faded. I watched the sea blacken, and then I walked on.

10

Serbia

There was no heat on the train from Budapest to Belgrade. It was New Year's Eve, it was bitterly cold, and it was snowing: three reasons why we should be allowed to travel for free, said the British journalist who invited me to join him in first class. The conductor was not amused. For a moment it looked as if he might even throw us off the train. Then he took our tickets with a loud grunt. The other passenger in our compartment, a sleeping, disheveled man, opened his eyes long enough to show the conductor his pass and closed them again.

One thing Paul Martin could say for the new world order: it made his business travel easier. Once a foreign correspondent in South Africa and the Middle East, now he commuted from his home in London to cover the war as an independent filmmaker. On this trip he would interview Ratko Mladić and finish a documentary about Kosovo. Croatia's return to fascism was his next project. I was surprised to hear that although he had not visited Croatia he knew the outlines of his story. The other passenger, who had awakened at the sound of Mladić's name, stared at the filmmaker.

"*Dober dan*," Paul said to him.

When the man did not answer, the filmmaker began to lecture me about the war. His opinions contradicted everything I knew. The Bread Line Massacre in Sarajevo, for example. Paul said the shell had come from a Muslim position. Nor did he believe the story of the captured Serbian soldier confessing to raping and murdering Muslim women. Look at his photograph, he insisted, as if he were showing me a copy of the month-old head shot printed in *The Times*. His fingernails are missing, his wrists are swollen. The Muslims probably tortured him until he was ready to admit to anything. What are you talking about? I said. But Paul had moved on to another subject: how Eagleburger's decision to call Milošević a war criminal had insured his reelection: it was just what he needed to rally support. The filmmaker could not decide if the secretary of state had done this by design or out of stupidity.

His monologue did not end until the train stopped at dusk in the Serbian border town of Subotica. Two guards took me to an office, where a policeman reluctantly gave me a visa. When I returned, the third man in our compartment was gone and Paul was arguing with a guard over the videocassettes marked **Kosovo** in his luggage. This is not normal, said the guard. You are supposed to take films *out* of Serbia not *into* Serbia. Twice Paul said he was returning from holiday to finish a documentary. The guard asked for his papers. All at once he smiled. He had made a great discovery.

"You are a filmmaker?" he said, relaxing into a conversational tone.

Paul nodded.

"Please," he said, handing his documents back to him. "May I have your card?"

"No," Paul replied, and hoisted his luggage onto the overhead rack.

The guard stiffened. "Why?" he said.

"Because you were not nice to me," Paul said contemptuously.

My heart sank. The guard pointed at both of us. "Come with me," he ordered. Paul asked him, in all seriousness, for help with our luggage. The guard shook his head. "Why?" said Paul. "Because I am the guard and you are the passengers," sneered the guard.

Night had fallen when we were escorted to an unheated warehouse, where two policemen were waiting. The guard told us to remove our overcoats and unbutton our shirts. He took his time, in the freezing air, opening our bags, and when our belongings were spread across the concrete floor the questioning began. Why are you bringing film into Serbia? the guard asked again. I am returning from the Christmas holiday to finish a project, Paul explained again.

"You are Catholic?" said the guard.

"Protestant, not Catholic," Paul said emphatically.

The guard was pleased to hear this. "Yes," he said with a smile. "Orthodox and Protestant celebrate Christmas the same."

Yet the guard could not understand why Paul had already taken his Christmas vacation. The Orthodox holiday was a week away, coinciding with Epiphany in Western Christendom. Do you have a different calendar? he wondered.

I let out a nervous laugh.

"What's so funny?" the guard demanded.

"Nothing," I said, shaking with cold.

"Then get out of here," he commanded.

We threw our clothes into our bags and hurried to the train.

"And they wonder why they've lost the propaganda war," said Paul.

The first person I met in Belgrade was Radovan Karadžić. He and his wife (and two bodyguards) were in the lobby of the Hotel Intercontinental, when Paul and

I checked in. The filmmaker, hugging Karadžić's wife, introduced me to the Bosnian Serb leader, as a fellow poet.

"He thinks the way we do," Paul told the president.

Karadžić was a little drunk, so he may not have noticed the contortions I went through to mask my horror. Or perhaps he did: he was a psychiatrist, after all (his former patients living under a siege of his devising liked to recall that paranoia was one of his specialties). And he was a poet destined to be charged with crimes against humanity. He was already guilty of what in another context Czesław Miłosz called a grave sin: "though much can be forgiven a poet, he must not become a seducer, not use his gifts to make his reader into a believer in some inhuman ideology." Karadžić would not be forgiven for preaching—and practicing—ethnic cleansing. He was a bad poet (though he had gone to New York City on a Fulbright fellowship to study underground poetry), but it was foolish to deny that in one of his couplets he had foreseen the future: "Take no pity let's go/ kill the scum down in the city"—a future he himself had invented.

He was never comfortable in Sarajevo. Born in Montenegro, he moved to the Bosnian capital as a teenager; and though he went on to establish a successful practice (he was the soccer club's psychiatrist), the literati viewed him as an ambitious, untalented boor. A criminal conviction in the 1980s for fraud further diminished his standing among the writers, who were not impressed by his ancestry: he was related to one of Serbia's most fabled literary figures, Vuk Karadžić (1787–1864), the philologist, collector of folk epics, and codifier of the Serbian language. A tragic mistake: the siege of Sarajevo was his revenge for the slights he suffered at the hands of the city's elites and the centerpiece of his strategy to become the father of a Bosnian Serb homeland.

Strange to think what Vuk Karadžić's descendant wrought. It was the elder Karadžić who created a literary language out of the Serbian vernacular, writing down the epics of the Battle of Kosovo. He introduced phonetic spelling, completed the Cyrillic alphabet, produced the first Serbian grammar, and translated the New Testament into Serbian. Collecting poems, riddles, proverbs, spells, curses, and songs from Bosnia, Croatia, Dalmatia, Montenegro, and Serbia proper, the folklorist concluded, from his office in Vienna, that all South Slavs were Serbian. In 1849, he published "Serbs All and Everywhere," the first modern articulation of Greater Serbia, an idea at odds with the ideals of Ljudevit Gaj's Illyrian Movement. Yet because Gaj promoted the same form of the vernacular as Karadžić, within the year Croatian and Serbian writers signed a literary agreement in Vienna, uniting Karadžić's linguistic work with Gaj's reforms to create a new literary language, Serbo-Croatian. And this was still the language of Bosnians, Croats, and Serbs (allowing for regional differences), despite efforts on all sides to separate it into three distinct languages, Bosnian, Croatian, and Serbian.

Radovan Karadžić's rise to power, rooted in his ancestor's romantic nationalism, was orchestrated by the novelist Dobrica Ćosić, and together they gave new meaning to literary politics. In this marriage of the pen and sword Karadžić, armed with a *gusle* and newly discovered Orthodox faith, took to visiting the countryside—the "source" of the pure Serbian spirit—to whip up hatred for the Muslims. The sharp division in the Balkans between rural and urban life, a charming feature of Slovenia, was lethal in Bosnia, where Muslims tended to congregate in cities surrounded by Serbian villages. Karadžić's troops aimed their guns at his former neighbors in Sarajevo, singing a folk lyric:

> Oh, beautiful Turkish daughter,
> Our monks will baptize you.
> Sarajevo, in the valley,
> The Serbs encircle you.

Kitsch on a grand scale. And war was a logical outcome of a literature and politics founded in kitsch, which is immune to criticism, to the wisdom, for example, embodied in three of my favorite proverbs collected by Vuk Karadžić: "A lie has short legs." And: "The devil rides on a man who lies." And: "You need a lot of spades to bury the truth."

The devil was riding hard on Radovan Karadžić's brawny shoulders. Among his boldest lies: Sarajevo was not under siege: the armored forces circling the city were merely "protecting" the Serbs from the Muslims, who had destroyed the National Library, National Museum, and Oriental Institute. Muslim snipers were shooting Muslim children, et cetera, et cetera. Perhaps Karadžić even believed himself. Delusions of grandeur are not uncommon in the literary trade, to say nothing of the political class. And certainly his followers trusted him, along with his apologists in the media, here and abroad. "I am glowing like a cigarette/ On a neurotic's lip," he wrote in 1983, in a poem prophetically titled "A Morning Hand Grenade." It was a faithful self-portrait, I decided, when he told me that if NATO launched air strikes on Serbian positions his gunners would shoot down every plane.

"I have thousands of Russians who want to come to our defense," he boasted, running his hand through his mop of grey hair. "If the West intervenes, Yeltsin will be gone in twenty-four hours and Russian troops will be everywhere."

He was playing on a genuine fear in diplomatic circles: that in the name of Orthodoxy and Pan-Slavism Russia would side with Serbia in any confrontation with the West, as it had in the past—whenever it suited its purposes. Russia and Serbia were not, strictly speaking, historical allies. ("When push comes to shove," writes one historian, "Russia has always abandoned the 'savages.'") But they had joined forces often enough to lend some credibility to Karadžić's threats, providing timid Western politicians (mindful of the unsettled political climate in Moscow and in the newly independent Soviet states) with another excuse not to take decisive action in the Balkans.

From his pocket he withdrew a brass figurine, supposedly taken from a dead Bosnian soldier, bearing the trademark, **MADE IN PAKISTAN**. Paul nodded. Proof, said Karadžić, that *Mujahedeen* (literally, holy warriors) veterans of the war in Afghanistan were fighting on the Bosnian side. It made no difference. Indian Hindus were coming to his aid.

Hindus? I said, amazed. I could not tell if Karadžić, a gambler rumored to have lost hundreds of thousands of dollars in Belgrade casinos, was bluffing. I knew that mercenaries were fighting on all sides in Bosnia, but what struck me was Karadžić's eagerness to impress. It was hard to believe that such an ordinary braggart could keep the international community at bay.

Hindus, he repeated.

What a heady feeling to travel to Geneva for peace talks called to stop a war you began!

Just the start of a long process, Karadžić said with pride. We're down to maps, but the Muslims want to keep fighting. Paul nodded once again. As for Eagleburger's claim to have enough information on the chain of political and military responsibility from Milošević to Karadžić to Mladić to charge them with war crimes for the atrocities committed in Bosnia?

"Larry's fishing for a new job now," the president said smugly.

The cab driver took an instant dislike to us. Climbing into his Mercedes, I broke off part of the plastic protector between the front and back seats. Then Paul lost the address to the peace vigil he wanted to cover. So we drove up and down Belgrade's snow-covered streets, searching for protesters. Finally, Paul motioned the cab driver to pull over and went into a skyscraper to get directions.

An hour passed without any sign of him. The cab driver was irate. His meter was broken, he was using up gas, and the next move he was plotting did not bode well for me. Across the street was a parked police car with its lights flashing (a drunk driver was headed for jail), and every minute or so the cab driver would scream at me, pointing first at his meter then at the policemen. In Serbia for less than six hours, I was on my way to another ugly encounter with the authorities. I got out of the cab before the clock struck twelve (no sense in kissing the cab driver), pretending to look for Paul. In fact, I was making my New Year's resolution: to quit the Balkans as soon as possible.

The filmmaker reappeared a few minutes after midnight, with three young cameramen from Sarajevo. The cab driver had not even shifted into gear when another car, skidding out of control, plowed into us from behind. Paul left a wad of dinars on the front seat and we set out on foot.

The peace vigil was winding down when we arrived. About a hundred young men and women milled around the snowbound square, ringed with candles, between City Hall and the Presidency building. "Nobody wants to put on a uniform," said one cameraman, refuting, perhaps unintentionally, the popular Croatian and Slovenian expression: namely, that only Serbs liked the military.

An ill-fitting overcoat, a mouthful of rotten teeth—Boske Perišić, the cameraman, had a haggard look. At the start of the war he had fled with his Serbian father to Belgrade, and he had not heard from his Muslim mother in three months, not since receiving word from Sarajevo that a Serbian grenade had killed his sister. But Boske was quick to defend the Serbs. "They're always fighting for their freedom," he said. "From the Turks, the Austrians, in World War One, World War Two, and now again. Look at what they're up against. The Croats went after the Serbs for a long time before they retaliated, and they have the Vatican behind them. Now they've wrecked the country."

It was a common Serbian complaint—that the whole world (Germany, the Vatican) had turned on them. Boske lit a candle for his sister and planted it in the snow. "I want to leave Yugoslavia," he said mournfully, "but where can I go? No one's taking in Serbs anymore."

"Believe me," said Boris Kostov, an independent television producer. "Serbs built their house right in the middle of the crossroads between the East and West, and then they fucked the world!"

His New Year's Eve party was still going strong when we returned from the peace vigil. Television monitors and lights flashed, and from the nineteenth floor we watched snow falling over Belgrade (which means White City). Boris had set out roasted lamb, cakes, and beer for his employees, who kept saying to Paul, "The bombs must be coming!" Boris nodded enthusiastically. A short nervous man with bulbous eyes, he had a sort of canine eagerness, which I attributed to his membership in Serbia's small Bulgarian minority: things might turn precarious for him in Greater Serbia. Not wishing to miss out on any conversation, he prefaced each remark with an earnest plea. "Believe me," he would say, "you can never defeat Serbia. It will keep fighting for hundreds of years, but it will lose every peace." Or: "Believe me, all the daughters of Russia's high commanders are strippers and whores in the West. These men will get their revenge some day."

Boris had unwisely hitched his star to the fortunes of Jezdimir Vasiljević, the flamboyant banker and host of the sanctions-breaking chess match between Bobby Fischer and Boris Spassky. Vasiljević had run for president, too (Boris had produced his television advertisements during the campaign), and, though all his neighbors, including Boris, had voted for him last week, when the election results were announced, Vasiljević did not receive a single vote from his own district.

"How do you explain it?" I said.

"Believe me," said Boris. "The opposition parties got at least a quarter of the total vote, but Billy"—his nickname for Milošević—"Billy made sure they only got 2 percent!"

Rigging elections came naturally to Milošević. His first presidential campaign made a mockery of the OSCE document Yugoslavia had signed in 1990, affirming that fair elections and freedom of expression were the safeguards of democracy. Bowing to pressure to give his opponents air time on state TV, Milošević granted them one night to present their cases. Thirty presidential candidates, including what Warren Zimmerman called "drunks, religious freaks, rock stars, and stumblebums" recruited for the occasion, spoke one after another into the early morning. Even the serious candidates looked foolish. Milošević went on to win by a wide margin. And in this election he had taken no chances, stealing enough votes to insure a decisive victory over his main rival, Milan Panić. When Vasiljević's wife went to the polling station in her neighborhood she was told her name had been expunged from the rolls.

"Believe me," Boris said with a leer. "Billy is the greatest director of all."

"The banality of evil" was not what came to mind when I saw Ratko Mladić on New Year's morning in the lobby of the Intercontinental. His icy blue gaze I recognized from Paul Celan's poem on Nazi executioners: "Death comes with eyes that are blue, with a bullet of lead he will hit in the mark." And Mladić had made his mark on this war. The son of a Partisan killed in a raid on the village Ante Pavelić came from, he commanded the JNA in Krajina with a singleminded ferocity, ruthlessly expelling and exterminating Croats; and when he took over the Bosnian Serb army he moved quickly to occupy 70 percent of Bosnia. The refugees, the death camps, the siege of Sarajevo, these were his work. He had the look of someone who held the high ground. I did not know what to say to him. Paul Martin did. The general, who was on his way to Geneva for the peace talks, told the filmmaker he would retaliate against any country that shot down Serbian planes. And he meant it.

Everyone was a criminal. You heard that everywhere. The UN sanctions were to blame for the black market. It was the West's fault the dollar fetched more than three times the official exchange rate, which was revised every hour. German industrialists were out to destroy the Serbian economy, and so on. The economy was indeed in a free fall: hyperinflation had set in (20,000 percent annually—a rate not seen in Europe since the crash of the Weimar Republic—and soaring to 300 million percent per *month* by summer); industrial output was down by a quarter; hundreds of companies had gone bankrupt, idling more than a half million workers; and the average wage for those still employed had declined from $750 a month to $75. Serbs were living on their savings, the value of which fell by the minute. And the sanctions made things worse. Gasoline cost almost $14 a gallon, cars were parked for hours in front of the few service stations still open, black marketeers selling fuel in plastic containers lined the streets. A shortage of heating oil left people freezing in their apartments. At any moment the entire electrical system might crash, because of all the space heaters in use.

I went to the open-air market to change money. In the stalls were slaughtered pigs, heaps of fresh vegetables, wine in used bottles. Old women hawked smuggled cigarettes. Pensioners sold their kitchenware. The sanctions worked to Milošević's advantage: he could blame the failure of his command-style economic policies on the international community. And he was not alone in profiting from the hardship. There were long lines outside the private banks, where thousands of Serbs, lured by promises of 14 percent monthly returns, were investing their savings in pyramid schemes; the hard currency the banks took in fueled the war machine. But the gold rush atmosphere would last only until the banks ran out of investors and the pyramids collapsed.

The Slovenian economist Jurij Bajec also blamed the West for the chaos. He was still teaching at Belgrade University, having found a new subject—the deleterious effects of the sanctions—instead of moving back to his homeland. Like Serbs, Bajec could not understand why George Bush had refused Milan Panić's request to ease the sanctions, a goodwill gesture that might have helped him to defeat Milošević. Here was proof, of the same order as Eagleburger's labeling Milošević a war criminal, that the West wanted the strongman to stay in power. Belgrade's elites were indeed no less prone to paranoia than the masses. People behave differently in a locked system, Bajec argued. They're illegal actors in an illegal economy. I wished I had asked Bajec to discuss the differences between the locked systems of Belgrade and Sarajevo. No one was besieging Belgrade.

The money-changers (who worked for the banks) were in the unlit tunnel next to the market.

"*Devize, devize,*" they cried. Hard currency, hard currency.

Some were women, some were children. It was hard to tell who had the best rate.

The night clerk was shaking with anger. "Hey, mister," he said, pounding the counter. "Are you a journalist? Why the West, why the United States want war with Serbia?"

How I regretted moving from the Intercontinental to the Slavija, a bleak hotel near Saint Sava Temple, in order to save money. The phone and television in my room did not work, the dining room was given over to armed soldiers and black marketeers, and the clerk refused to return my passport.

"What's the matter with Americans?" he cried. "Why don't they like Serbs?"

Not all Americans feel that way, I told him.

"All journalists are liars," he replied. "The only ones telling the truth are from India or China."

I was tempted to ask him when he had learned to read Hindi and Chinese, but the bellhops were closing in around me. I said I was here to tell the truth.

"I won't believe it until I see it," he said. The bellhops, nodding in agreement, tightened their circle. "Maybe tomorrow you will get your passport," the clerk snickered.

Kalemegdan, the ancient fortress overlooking the confluence of the Sava and Danube rivers, marks the border between Central Europe and the Balkans. Celtic fortifications gave way to the Roman legionary camp of Singidunum destroyed by the Huns in 441, rebuilt by Justinian, then razed again by Avars and Slavs. Belgrade was an important citadel by the ninth century, the site of many battles in Byzantium's ongoing war with the Magyars. Frequently leveled, the citadel fell to the Magyars soon after Constantinople recognized Stefan Nemanja as Serbia's grand *župan* in 1159. His sons, Sava, founder of the Serbian Orthodox Church, and Stefan the First-crowned, established the Serbian kingdom, which at one time covered most of the Balkan Peninsula. The kingdom endured for precisely three centuries: in 1459, at Smederevo, a fortress just downriver from Kalemegdan, the Ottomans concluded their century-long campaign to conquer the Serbs.

It was Sava who coined the phrase, *Only Unity Saves the Serbs*, advice ignored often enough to make disunity a major theme of the conquests, uprootings, and uprisings known as Serbian history. This was true even before the Battle of Kosovo in 1389, when Prince Lazar's defeat gave rise to a mythology, rooted in loss, of Serbs betrayed by their brethren. Indeed Serbs fought against Serbs in the Sultan's army (Balkan warfare is nothing if not fratricidal), but the epic poets, writing two hundred years after the fact, wrongly accused the Turtko or govenor of Bosnia, Vuk Branković (who was also Lazar's son-in-law), of treachery. Nor is it clear that the Serbs even *lost* the battle. The Turks retreated to Edirne after suffering heavy casualties, comparable to those of the Serbs, and it was another seventy years before the Serbian kingdom fell. Because Turkish rule was harsher in Serbia than in other provinces (the nobility was put to death, the peasants were worked to death), unity among the Orthodox usually took the form of flight (notably the 1690 exodus from Kosovo of 30,000 Serbs who settled in Krajina) or belief in the resurrection of the Serbian state, the litany for which was provided by the Kosovo epic. The two were not mutually exclusive.

The legend of Lazar's refusal to live in shame inspired Serbian nationalists to free the Balkans of the Ottoman yoke. In 1804, a peasant called Karadjordje or Black

George ("Future king of the paupers," Vasko Popa wrote) led the first Serbian uprising, liberating Belgrade and upholding the tradition of internecine battle. Black George, a ruthless illiterate (he had already killed his stepfather), founded his dynasty with a certain flair, hanging his brother for rape and then leaving the corpse dangling from the front gate for his fellow rebels to see when they came to dinner. The king also met a violent end (murdered by Miloš Obrenović, begetter of a rival royal line), after inaugurating "the Serbian problem": regularly stirring up trouble and drawing the Great Powers into the Balkan arena. Under the banner of Pan-Slavism Serbs fomented rebellion in Bosnia, fought Turkey and Bulgaria, started the first Balkan War, caused the second, and set off the Great War. In short, Serbian nationalism drove the liberation of the Balkans from the Ottoman and Austro-Hungarian empires. Milošević was not exaggerating when he said that Serbs may not know how to work but they do know how to fight.

One bleak afternoon, I walked through the spacious park surrounding Kalemegdan and along the serrated walls of the fortress, an extensive fortification of stone turrets and towers and gates. The sky was grey, the snowpack underfoot hard as rock. Trawlers broke through the ice on the rivers, which boomed and cracked. Old men helped their wives down the icy stone steps leading to a chapel and restaurant. Seeking warmth, I ducked into the Military Museum. Every room was taken up with schoolchildren ogling over maps, lances, hatchets, helmets, scabbards, swords, and guns. Here in engravings, paintings, and photographs was an illustrated history—military and migratory—of the Serbian people. The assassination of King Alexander in 1903. Victories in the Balkan Wars over Turkey and then Bulgaria. Gavrilo Princip and his fellow conspirators. I could not take my eyes off a drawing, from the 1915 Serbian retreat to Albania (an exodus on the scale of the 1690 migration), of women kissing the feet of hanged men. At the conclusion of the World War One exhibit three bearded men who called themselves Četniks led me out into the cold.

Ah, the history lesson continued outside, in front of the walls, where children were climbing over cannons, tanks, and missiles dating from World War Two. Yugoslavia found its soul, according to Winston Churchill, on 27 March 1941, when army and air force officers mounted a *coup d'etat* in Belgrade, deposing Prince Paul who had just signed the Tripartite Pact. This open defiance of the Nazis (Serbian crowds attacked the German tourist board, chanting "Better war than pact! Better graves than slaves!"), a heroic and suicidal act in the spirit of Kosovo, precipitated Hitler's invasion of Yugoslavia. And while mythology held that it forced the Nazi leader to postpone his Russian campaign, in truth it ushered in the Ustaša terror, civil war, Partisan victory, and the demise of the Serbian monarchy. I remembered what Boris Kostov had said: "Believe me, life means very little to a Serb. If he loses his family, he'll fight you to the death because he has nothing left to live for."

Zoran Milutinović demanded to know who had shut off his e-mail. I steeled myself for a lecture on the machinations of the West to deny him and his friends news of the outside world.

The scholar had a new subject: propaganda. A report in a German newspaper last summer of JNA planes bombing Jajce had so angered him that he considered joining the mercenaries fighting on the Bosnian side. But when he decided the report was false (though Serbian forces were shelling the town and would overrun it in the fall, putting to flight 40,000 Muslims and Croats) and saw no correction in the media, Zoran concluded that Western journalists, taking their cue from their governments, were writing on behalf of the Muslims. If the West wants to end the war, he told me, it should not support any nationalist—Izetbegović, Karadžić, or Milošević.

A light snow was falling outside the unheated restaurant. Inside, it was very cold, and though we were the only customers the waiter would not serve us more than one cup of tea. A skinhead, banging through the door, came to our table and asked for money. Zoran gave him 500 dinars.

"That's normal now," he said.

With an exemption from military service (curvature of the spine) and a university lectureship, Zoran was better off than some. It was a good thing he taught critical theory in addition to Slovenian literature, which now appeared to be a doomed subject. Even Zoran took a dim view of it. Like most Serbs, he felt doubly betrayed by Slovenia: first, for defending Albanians in the constitutional crisis that led to the revocation of Kosovo's autonomous status, then for waging what everyone described as a "dirty war." Slovenians, so the argument went, seemed to forget that Serbia had sheltered thousands of Slovenian children from the Nazis. And they did not realize how good they had it in Yugoslavia, which offered them a chance to preserve their language and identity, raw materials at a fraction of their value, and a market for their inferior finished goods, including their literature, which, according to Zoran, was unfairly promoted, here and abroad. Perhaps now it would sink back into its rightful place.

Only Tomaž Šalamun was spared his withering scorn.

"A great poet," he said with emotion, then returned to his theme of Western propaganda.

But he had inoculated himself against a difficult truth: Serbian propagandists had set the pace in Yugoslavia, at Milošević's behest. "A campaign of intense propaganda was needed to mobilize the population, to make war thinkable in Yugoslavia, let alone inevitable," Mark Thompson writes in *Forging War: The Media in Serbia, Croatia and Bosnia-Hercegovina*. Kosovo was where Serbian journalists perfected the art of manipulating public opinion against other ethnic groups: "a model," Thompson suggests, "which identified and stigmatized a national enemy, homogenized Serbs against this threat, and called for resistance." In their news accounts the "enemy" was always holding Serbian children hostage before killing them in a gruesome manner, unarmed Serbs defended themselves against hostile forces, and so on. Kosovar Albanians were the hatemongers' first victims, followed by Slovenians, Croats, and Bosnian Muslims. Some typical headlines: "1941 started with the same methods," "The whole Serb people is attacked," "Genocide must not be repeated." All madness, of course, designed to inflame the Serbs' hatred of the Other. Likewise the accompanying attacks on the Vatican, Germany, America, the CIA, Free Masons, and Jews. But Zoran would hear none of this. Anything I said was by definition tainted by my "bias" against Serbia.

"One day people will try to find out the truth of our times," he said. "But by then it will be an academic question. Nationalists destroyed the country, with the help of the West, and now all the connections are gone. What comes next will be the breaking of communications with Europe. Milošević doesn't care. He doesn't even communicate with his wife."

Not true. Close political partners, Slobodan Milošević and Mirjana Marković had a marriage in the style of the Ceauşecus: the pursuit of power would hold them together until death did them part (which might take the same violent form as the executions of the Romanian First Couple). Milošević and his future wife grew up together in the provincial city of Požarevac, under separate clouds: his parents were suicides; her Partisan mother, denounced as a traitor, was killed in the war. But at the university Milošević displayed a certain zeal for Marxist doctrine, and since Marković's other Partisan connections were exemplary (her father was a political commissar, her aunt was Tito's lover) the couple advanced through the party ranks. He became a bank director, she taught Marxism at Belgrade University. They sold out one ally after another in their bid for power, yet they remained true to their origins, despite Milošević's opportunistic embrace of nationalism. Serbs called him the "Red Bandit." She was a founding member of the League of Communists. They knew the value of curtailing the flow of information. Odds are, Zoran's own government cut off his e-mail. And I felt for him in his isolation, because sometimes his wit allowed him to see through his countrymen's paranoia.

"When Vukovar was destroyed," he was saying, "our media told us it was liberated. Now my friends want to move to San Francisco and liberate a house."

His friends had in fact scattered to Amsterdam, Johannesburg, and London, and though he himself had applied for a Canadian visa he was resigned to staying here. His grandfather's story—how the people from his village would send one son to the Partisans and one to the Četniks because they never knew who would win—no longer seemed so distant.

"I won't need foreign languages anymore," he sighed. Then he brightened. "Do you know the joke about Karadžić's assistant telling him an American reporter wants to see a concentration camp? 'Show him one where the people are well-dressed and well-fed,' says Karadžić. 'We don't have one,' says the assistant. 'Then show him any camp and let him spread more lies about us!'"

Bogdan Bogdanović had long warned of the threats to cities. The former mayor of Belgrade, one of Yugoslavia's leading architects, had predicted the destruction not only of Vukovar and Sarajevo but also of the Serbian capital. For cities can "fall spiritually, from within." Here the damage was invisible: the buildings were intact, but the life had gone out of them. A civilization was dying. Bogdanović's monument in Vukovar to Ustaša victims illustrated why. In 1977, in a series of prophetic sketches for the memorial, he drew buildings and cities, then covered them with images of flames and ashes, leaving visible only the tops of Gothic spires and cupolas. Of the completed monument, a granite and copper dome, the architect said, "you had every right to visualize an entire city buried beneath it." Now Vukovar, including his memorial, lay in ruins similar to what he had envisioned in his sketches.

Bogdanović, who was also a gifted writer (he compared the siege of Dubrovnik to a madman throwing acid in a beautiful woman's face, promising her another one in return), likened cities to novels: both explored and celebrated civilization's highest values. Thus he believed everyone should be taught how to read a city, an art lost on his countrymen. "We—we Serbs—shall be remembered as despoilers of cities, latter-day Huns," he wrote. And if it was true, as he suggested, that the rise and fall of civilizations was a function of the Manichaean (his word) war between city builders and city destroyers, then the Serbs were indeed contemporary barbarians. "I am with you in your troubled, sleepless nights," Bogdanović wrote in an open letter to his friends in Sarajevo. "Defending the city is the only valid moral paradigm for the future." In despair, the aging architect had gone into exile.

It is said that Belgrade has been destroyed forty times. One day on a walk through the Upper Town, near the vacated American Center (still advertising an exhibit titled *Seeds for Change*, above which was scrawled, in English, **Gringos Go Home**), I wondered how Serbs would breathe life into their capital again. Where a row of handsome Secession-style buildings gave way to bland examples of the International school, an evangelist for Saint Sava pinned a gold cross on my lapel, demanding 3,000 dinars. When I refused, he yanked the pin off, tearing my coat, and shouted at me until I was out of earshot. I stopped at a kiosk to look over the erotic magazines, some of them more than a year old. The lighting was poor, the sex rough. I was thinking about the significance Serbs found in the date of Bosnia's recognition by the EC: 6 April 1992, the fifty-first anniversary of Hitler's Operation Punishment, an Easter Sunday bombing of Belgrade that killed and wounded thousands of people, leveled much of the city, and destroyed the National Library. But here was another coincidence of dates: the JNA bombardment of Sarajevo, which began on recognition day, marked the forty-seventh anniversary of the Bosnian capital's liberation from the Nazis. "The destruction of Vukovar and Sarajevo will not be forgiven the Serbs," Charles Simic writes. "Whatever moral credit they had as a result of their history they have squandered in these two idiotic acts. The suicidal and abysmal idiocy of nationalism is revealed here better than anywhere else. No human being or group of people has the right to pass a death sentence on a city."

The interview with Daniel Schiffer took up the last page of *The International Weekly,* an English supplement published by *Politika*, Serbia's largest media organization. The French philosopher explained his presence in Belgrade as the continuation of a peace mission begun by Elie Wiesel. After working with the Nobel laureate to secure the release of prisoners from the Serbian concentration camp at Manjača, Schiffer now called himself a spokesman for truth and justice, condemning atrocities on all sides. He deplored the foreign media's "satanizing" of the Serbs, and in an open letter to European intellectuals he said the Serbs were fighting against resurgent fascism and Muslim fundamentalism. His invective against Croats and Bosnian Muslims was severe. Serbian political figures were spared any criticism. Indeed Schiffer had nothing but praise for Dobrica Ćosić, whom he described as "a highly cultured and cultivated man." The Serbian president's humanity and profundity so impressed the scholar that he likened him to Václav Havel.

A tragic error. Schiffer was interested in the relationship between intellectuals and society—he had published a book titled *The Discrediting of Intellectuals*—but it was power that fascinated him. And the intellectual's relationship to power defined an essential difference between Ćosić and Havel. The Serbian novelist lusted after it. The Czech playwright attempted to "live in truth," to recognize the moral dimensions of every problem—personal, political, aesthetic—and act accordingly. Ćosić thus divided the world into Serbs and their enemies. Havel cultivated goodwill even among his opponents. Ćosić made a fetish of the Serbian Question. In his reflections on Yugoslavia's imminent breakup, he warned Serbs—in prose as turgid as Havel's was electric—that they risked annihilation and must be prepared to defend themselves. The playwright appealed to what was best in Czechs and Slovaks alike. The novelist played on Serbian fears. Czechoslovakia's peaceful dissolution was a testament to what Havel called "the art of the possible." No wonder he was leading his country back into the European fold, while Ćosić was the titular head of a pariah state engaged in a genocidal war.

Ćosić traded one collective identity (Partisan general, Communist novelist) for another, a change announced in his magnum opus, *This Land, This Time,* a quartet of novels about the Serbian experience of World War One. "It began with an attack on Serbia," Ćosić declared, introducing the paranoid tone destined to drown out all others in Serbian life and letters:

> The Austro-Hungarian Empire sent a punitive expedition (*Strafexpedition*) against the Kingdom of Serbia, to destroy it and open up to Germanic conquest the way to the Bosporus and the East. The first bullets of World War I were fired at Serbian soldiers; the first man killed in the war died on the right bank of the Drina River; the first artillery shell was hurled against Belgrade, and the first house was destroyed there—and then more houses than anybody could remember. August 1914 witnessed the first cases of women and civilians being hanged by the victors; the first murders of old men and children, raping of women, and looting; the first destruction of Orthodox churches and poisoning of village wells. All this was on Serbian territory occupied by the armies of the Austro-Hungarian monarchy.

Ćosić's circle produced the likes of Vojislav Šešelj, a brilliant sociologist-turned-murderer and mad leader of the White Eagles, a paramilitary unit modeled after the Četniks. Šešelj, a self-styled duke from Sarajevo, promised to butcher Croats with rusty spoons (his White Eagles were responsible for much of the mayhem in Vukovar) and extend Greater Serbia to Thessaloníki. In Vuk Drašković he found an ally, a journalist whose long hair and flowing beard gave him a Christ-like appearance. And Drašković had a messianic streak: Serbs must be delivered from their enemies, was the message of his 1982 novel, *The Knife*. Its graphic descriptions of Bosnian Muslim atrocities committed against Serbs in World War Two boosted him into the public arena. Soon he was a leader of the opposition, with his own militia, the Serbian Guard. To his credit he disbanded his militia after the war in Croatia, distanced himself from Šešelj, and began to talk of democracy—without, of course, renouncing his nationalism. Saving Serbs from the fate they had suffered in World War Two was the first article of faith of every Serbian politician, and some in the West. Even Elie Wiesel had briefly succumbed to the temptation to view the present conflict through the bloody lens of the last war, perhaps because on his journey to the region in November he had spent so much time among the Serbs. During a half-hour visit to the Bosnian capital he reprimanded the besieged citizens for committing atrocities comparable to those of their enemies, prompting one Sarajevo journalist to ask, "where are the strands of sanity we are supposed to hold onto? Can anyone take it anymore?" But soon Wiesel would speak out forcefully against the siege. Not Daniel Schiffer. He was staying in Belgrade as a guest of Stankom (a company run by one of Milošević's associates) to offer Serbs his moral support. Unfortunately, he played right into the hands of the warmongers: at last a European intellectual is speaking the truth! The Serbian media loved him.

Srdjan Rajković slept in a different house each night, avoiding the military authorities anxious to press him into service. Tens of thousands of men had fled the country for the same reason; those of draft age who stayed were divided, roughly speaking, into war supporters (soldiers, profiteers) and opponents, like Srdjan, who were too poor to leave. Srdjan, a tall nearsighted student with a ponytail and a cold (courtesy of the unheated classrooms at the university, shut down now until the weather warmed up), was ill at ease in the Hotel Slavija dining room. Ten soldiers were drinking *šljivovica* at the next table, and whenever Srdjan's girlfriend raised her voice he begged her to be quiet—"or else we'll all get killed."

Jelena Subotić ignored him, confident the soldiers could not understand English, which she spoke with a British accent. The soldiers could not take their eyes off this high-spirited woman, who ignored them. She held her countrymen in contempt, blaming the war on primitives from the provinces, and saw no contradiction in writing on cultural matters for *Politika*, the newspaper bearing the most responsibility for fanning the fires of war. If, as George Steiner argues, "The genius of the age is that of journalism," *Politika's* contribution to post-Tito Yugoslavia, which included reprints of World War One propaganda posters ("Serbia Needs Your Help"), was to foreground the Serbian Question. War, according to Mark Thompson, merely "substantiate[d] the fantasies of national beleaguerment and solidarity which the media had been conjuring." And *Politika* was the print medium's dream factory. Jelena described its politics as neutral or even in the opposition.

"We're living hysterically now," she said in an animated voice. "Theater tickets sell out months in advance. *Richard III* is very popular. People are sick of the news. They want to run away. We can't make plans. We could be bombed. Our borders could be closed."

Srdjan tried to interject, but she cut him off.

"So we live quickly. We make parties, we see friends, we smoke and drink and look at MTV. The problem is, too many artists and writers work for the government. There are no people in the middle anymore. We have our writers, and they have theirs. Families are breaking up over this."

"I'm not speaking to my grandfather," Srdjan confided, "because he's stupid and ignorant. At a family lunch he raised his glass for a toast: 'Long live Serbia and God destroy America!'"

"We know Western fashions, music, art, film," said Jelena, reaching for her cigarettes. "We were up to date until the blockade. We need to feel up to date."

Srdjan handed her his lighter. "My grandfather, for instance, knows the population of every European country, and now that the Soviet Union's falling apart he's going crazy. How many people do they have now? he wants to know."

The soldiers stood up and walked out of the dining room. Srdjan sighed with relief.

"We're a lost generation," said Jelena. "We can watch the worst mutilated bodies on television and not even wink."

"We know it could happen to us," said Srdjan, "but we don't panic. We laugh."

"One day we'll be able to express ourselves," she said.

"Provided we don't get killed," he interrupted.

"My American friends were amazed that we could name all fifty-two states," she said.

"Fifty-two?" I said.

"Yes," she said. "Serbia and Montenegro want to be your fifty-first and fifty-second states."

"There are hundreds more where they came from," the general director assured me. The headquarters for his soccer club were in a heated office overlooking a new stadium with seats for 100,000 fans—part of Red Star Belgrade's unfinished sports complex; construction was stalled on the practice fields, indoor tennis center, and hotel, because the sanctions, which prevented the club from playing outside its borders, had cost it $20 million in revenues. One year after winning both the European Cup and the Coca-Cola Cup against the champions of South America, Red Star Belgrade, the world's best soccer club in 1991, had sold off its first eight players to foreign clubs just to stay afloat.

"Before the war," said Vladimir Cvetković, "our team was made up of players from everywhere. When we beat Bayern Munich, a Romanian took the ball from a German and passed it to a Croat, who passed it to a Serb, who passed it to a Macedonian, who scored the winning goal!"

The quintessential Yugoslav institution: founded in 1945, Red Star Belgrade reached its height of glory in the same year the country split apart. Cvetković, a former star of the Yugoslav national basketball team, said his club's success transcended ethnicity.

Soccer fans were another matter. Cvetković would not talk about the infamous match, in April 1990, between Red Star and Zagreb Dinamo. The Serbian *Delije* fan club charged the Croatian Bad Blue Boys, partly in retaliation for the Croatian elections that swept Tuđman into office, and in the ensuing riot scores of policemen and spectators were injured. Millions of Yugoslavs watching the mayhem on television got a glimpse of the future. For the notorious paramilitary leader, Željko Ražnjatović, otherwise known as Arkan, trained his murderous followers in the *Delije* fan club.

Arkan was under investigation in several European countries for drug trafficking and wanted on murder charges in Sweden, the Yugoslav secret police having hired him in the 1970s to "liquidate" Ustaše living abroad. But it was in the Yugoslav wars of succession that Red Star's best-known fan found his métier. Backed by Milošević, Arkan's Tigers and Vojislav Šešelj's White Eagles led the first "ethnic cleansing" operations in Croatia and Bosnia. In ski masks or with their faces painted, they showed off their prowess to reporters by slitting pigs' throats. And they spared no one—women, children, Serbs who objected to their methods—in the grisly trail they left from Vukovar to Zvornik.

Cvetković rose to his feet: how could he control Red Star's fans? He gave me a medallion of the cup-winning team and showed off a photograph of the trophy.

"That's ours forever," he said. "No sanctions can take that away."

Radio B-92's operating license had long since expired. But the independent station, occupying four small rooms in a dilapidated building, remained on the air. According to one argument, Milošević, unlike Tuđman, knew the value of allowing limited press freedom. As long as he retained control of state TV, independent outlets (whose range was limited to the capital) like TV Studio B, the weekly news magazine *Vreme,* and B-92 could siphon off some of the hostility toward his regime. For example, B-92 (the call signal came from the emergency telephone number for the Belgrade police) invited listeners on the night of the presidential election to call in and describe how Milošević should be executed. It was a fine imaginative exercise. B-92's listeners revealed a macabre streak, and although the station organized peace rallies, encouraged draft-dodging, and mounted drives to send clothing and medicines to Sarajevo, Milošević showed no sign of losing his grip on power.

Yet he was reviled by countless Serbs. In May, thousands had unfurled a black band of mourning down the length of Belgrade's main thoroughfares to protest the shelling of Sarajevo. Ivana Balen, an animated leader of the anti-war movement, blamed Milošević's reelection on his skill at teaching Serbs to elevate their patriotic feelings above their common sense. "Yes, we live miserably," she told me at B-92, "but we have our flag." And the international response reminded her of the Balkan Father Syndrome, whereby a man comes home to find his family embroiled in an argument. Instead of asking what happened he hits everyone. Such was the effect of the sanctions.

The war had plunged Ivana into an identity crisis. Born of a mixed marriage, raised in Zagreb, and living now in Belgrade, in her hometown she was called a Četnik, in the Serbian capital, Ustaša. But there are no differences between Serbs and Croats, she argued. What can the nationalists do with that? They introduce new words into the language, they teach people how to hate.

"Why is Sarajevo hit the hardest?" she said. "Because that's where the most people lived most peacefully. We destroyed something good. Slovenes wanted to be richer and they're poorer. Croats wanted their own land and lost a third of it. Serbs wanted their flag flying everywhere, and I know a women who had her baby in Paris because she didn't want the birth certificate to read *Belgrade*."

We walked from the station to the Presidency, where a group of women dressed in black held a weekly peace vigil. A solemn procession. Four inches of new snow lay on the ground. A cold wind swirled. "Three hundred thousand people protested here in May," Ivana said. "You see, not every Serb wants war. But what can we do? The smartest young people left, others were displaced, and the rest were killed. What's left? Only people who know how to hate."

The archbishop had lost his place in the service. He consulted his notes, with the cameras running: the Christmas service was being broadcast on state TV. Better to show this than footage of a ragtag group of Bosnian defenders retaking Srebrenica, a mining town near the Serbian border—though before long Ratko Mladić would besiege the eastern enclave again, driving Bosnians into the hills, with the blessings of the Orthodox clergy. They were quick to charge Muslims with carrying out genocide against Serbs. Confronted with irrefutable evidence of Serbian concentration camps, however, the Holy Episcopal Synod issued a remarkable statement: "In the name of God's truth and on the testimony from our brother bishops from Bosnia-Herzegovina and from other trustworthy witnesses, we declare, taking full moral responsibility, that such camps neither have existed nor exist in the Serbian Republic of Bosnia-Herzegovina." Serbs returning to their faith took the Church at its word. Signs of the Orthodox resurgence were everywhere. A new cathedral was rising from the spot where the Turks had burned Saint Sava's bones; the mural on the wooden fence around the construction site was an essay in martial imagery: haloed figures clutching swords and scepters. During the Christmas mass, I experienced a strange mixture of feelings. The chanting, icons, candles—such a beautiful liturgy in the service of such mayhem! Boris Kostov had said to me. "Do you know what happened to Croats when they lived with Serbs? They stopped being so fanatical about going to church. Believe me, Serbs only go to church once or twice a year." He gave me a quizzical look. "The strange thing is, every fifty years or so the whole region goes crazy."

Peacocks wandered down the unplowed paths of the zoo in Kalemegdan. I saw zebras in unmucked stalls, lions circling snow-covered concrete pens, two elephants in a tiny, unheated building. It was hard to tell if the alligator sprawling in an empty pool was asleep or dead. "Out of the nightmare of smoke and fire came the maddened animals released from their shattered cages in the zoological gardens," Churchill wrote after the Nazi bombing of Belgrade. "A bear, dazed and uncomprehending, shuffled through the inferno with slow and awkward gait down towards the Danube. He was not the only bear who did not understand." The scrawny American buffalo taunted on this winter afternoon by a young man and two women must have felt the same confusion when the man, egged on by his friends, began to swat it with a stick. The women laughed. The bison twitched. Muslims had fed Serbian babies to the lions in the Sarajevo Zoo, according to a recent report on state TV, which was why Bosnian Serb forces had to kill all the animals. Wolves were howling in the Belgrade Zoo.

"What to do?" Milan Djordjević kept saying. And: "Catastrophe!"

The poet stopped in the middle of the street and shook his head. There were better places for an American to run out of money than Belgrade. No bank would cash travelers checks or give cash advances on a credit card. What a foolish decision to change my last $100 before I needed the dinars! I made a quick calculation: I had just enough left for a train ticket to Budapest.

Milan escorted me to the station, keeping an eye out for the military authorities. A bearded, bespectacled father of two, he had his "shelters" to sleep in at night, but daytime was a different matter. Still, he insisted on helping me, though out of nervousness he began to dream aloud about countries to emigrate to—Australia, Canada, Switzerland. Then he slipped on some ice and came to his senses. He took a kind of rueful pleasure in Belgrade's growing dinginess. "It isn't very white anymore, physically or spiritually," he sighed. "But it used to be the only cosmopolitan city in Yugoslavia. Now it's going to be pure. That's no improvement." *Desert,* the title of his new book, was Milan's metaphor for the White City, and it was from him that I first heard what was to become a common prophesy: that when the war ended the Serbs would begin a civil war among themselves.

"We have a saying: *when poverty comes through the door, love goes out the window*," he told me when I purchased my ticket, then he hurried away.

Belgrade, Novi Sad, Subotica. Like a conductor, I silently ticked off the names of the cities on the way north. I had in mind a certain inspector of the Royal Hungarian Railways, Eduard Scham, the mad father at the heart of Danilo Kiš's *Garden, Ashes*. Long before I ever thought of traveling to the Balkans, Kiš's semiautobiographical novel of a boy's coming of age during World War Two was working its magic on me, presaging this train ride. For the madness of the narrator's father takes a peculiar form: Eduard Scham is composing an imaginary timetable for the entire world, linking the remotest cities and islands "in an ideal line"—what his son calls "an apocryphal, sacral bible in which the miracle of genesis was repeated, yet in which all divine injustices and the impotence of man were rectified." The marvelous timetable (ideal for literary travelers), crammed with pictures and ideograms, is, of course, unfinished, and Scham, unfortunately, is under contract to produce a travel guide; the publisher's rejection of his "masterpiece" spells the beginning of his doom, for his madness mirrors the larger disorders afflicting the world. He moves his family to a hovel next to a railroad embankment, where the passing trains leave them in a constant state of anxiety. It is a mark of Kiš's genius that he leaves unspoken the central fact of *Garden, Ashes*: that Eduard Scham, a comic figure "guided by his star" (six-pointed and pinned to his greasy black frock coat), will board a cattle car to a concentration camp. And Kiš renders him in such loving and laughable detail that long after the book is closed he hovers in the imagination, like a ghost.

Kiš was similarly haunted. He was born in 1935 in Subotica to a Hungarian Jewish railway inspector (whose *Bus, Ship, Rail, and Air Travel Guide* enjoyed another run after the publication of *Garden, Ashes*) and his Montenegrin wife. His father and everyone on that side of the family perished at Auschwitz, and he was saved only because he was baptized in his mother's Orthodox faith when anti-Jewish laws were passed in Hungary. After the war, his mother took him to Cetinje, Montenegro (her warrior relatives included an "Amazon" who beheaded a Turkish despot), where he studied violin (Gypsy music was his specialty) and trained to be a poet, translating French, Hungarian, and Russian poetry. Indeed Kiš's prose, Joseph Brodsky suggests, "is essentially a poetic operation," absorbing emotions linguistically instead of expressing them. Hence my enthusiasm for *Garden, Ashes* and *A Tomb for Boris Davidovich*, his "suite of fictional case histories of the Stalinist Terror" (Susan Sontag's phrase) which touched off a firestorm in Serbian literary circles and caused him to go into exile. Kiš's crime was to follow up his masterpiece on the Holocaust with an equally devastating portrait of the Soviet horror. His gallery of ill-starred Poles, Russians, Romanians, and Hungarians (no Yugoslavs) offended the literary collectivists (soon to change their allegiance to nationalism). Kiš's allegiance was to his art; and as "a contemporary of two systems of oppression," he explained in *The Anatomy Lesson*, a book-length reply to his detractors, "two bloody historical truths, two networks of camps for the annihilation of body and soul, and to speak of only one (fascism) in my books, overlooking the other (Stalinism) with the aid of a psychological blind spot—this intellectual obsession, this moral and moralistic nightmare recently began to put such strong pressure on me that I could only resort to 'letting the lyrical blood' from my artery." For this he was vilified. More irony: since his death in Paris in 1989 Serbian nationalists had claimed him as one of their own.

Kiš would have been mortified. In an essay published almost twenty years before Yugoslavia broke up, he described nationalism as "collective and individual paranoia." Here was an antidote—never administered—to the venom already poisoning the culture. "The nationalist is by definition an ignoramus," said Kiš, which did not endear him to his fellow writers. And we can only imagine what might have happened if Serbian scholars had added his essay to the curriculum instead of promoting only Serbian writers. What the nationalist fears, said Kiš, is his brother: an uncannily accurate portrait of the Yugoslav wars of succession. The writer understood how the Other—the Jew, the Gypsy, the individual—threatens the collective. "Here is something we can all count on," Charles Simic writes. "Sooner or later our tribe always comes to ask us to agree to murder." But lyric poets "perpetuate the oldest values on earth. They assert the individual's experience against that of the tribe." These were Kiš's values. And I mourned his premature death, not least because it prevented him from writing another masterpiece, this time on the bloody historical truth of nationalism. Nor would his countrymen's barbarity have surprised him. As he said in one of his last interviews, "I am convinced that history is the history of misfortune, that its worst aspects recur endlessly, over and over."

On 4 November 1956, three days after Prime Minister Imre Nagy announced Hungary's unilateral withdrawal from the Warsaw Pact, Soviet armored divisions invaded the country, laying waste to Budapest and killing nearly 3,000 people. Hundreds of thousands fled to the West, hundreds of freedom fighters were executed, Nagy was condemned. Why did the Soviets wait for three days? Nikita Khrushchev was on the island of Brioni, convincing Tito the invasion was necessary; rainstorms grounded their planes. And it was on this same island, thirty-five years later, that EC diplomats negotiated the JNA's withdrawal from Slovenia, granting independence to the republic and dooming Bosnia and Croatia to war. Symmetry is indeed a guiding principle of Balkan history. "O Europe is so many borders," wrote the Hungarian poet Attila Jósef, "on every border, murderers."

Theories about the causes of the war abounded. And the language of culpability was easy to master. Serbian queries into the derivation of my English surname, for example, inevitably led to diatribes against Germany, Exhibit A being Warren Zimmerman: his name was German, therefore America had plotted with Germany to break up Yugoslavia. In Budapest I heard more balanced views. The soft-spoken man who rented me a room in his house pinned the blame on the West's misunderstanding of the Balkans. A lexicographer told me the Communists' systematic lying had created an ethical vacuum filled by war. The poet Peter Kantor took a cultural view. He was a short man who lived in a sparsely furnished, high-ceilinged flat; his voice echoed off the walls, as if in a canyon. He said too many Yugoslav writers refused to explore the middle ground of experience, the complexity of the human condition. Either they were trapped in their myths and folklore or they employed a highly abstract form of reasoning drained of emotion.

"I don't know that they connect," he said.

I suggested an analogue in the sharp division between rural and urban sensibilities, but the poet had moved on to another subject, the Hungarians' distrust of their government and media, which he compared to a line of traffic. In America, he said, no one passes a slow car because you assume the driver has a reason for going that speed. But if you're stuck behind a car in Hungary you're right to pass it, because you discover the other driver is stupid or likes going slow. But I like to drive fast. Of course, sometimes we have fiery crashes. I suppose you could say we have a healthy skepticism, which can degenerate into chaos when no one follows any rules.

"Budapest isn't a bad place to think in," George Konrád believed. "Without a little danger, thinking loses its edge." I needed space to think in before returning to Serbia. Clarity, too. So I called Konrád at the headquarters of the Social Democratic Party. Since the fall of Communism this writer had become a leading political figure (and president of PEN International), yet he remained an intellectual gadfly. Like Havel, he used the compass of civic society to chart a course between the shoals of Communism and nationalism. The former dissident imagined that Central Europe as a governing idea was more relevant than ever: "Our region requires an ability to get along with others," he wrote.

His was a fading vision. In Geneva, Cyrus Vance and Lord David Owen had just unveiled their peace plan, modeled after the Swiss federation, proposing to divide Bosnia into ten cantons, nine of them defined in ethnic terms: three each with Serbian and Muslim majorities, two with a Croatian majority, one with a mixed Muslim and Croatian population; only Sarajevo would remain multiethnic. But the Bosnian Serbs, who would have to return a substantial portion of their winnings, did not want to live in isolated pockets of territory without secure landlinks to Serbia proper. And Bosnians refused to ratify the ethnic partition of their country. Only the Bosnian Croats liked the proposed maps, which gave them large swathes of territory contiguous to Croatia.

This did not surprise Konrád. He thought the atrocities committed in Bosnia and Croatia were a logical outcome of the desire to redraw borders along ethnic lines. "You cannot turn from a political to an ethnic understanding of a nation," he told me. "Maybe Yugoslavia was an artificial creation, but so were all the republics. And if you question Yugoslavia's right to exist, you must expect what followed. Now all ethnic nations will have to get their frontiers. They need less unjust borders.

"Impossible."

That politicians were adopting a tribal view of culture and society was to be expected: votes lay in that direction. But as Konrád said in a speech to Hungary's Serbs and Croats, "the politicians involved have addressed lesser evils by resorting to greater evils, and the simpleminded have followed them." His admonishment is worth quoting at some length:

> The South Slavs will go on living together for centuries—if not in the same village, then in the same region. With whom are they to work if not one another? With whom are they to trade if not one another? And they did so. But now they have surrounded themselves with walls of corpses and bred visions of mistrust, revenge, and terror.

Subotica, again. Tonight the border guards reserved their invective for the Gypsy family in the next compartment, angrily throwing them off the train. The threadbare parents and their two small children stood forlornly in the snow, repacking their luggage, the red-white-and-blue plastic bags torn open by the guards. A refugee from Mostar and I took over the empty compartment. He methodically tossed the rest of the Gypsies' belongings out the window. A refugee from Dubrovnik joined us, and while the Serbs talked over their losses (houses, cars, friends), sadly recalling Yugoslavia's pre-war harmony, I watched a guard pocket the headphones of the Gypsy woman's Walkman. It was not hard to picture Serbia's 150,000 Gypsies as the next target of the nationalists. Somewhere Danilo Kiš was tuning his violin to play the songs of the *lungo drom*, of the long road traveled by the Gypsies, of the rootlessness that terrifies true believers.

"Do you believe in God?" I asked the poet Ivan Lalić.

"Don't you?" he replied.

A thick coat of soot covered the floor and walls of the living room, the residue of a kitchen fire two days earlier which had destroyed part of the flat; only the quick wits of Ivan's wife, Branka, had prevented the fire from burning down the whole building. Candles lit the room, which would have resembled a prehistoric cave if not for the electrical wires dangling from the sockets. A wood stove provided the only heat. Branka, whose hands were stained with soot from cleaning, set out a plate of cold beef and bread. Ivan poured me a glass of red wine from Macedonia and sighed. He was not allowed to drink (recurring pancreatitis), a prohibition he meant to have his doctor lift by summer.

Ivan sketched a jagged line in the air: a map of Europe. He was explaining the schism between Eastern and Western Christianity: how the Roman heresy, papal authority, divided the Continent; how Byzantium became the South Slavs' spiritual home; how Serbia was Christendom's last bulwark against Islam. "Geography," he wrote, "very often overlooks the truth of legends. That's what makes traveling exciting: the constant uncertainty of outcome. In any case, there's no true map."

Branka turned to me. "He knows everything," she said. "It's so irritating."

"She's my muse," Ivan said playfully.

But an air of tragedy hung over the flat, apart from the fire damage: their son, a talented jazz pianist, had died three years earlier, in a storm on the Adriatic. When his handmade boat was swamped, he treaded water for seventeen hours—the two friends with him wore the life jackets—before his heart gave out; his fiancée lived on in his bedroom. Branka showed me photographs of him. Ivan spoke of turning his son's tapes into a CD. But they could never cover the production costs, not with the war, the sanctions, and now the fire. They could not even meet their living expenses.

All at once Ivan was castigating Slovenians. "They were unduly promoted for seventy years, during which we had to buy their awful products and read so much bad literature. Then they wanted to be on their own. Good. It's better for us. Now let's see how their goods compete in the West."

He waved away my protest that Slovenia had legitimate grievances. Indeed he was proud to have supported Milošević in his rise to power. The war he blamed on Germany, an assertion which brought to mind the concluding lines of his "Concert of Byzantine Music": "Whoever forgets the cause is condemned/ To celebrate unclear effects."

"And mind you," he said, "I love the German spirit. I've translated Hölderlin, and Rilke is as close to me as any poet. But Germany must not be that strong."

He had also translated a number of American poets, including Allen Ginsberg, who once visited him in Belgrade during a reading tour. Ginsberg was interested in only one word of Ivan's translation: *cocksucker*. Serbian had no such word, and Ivan was pleased with his coinage. But language has a life of its own: by the end of Ginsberg's tour Serbian slang had a new phrase for cocksucker: *Allen Ginsberg*. Ivan roared with laughter.

Melancholy was the dominant tone of the evening, though. "The poets, at times, have an intimation of what endures," Ivan wrote. And even before I left I knew that what would endure in my memory of this visit was a sadness as old as Byzantium.

Sonja Karadžić, director of the International Press Center in Pale, decided if and when journalists could travel into Bosnian Serb territory. Sonja did not help her father's cause, as I discovered when I applied to visit the capital of Republika Srpska. Duly accredited to the Houston *Chronicle*, for several days running I filled out requests, in a dingy Belgrade office, to go to Pale—a four-letter word, journalists said—only to have Sonja turn each one down without explanation. She rejected most travel requests, I learned, and even when there were stories that might have burnished the Bosnian Serb image she did little for foreign journalists. Under her guidance the press center had indeed become one of Republika Srpska's most corrupt institutions, which, for example, routinely auctioned off television footage to foreign networks—filmed with camera equipment stolen from the BBC. So much for telling the Bosnian Serb side of the story.

History, according to Hubert Butler, is against the pacifist in the Balkans, which made Vesna Pešić's presence in Serbian politics all the more remarkable. After waging a twenty-year campaign for democracy and human rights, she had founded Civic Alliance, the only political party not to endorse Greater Serbia. A slight, short-haired woman who spoke quickly, though not always clearly, in her drab sweater and skirt she carried herself more like the academic she was than a politician. One evening she received me in her bedroom-study, and as her aging mother served coffee Vesna sat on the narrow bed, adjusting the space heater. The blame for the war she placed squarely on the Serbian people, who were behaving like an abandoned lover, vengeful and irrational.

"I didn't know Serbs were such a stupid people," she said.

There was no rancor in her voice. Despite death threats, she continued to advocate Bosnian independence and respect for minority rights. She organized peace vigils, condemned war criminals, worked with refugees. Yet I felt an unbearable sadness in her dreary apartment. I could not imagine such a retiring figure leading anyone to freedom, to say nothing of a people in the grip of a warlord. She said herself, "If people want to live in a catastrophe, you can't stop them."

In the Military Museum was a replica of the infamous Tower of Skulls in Niš, a four-sided tower that originally contained nearly a thousand skulls, bleached white by the sun and rain. In 1809, during the First Serbian Uprising, when the Ottomans overran the Serbs at the Battle of Čegar, the Serbian commander fired his pistol into an ammunition dump, blowing up his outnumbered forces along with thousands of Turks. Better to commit suicide (and kill off some of your enemies) than risk impalement. The Turkish leader responded in kind, building the Tower of Skulls as a warning to the rebellious Serbs. On a visit to Niš twenty years later, the French poet Lamartine, taking refuge in the shade of what he thought was a marble tower, saw instead a ghastly triumphal arch. "In some places portions of hair were still hanging and waved, like lichen or moss, with every breath of wind," he wrote. "The mountain breeze, which was then blowing fresh, penetrated the innumerable cavities of the skulls, and sounded like mournful and plaintive sighs." The Tower of Skulls became a shrine after Niš was liberated in 1878 (though by then most of the skulls had vanished), and in the decade before the breakup of Yugoslavia thousands of schoolchildren made pilgrimages to the site, obeying Lamartine's imperative carved into a plaque: "This monument must remain! It will teach their children the value of independence to a people, showing them what price their fathers paid for it."

Perhaps Vasko Popa best understood the monument, which he memorialized in *Earth Erect.* His most important collection of poems explores the Serbs' place in history through their heroes, shrines, and myths: Saint Sava, the Orthodox monasteries, Blackbird's Field in Kosovo, the wolf (the Serbs' totem animal), Belgrade ("White bone among the clouds"). A celebration of Serbian history, daring in its time (he began composing it in 1950, when any expression of national pride was taboo), *Earth Erect* was my introduction to the Balkans. When I first read Popa as a graduate student, I could not imagine anything more terrible than the vision enunciated in his Tower of Skulls cycle: "From the eye-sockets/ Black clairvoyance looks/ To the end of the world." But as Octavio Paz wrote in his homage to the late poet, "Vasko's rifle does not kill,/ it is a maker of images."

"The Balkan horror," was what Milan Djordjević called the Tower of Skulls when I revisited the Military Museum with him. "So much history for such a small place," said the poet. "For us it's too complicated." He studied a drawing, from 1917, executed by a Croatian artist, of Serbs hanged by the occupying Austrians; his grandfathers, both Serbs, had fought on opposing sides in the Great War. "Serbs are great warriors," Milan admitted, "but it's better to be something else."

The weather remained cold and overcast. All the world is against us, said the Serbs. The plotters included Islamic fundamentalists, Austro-Hungarians, Germany, the Vatican, the United States, and journalists. Even as many discussions in Belgrade revolved around the threat of Western intervention, Serbian paramilitaries in Sarajevo were stopping a UN armored personnel carrier just outside the airport to shoot, point-blank, Hakija Turajlić, Bosnia's deputy prime minister, ostensibly because he was delivering a contingent of Turkish soldiers to fight alongside the Bosnians. There was only one problem with this scenario: Turajlić was traveling alone. The calls for air strikes grew shriller yet.

The small island at the confluence of the Sava and Danube rivers stymied city planners. But the novelist Dragan Velikić had dreamed up a creative use for it. He proposed to erect breakwaters to keep the so-called "Island of War" from flooding over every spring, then clear it of woods and build for Milošević and his friends a house, hotel, and runway. Now *that* was the cleansing Serbia needed.

"If all the crazies moved out there," he said, "this war would stop!"

We were in the Little Moscow café—"I prefer Moscow to be little," Dragan smiled. He took more ironic pleasure, and not a little pride, that the war allowed him to emulate Hemingway and Fitzgerald, at least in a financial sense: on the black market, the money he earned in hard currency for an article published in Vienna could support his family for a couple of months. And in his writings for *Vreme*, the news weekly that Mark Thompson called "a beacon of independence and sane resistance to the nationalist 'logic' of war," he could ridicule the Greater Serbians. But with a circulation of 25,000 *Vreme*'s influence was limited to Belgrade. I recalled the angry reaction of my Slovenian friends to Dragan's phone call during the Ten-Day War. Asking if they had read his new novel was indeed a testament to either appalling manners or gross ignorance. Yet he was much closer to them in his thinking than they knew. He had no illusions about his countrymen's barbarism—he joked that anyone could say anything in Serbia, because it made no difference—and he had a fine metaphor for the Balkan crisis: that Tuđman and Milošević were playing the four-handed piano of war.

An exhibition of photographs at the Museum of Applied Arts. If the inspiration for *Genocide against the Serbs* was the discovery, in 1990, of mass graves in Croatia and Bosnia—pits and caves containing the remains of hundreds, perhaps thousands, of Serbs killed in World War Two—this show insisted on historical continuity between the depravity of the Ustaše and the designs of the current Croatian leadership. Divided into two parts, 1941–1945 and 1991–1992, in the first were gruesome photographs documenting Ustaša atrocities—the special knife, produced in Germany, worked into a glove for "quick and efficient throat-cutting"; the corpse of a man with his head sawed off; a makeshift gallows with more than twenty dead men hanging one against another, like coats. Of the more recent photographs (corpses floating down the Danube, damaged churches in Slavonia, Tuđman, Pope John Paul II) the most powerful image, from the Ravni Dolac pit near Livno, showed no sign of humanity. What in another context might have passed for the work of a landscape photographer was now an emblem of horror. The view was from the bottom of the pit—dark, slippery walls, a splash of reflected sunlight in the upper right-hand corner. In 1941, more than two hundred women and children were thrown into the pit; a handful survived, without food or water, for forty-five days. The lavish two-volume catalogue of the exhibition included a forward by the Serbian Patriarch, who described, in poetic terms, a baby born in that abyss: "It went from the darkness of its mother's womb through the darkness of the damp pit and into the darkness of death." A German journalist struck up a conversation with me. "Once I traveled to Houston to write an article about dog fighting," he said with a weary smile. "And now dog fighting has come to Central Europe."

No one could believe what showed on state TV after midnight. The prohibition on pornography had been lifted in 1991, and war profiteers were creating a sex industry—prostitution rings, escort services, hard-core magazines. The first broadcasts of pornography coincided with news footage from the front. And the news production values were only slightly higher than those of the pornographers, who skipped over small talk and foreplay. Whether filming carnage or carnality, the new Serbian style was brutally simple. Now you could watch mutilated bodies during dinner and grainy love scenes at bedtime—anything to divert the people.

"Why am I voting for Vasiljević?" said a man in the street. "Because we love the same three things: chess, horses, and chaos. And if he is elected president we will have chaos!"

Some forms of chaos, though, were not welcome at Vasiljević's office near the Hotel Slavija. Beyond the metal-detection unit at the door were two bodyguards with handheld detectors who frisked me then carefully examined everything in my satchel. Boris Kostov was waiting in the lobby. Embarrassed to see me, the television producer chattered nervously about the weather.

"Believe me," he said. "It's because of the war. Otherwise it would be much warmer!"

Vasiljević was on the phone with Bobby Fisher when a bodyguard showed me into his office. After his chess match with Boris Spassky, Fischer had gone into exile to avoid the wrath of the U.S. Justice Department, and now he was teaming up with his sponsor to mass produce the clock he had invented for the match. Vasiljević wanted to close the deal by nightfall, but Fischer was still working on his design, on an island off the coast of Montenegro.

"These creative types are never satisfied," Vasiljević said in exasperation, handing the phone to the bodyguard. "Now Bobby wants to add a jigsaw puzzle. Soon it will be a computer!"

The bodyguard barked into the phone then slammed it down and left.

"That's the only way to get things done here," said Vasiljević.

Money usually did his talking. He had organized the Fischer-Spassky extravaganza as a publicity stunt, and Fischer's first match since winning the world championship from his Russian counterpart twenty years before had drawn considerable international criticism. Fischer earned $3.35 million for his victory in November—and a criminal indictment in Washington. He was facing a ten-year prison term and a $250,000 fine. Vasiljević dismissed the charge, saying the money I spent in Serbia made me a war criminal, too. A Vlach by birth (Serbs could not identify his facial features or accent), in his blue leisure suit he had a huckster's confidence. He predicted that within six months Serbia would be under new leadership. The banker had formed a coalition of businessmen to "absorb" (his word) the economy and the political class, buying up the country piece by piece. Nor did he rule out assassinations, a threat to take seriously from a man thought to be backing Arkan. The main thing was to get rid of Milošević, he said, the elections having demonstrated the futility of working for democratic change. He had even boycotted them, refusing to vote for himself once he learned that his wife's name had been removed from the rolls. His reasons for running for office were hardly noble—to see himself in the papers, to "verify" the democratic process, to see who wanted to work—but it was by no means clear that his election would have raised the general level of chaos. Serbia was a criminal's carnival, with journalists serving as barkers and black marketeers taking tickets.

And Vasiljević was an ambitious criminal. It was said that after losing his job as a clerk at the Hotel Slavija he figured, why steal a little from a hotel when you can steal so much more from a bank? His employment history should have given pause to those investing their savings in his Jugoskandic Bank. His pyramid schemes offered enormous returns, but it was too good to be true. In two months he would flee to Israel, complaining of government pressure on private enterprise. In fact, Belgrade was a dangerous place for rich criminals. Vasiljević's talk of ousting Milošević was just talk. For even as he put together a deal to manufacture chess clocks he was planning his boldest, and most successful, move yet: to make off with more than seven million dollars of his investors' money.

An inevitable subject at this hour, said Zoran Milutinović.

The professor had invited me out for a night on the town, which in Belgrade's present circumstances meant retiring to a friend's apartment for rounds of champagne and raisins interspersed with cups of coffee or tea. The air was full of smoke, the revelry was frantic. Zoran's friends were trying hard to have fun in a city whose famous night life had vanished with the war. Lacking the money to go to the bars, jazz clubs, and restaurants which once stayed open until dawn, determined to avoid the war profiteers and soldiers who made a habit of shooting up cafés, they visited each other's flats for entertainment. At three in the morning, the party showed no sign of ending.

A logical place for our conversation to go, said the high school teacher hosting us. Mirrors surrounded the living room, and she was sitting in front of the tallest one, which made her slender figure look even more delicate. Naturally, the discussion had turned to Arabic grammar. It was over tense changes that another woman, Alica, began to argue with Stojan, a graduate student in English.

Zoran rolled his eyes. "I'm tired of living in an interesting country during interesting times," he said. "A sense of humor would help, but I can't laugh. I'm desperate."

Still, he was in a playful mood. More than once tonight he had proposed to Alica. A court reporter and translator of Calvino, Alica had job offers in Milan and Lugano, but she wanted to watch the final act of the war from Belgrade, which she thought had begun to resemble one of Calvino's fantastic excursions. Just this afternoon she had passed a bookstore with a sign in its window advertising guns. "A true entrepreneurial spirit," she said. Zoran fashioned a wedding ring for her out of the wire from a bottle of champagne and slumped to the floor in front of her.

"You're supposed to be on your knees, not your bottom," she giggled.

He made a face.

"What's the matter?" she said.

"Nothing," he said. "I'm just sad."

"Like a Slovene," she teased.

"I'm not pathetic," he said with a laugh.

The conversation took a serious turn. What Serbia needed, everyone agreed, was a purge. The nationalists have to go, and so forth.

But how will you achieve that? I said.

They had no idea.

"The Nazis were defeated decisively," Stojan offered halfheartedly.

But they knew Serbia would not be conquered.

When the party broke up at dawn, Zoran and I boarded a bus crowded with men and women commuting to work. We said little during the trip across the city, arranging only to meet later in the day. The pale sun was burning through the clouds as I walked back to my hotel, my head reeling. As it happened, I did not see Zoran again. That afternoon he phoned to say that for the next week he would spend his free time with his aunt, his cousin in Sarajevo having taken sick and died. Probably he would have survived, said Zoran, if not for the war.

11

Montenegro

The morning television programming was devoted to America. The lieutenant colonel translated what was on the screen—first, "The Making of a Continent," with aerial views of the Grand Canyon, then an animated history of the United States. Revolutionary War, he said, summing up a succession of images beginning with the Boston Tea Party. The officer, Miroslav, did not know what to call a cotton gin, but a smile of recognition crossed his face when a shirtless black man appeared on an auction block.

"Nigger," he said with solemn authority.

"Slave," I corrected him.

I had spent the morning with Miroslav, arranging to interview General Radomir Đamjanović, commander of the Montenegrin forces. Đamjanović was a good soldier, according to the EC monitor in Dubrovnik who had offered to drive me to Herceg-Novi, the resort city on the Montenegrin side of the border, to meet the general. When Croatian authorities would not let me cross the border, I decided to travel the extra 2,000 kilometers (Dubrovnik-Split-Zagreb-Budapest-Belgrade-Podgorica) to talk to Đamjanović. Perhaps he could explain why his forces had wreaked such havoc in Dalmatia.

I had other reasons for coming to Montenegro (Crna Gora in Serbian: Black Mountain), chief among them the wish to explore the truth of a statement made by a native son, Milovan Djilas: "The great myths—a doomed Serbia, the flight of its nobility into the Montenegrin mountains after the fall of the empire at Kosovo, the duty to avenge Kosovo, Miloš Obilić's sacrifice [Sultan Murad's assassin at the Battle of Kosovo, who was then beheaded by the Turks], the irreconcilable struggle between Cross and Crescent, the Turks as an absolute evil pervading the entire Serbian nation—all these myths took on their sharpest and most implacable aspects here." Serbian nationalism was rooted in this small (population 620,000) mountainous republic—Vuk Karadžić was its first historian, Slobodan Milošević's parents and Radovan Karadžić grew up here—but a growing number of Montenegrins no longer thought of themselves as Serbs. They led the increasingly vocal opposition to Montenegro's political leadership, the Communists-cum-Serbian nationalists loyal to Milošević. "I feel sorry for the Montenegrins," the EC monitor had told me. "They're stuck with the Serbs, for better or worse. But they also committed their share of atrocities along this coast."

"The man who celebrates peace"—that was how Miroslav translated his name, reserving his scorn for the "stupid politicians" responsible for the war, which had cost him 25,000 deutsche marks on deposit in a Slovenian bank. Once based in Croatia, where old friends of his had turned into enemies overnight, at least he was spared the shame of some of his fellow officers whose wives, he said, worked in Croatian munitions factories while their sons fought in the HVO.

"I know of a Croat who killed his wife and son because they were Serbs," he declared.

When I gave him a dubious look, he switched off the TV and showed me two items from the JNA newspaper—how effectively the army would respond to NATO intervention and how many journalists had been killed in the war. Then he returned to the article he was slowly typing up for the paper. I was frustrated, having already devoted two days in Belgrade to rearranging this interview. In the lobby soldiers milled about a line of young men waiting to be inducted. No wonder the JNA remained Yugoslavia's most entrenched institution. An army can survive for a long time on inertia.

"What to do?" the lieutenant colonel said with a shrug.

The phone did not ring until noon. Miroslav answered it and listened for several minutes without saying a word. When he hung up he told me to write out ten questions.

"Quickly," he urged. "We will fax them to the general. He must fly to Belgrade now."

"Then when will I meet him?" I said in despair.

But the lieutenant colonel had something else on his mind: finding the right translator for the interview. "Only a beautiful woman will do," he said. "And in matters of love age means nothing."

Podgorica's heyday, if it had ever experienced such a thing, must have come during its brief term as the capital of the Serbian empire, nine hundred years ago. In those days it was called Ribnica, after the river that flows into the Morača, a minor emerald deity meandering through the town. In fact, the Morača is about all there is to recommend in Podgorica (literally, Under Gorica Mountain). The city lies in the Zeta plain, a fertile agricultural region from which the medieval Serbian principality encompassing much of present-day Montenegro took its name. The Serbian noblemen defeated at Kosovo found refuge in Zeta, which at the time was relatively independent. Ottoman rule over the principality, from the fifteenth century until the Congress of Berlin in 1878, when Montenegro was formally recognized, was nominal. Hence the lack of Turkish influence on Podgorica, a charmless modern town "built without eloquence," Rebecca West wrote. All but destroyed in World War Two, the town was rebuilt in uninspired socialist architectural terms, and renamed Titograd, which it was still called in the only tourist brochure available—a German-language edition with photographs, from the coastal resorts, of nude sunbathers. By any name the capital of Montenegro was a dismal affair.

I had taken a room in a hotel along the Morača, and on my return from JNA headquarters I halfheartedly asked the clerk if he could put me in touch with Slavko Perović, the poet who headed the Liberal Alliance, the main opposition party.

"Let me give him a call," said a large, well-dressed man who had dropped into the hotel to use the telephone. Within minutes he had set up an interview with Perović, his cousin. "Shall we go?"

What a change from my dealings with the JNA, I told him when we drove off.

"Serbs are impossible to talk to," he said. "No good can come if we stay with them."

Two-thirds of his countrymen felt the same, according to Borislav Vukotić. He was a proud Montenegrin, a gregarious sea captain who had named his sons after his country's leading historical figures: Petar Petrović II, the poet-prince known as Njegoš, and Nikola Petrović, Montenegro's only monarch, who won independence from the Turks. Freedom from Serbia was what Montenegro needed, said Borislav, who supported not only the UN sanctions that prevented him from working but also NATO air strikes against Bosnian Serb military positions.

"They're warriors in the East," he muttered with a dismissive backstroke, a common gesture here. "Only intervention will stop them from spreading their disease all over Europe." He used the same gesture when he showed me the draft notice he was bringing to his older son, Petar. Once already Borislav had told an induction officer to fuck off, when Petar had been drafted at the beginning of the war. He promised to do so again if the JNA tried to pick up his son.

Down a street of drab façades we drove to a boutique, where we found his older son arguing, on a portable phone, with a newspaper editor. Petar was demanding a retraction—the paper had called him a Serb—and when he hung up in anger he announced, "I am Montenegrin." So it seemed. On one wall was the national banner, on another, a painting of King Nikola. The racks were flush with designer clothes from Italy, which Petar had smuggled through Albania.

"Ninety percent of my customers are in the Liberal Party," he said, fingering a Crna Gora medallion. "I must keep them satisfied."

His next call was to his brother, Nikola, who arrived in minutes with the two prettiest women I met in the Balkans. The physical beauty of Montenegrins, the tallest Europeans, is legendary, and it was my good luck to have Maja Đomović offer to be my translator. (Eat your heart out, Miroslav!) Hers was a classical beauty—olive skin, high cheekbones, shoulder-length black hair—tinged by melancholy. Maja had almost completed her English studies at Sarajevo University when the war broke out; when Novi Sad University refused to transfer her credits she returned to Podgorica to work for Slavko Perović. Her brother had fought in Trebinje; every night on TV she expected to see his name listed among the dead; if he was called up again, her family would send him into hiding.

"My life has been cut into ribbons, one piece here, others over there," she said when we set out for the mountain town of Cetinje. Nikola was at the wheel, in designer sunglasses and sporting a revolver. "Everyone's armed in Montenegro," Maja explained.

The brilliant light and mild temperature, a welcome change for a traveler from the north, brought to mind the high deserts of New Mexico. And it grew brighter yet beyond the orchards, vineyards, and rounded hay piles at the edge of town, when we began to climb through the stark hills and stone outcroppings of the karst. It is said that when God created the earth He used the leftover rocks to make Montenegro. This was a fierce landscape of crumbling stone terraces, ruined sheepfolds, stunted oak trees. Hitchhikers lined the road; along it a peasant was leading a pair of horses laden with chopped wood. Maja repeated a Montenegrin expression: *Mountains make big people.* Indeed the old woman strolling near the summit outside Cetinje was over six feet tall.

Coal smoke shrouded the capital of Old Montenegro, a thirty-by-sixty kilometer patch of harsh territory that was the only Slavic land outside Russia to stay independent for most of its history. Three times the Turks destroyed Cetinje without conquering the Montenegrin tribes. Fabled for their fighting abilities, the tribes, thirty-six in all, organized the clans and zadrugas of this frontier society, in the same manner as the tribes in Albania, administering Old Testament justice. Tribal law did not give way to a civil code until the reign of Njegoš (1830–1851), though the paramilitaries who rampaged through Dalmatia last year revealed the code's fragile hold on society. Or were they honoring another Montenegrin tradition—the *hajduks* and bandits who specialized in raiding and plundering coastal towns and villages? The myth of independence cut both ways. "It was from Cetinje that the tribes were called to mutiny and rebellion," Djilas wrote, "and individuals to human freedom—for faith and fatherland."

But Montenegrins were also adept at playing the Great Powers off one another, a diplomatic inheritance spurned by the current leadership when Italy offered it a sizable aid package in 1991 in return for its support of Lord Carrington's peace plan. Carrington proposed a loose association of sovereign or independent republics, sufficient guarantees for minorities, and no border changes—in hindsight, the best plan of all—but breaking ranks with Milošević, who refused to grant to Albanians in Kosovo the same autonomy he demanded for Serbs in Bosnia and Croatia, proved to be impossible for Montenegrin President Momir Bulatović. The former Communist lacked the nerve and cunning of his illustrious predecessors, the men of the Petrović clan, who ruled here for two centuries.

A word about Njegoš, the Petrović looming largest in the Montenegrin imagination, and not simply because he was nearly seven feet tall. The bishop-prince was the greatest Serbian poet. All Montenegrin schoolchildren memorized his poems, which had lately enjoyed a revival in Serbia, too. The myth of Serbdom was his theme, and his masterpiece, *The Mountain Wreath*, a verse drama published in 1847, celebrates the Christmas Eve extermination of Muslims carried out by Serbian warriors dedicated to cleansing the Slavic lands of the Turkish "spitters on the cross," converts to Islam. Njegoš wanted to eradicate this "domestic evil," in life as well as poetry. To redeem the loss of the Serbian empire and restore the Montenegrin state, he justified evil as a defense against evil, "the adder in our breast." *The Mountain Wreath,* realpolitik in decasyllabics, reinforced the belief that Montenegrins were the purest Serbs. A distortion of our history, said Maja, though not until World War One did some Montenegrins distinguish themselves from Serbs. For Montenegrin heroism was founded on Lazar's defeat at Kosovo, which Njegoš mythologized in his bid to throw off the Ottoman yoke. Many Montenegrins were now challenging it in their drive to free themselves from the Serbs.

For even older than the myth of Serbdom was the memory of the Montenegrin tribes. And certain tribal habits endured: the spirit of lawlessness, for example, governing the recent invasion of Dalmatia. Or the quirky tradition of forgiveness embodied in *Izmirenje,* a ceremony in which a murderer would beg forgiveness of the victim's relatives to end a blood feud. Hubert Butler attended an *Izmirenje* in 1937, and what he witnessed was remarkable. In the churchyard by a monastery, a hundred bareheaded men from the murderer's family lined up opposite a hundred men in top hats from the victim's family. A priest read out the sentence—the victim's brother was to be the godfather of the next child baptized from the murderer's family—then called for everlasting friendship and respect between the two sides. The men embraced and kissed, then sat at separate tables to eat, in silence, a feast prepared by the murderer's family. Slavko Perović's call for Montenegrins to pay reparations to the citizens of Dubrovnik invoked this spirit of *Izmirenje.*

Perović was a charismatic figure, in some respects Njegoš's heir, minus the Kosovo trappings. As public prosecutor he had opposed the Greater Serbia propaganda, a decision that cost him and hundreds of his followers their jobs; his party, founded in 1990, was branded an outfit of anti-Serbian terrorists for its support of Marković's reforms. Indeed the liberals counseled change by democratic means. David and Goliath. The poet's sling was his belief in the political process; his foe, the JNA, which had joined with Milošević to preserve its privileges. Once he had envisioned a confederation of Yugoslav republics making a peaceful transition to democracy, but the blood spilled in Croatia and Bosnia ruined any possibility of the South Slavs reuniting. Thousands of Perović's supporters took to the streets when the war began. The day after he spoke out against the shelling of Dubrovnik, his death notice was plastered to buildings and trees around the country, prompting him to talk openly of independence, a message that resonated with Montenegrins.

The liberals' gains in the parliamentary elections emboldened Perović to imagine Montenegro as a custom-free zone, which might hasten Serbia's democratization. But his phone was tapped, his followers were harassed, and the MIG fighter jets swooping low over Cetinje were daily reminders of the state's power. Perović was working, with no money and an ailing heart, against propagandists who, in his words, made Goebbels look like an amateur. Two bodyguards were always with him.

This day, as it happened, he was battling pneumonia. His flat, on the top floor of a four-story apartment building, resembled a Greenwich Village loft (from one wall hung a guitar, banjo, and violin), and under the skylight this pudgy man with a boyish face seemed closer in spirit to a poet than to a prosecutor. He was the Václav Havel of the Balkans, blessed with the kind of wit the playwright had displayed during the Velvet Revolution. At tense moments in negotiations with the Communists over the transition of government, for example, Havel would ask for a minute of silence followed by a minute of laughter. Perović was telling me in mock seriousness that the earth is made up of two poems, "The Raven" by Edgar Allan Poe and Njegoš's only love poem, "Night Gathers the Age."

"If I recite Poe to my wife, she knows I've been drinking for four days," he said between coughs. "But if I read to her from Njegoš, she knows I've gone mad."

But true madness, he went on, is Serbian politics, a church organ Milošević plays like an evil genius; the intelligentsia supplies the score. Even Belgrade University was founded on the Kosovo myth, and for over a century Serbs have gone there to learn how ridiculous they are. They should be ashamed they lost that battle, but they turn it into a point of honor. And since we never had our own university many of us studied in Belgrade, which helped to plant the Greater Serbia idea here. The Serbs call us the purest Serbs, but they can't divide us from our memory of 950 years of freedom as Montenegrins. We didn't lose our battles. We weren't slaves. But now our young people are leaving, replaced by refugees. This is how Serbia will collapse. The father of this miracle is Dobrica Ćosić, who thinks he's Tolstoy. Karadžić is one of his sentences; others are Arkan, Mladić, Milošević.

He is that rare novelist who gets to see his books lived rather than read. Do you know why Yugoslavia disappeared? Because we didn't know the other cultures. Hate grew from ignorance. For example, Bosnia's Muslims were our noblest people—you could see it in their art, literature, music—and I can't describe the shame I feel over Serbs raping women and destroying everything sacred. It will take decades to decontaminate Serbian political space. For myself? Perhaps what I'm living through will be poetic material some day. My wife doesn't like to hear me called a poet-politician, because that's what Karadžić is. We have a special kind of Montenegrin chicken that should be bright—we call it pearl—but it must try at least thirty times before it can pick up a piece of corn. That's Karadžić.

Then he was coughing again. Something in his apology for being sick suggested that he considered his illness an affront to the national character. When I recalled the story of a Napoleonic colonel meeting a local family of seven living generations—the longevity of Montenegrins is well known—the poet gave me a weary smile.

On the drive out of town, the mountain light glowed as the sun set over Lake Skadar, which divides Montenegro from Albania. I asked Maja what people did for a living in Cetinje.

"I don't know," she said with a shrug. "I just know they don't work very hard and they live well. If not for Serbia they'd work even less and live much better."

"The Serbs like to joke," I began to say.

"Yes," she interrupted, flushing with anger. "How lazy we are. But Montenegrins know how to work. They just don't want to work for Serbs."

Nikola dropped us off at my hotel, and we went to the bar to translate Perović's "Swans Will Be Shot Again":

> Nothing makes you happy
> Not the sun nor the rain birds games
> You just disappear
>
> Dead smiling
> Noble
>
> You turn into a monument—the royal "we"
>
> No fish swim in your rivers
> No butterflies flit above your waters
>
> In your eyes I see
> Swans will be shot again

"Do you like it?" said Maja. "Slavko has an artist's soul. Not like Karadžić, who only wants power. Karadžić is a bad poet, though they say he's a good psychiatrist. Very calm. But it's the calm ones who are the most insane here."

Boxes of televisions, stereos, VCR's—the apartment resembled the showroom of an electronics store. And the young men in black leather jackets who brought me here to use a looted fax machine from Dalmatia (the hotel clerk had called them for help) had distinctly mercantile attitudes toward their holdings. They made an interesting pair. One had a wife and daughter in Banja Luka, a leveled house in Sarajevo, and a mistress in Podgorica; the other, a pot-bellied man with a boyish face, had a bad temper. When he switched on a TV set with the remote control and saw Bosnia's UN ambassador being interviewed, he cried, "Turkish!" and pretended to kick in the screen. Then he faxed my article on censorship in Croatia and Serbia to someone in Kosovo. Once he realized his mistake he raised his fist at me and laughed. "Only in the Balkans," Ivo Andrić wrote, "can anything happen anytime."

The black marketeers insisted I join them for dinner back at the hotel, and it seemed unwise to turn them down. Our host, a grizzled Slovenian with a gold medallion of Tito hanging from his neck, was in charge of their operation. He told the young men—his "ministers of finance"—to count their money. And when a platter of smoked fish arrived he sent it back and ordered another dish. The hotel clerk hovered around the table until the old man invited him to take a seat. The clerk toasted him, and then, wheeling around, pulled a revolver from his belt and made for the door on the run. He had just gained the terrace overlooking the river when he fired off four shots into the night—a variation on an old custom, as it turned out. Because many churches did not have bells, Montenegrins used to fire their rifles to signal the start of the service, a practice certain to catch God's attention.

Only now did I realize that everyone at the table was armed.

After the last course, the old man unwrapped the tin foil from around a marble-sized chunk of what I took to be hashish, worked it into the blade of his dinner knife, and, when he had got it lit, offered it to me. The honored guest, et cetera, et cetera.

"Hashish?" I asked with some trepidation, and inhaled. But the smoke burning my nostrils was like nothing I had ever tasted.

The ministers of finance were laughing. "Jesus Christ," they said in unison, making the sign of the cross. The clerk gave me an envious look. The old man smiled. "Opium," he said. I tried to pass the knife back to him. "For you," he said. My head began to spin.

"Are you cold?" said Uroš. "A monk must be strong."

We had come to the last room of the unheated museum adjacent to the Orthodox monastery, and I was anxious to leave. I had returned to Cetinje with an ambitious plan: to tour the monastery and Petrović palaces, then to hitch a ride up Mount Lovčen, Montenegro's tallest peak, to see the mausoleum, designed by

Meštrović, containing Njegoš's remains. But the monk's lecture had gone on all afternoon, forcing me to abandon the rest of my tour. Now I was in danger of missing my long-awaited interview with General Đamjanović back in Podgorica. Not that Uroš cared. His mind was on eternal matters. He was a study in black—black habit, long black beard, black ponytail; his eyes burned with belief. Two young men trailed us, and from time to time Uroš would point at something—icons, habits—and say to them, "*Naš, naš*." Ours, ours. Their interest in the monastery was perhaps conditioned by their desire to escape the draft. Uroš was testing their faith.

"If you want to be a monk," he said, "you have to be a warrior."

He had a point. The monastery, which from Cetinje's founding in 1484 served as the court of Old Montenegro, was the center for uprisings against the Turks. A symbol of defiance, rebuilt after each sacking of the town, it housed the first Serbian printing press (on which Njegoš published *The Mountain Wreath*) until the type was melted down and turned into bullets to use against the Turks. Next door was Montenegro's first prison, built by the poet-prince, whose strict sense of justice—Turkish prisoners were decapitated, their heads lined up above the monastery for all to see—permeated society. A truant needed only to bring a severed Turkish head to school to be excused for missing class. Nikola banned beheadings in 1876, but through World War One Montenegrin soldiers cut off the ears and noses of their enemies, a practice that Uroš did not condemn. Present-day atrocities he absolved in his quest to restore the theological underpinnings of the warrior society: the people had lost their way under the Communists. The monastery was closed after the war—there was even a plan to paint it red—and when the Patriarch came to reopen it in 1990 he was threatened with an axe. Twice the monastery was stoned. One day an old Partisan woman accosted the monk.

"I didn't destroy the churches and monasteries to live this long," she cried.

But the collections of Russian icons and ceremonial attire in this room were intact, though on one wall was a painting riddled by wormwood. "Tito's advisor was the Devil," Uroš explained, "so he made Yugoslavia into a demonic state. If you wanted to go to church, you went to jail."

"Then why did you become a monk?"

"You can make theories about a jar of honey and not know what it tastes like. Monasticism must be tasted. I wanted to get to the roots of our culture. And our culture was created by monks."

No monk was more important to Uroš than Njegoš's uncle, Petar I, Montenegro's only saint, remembered for leading his men into battle crying, "If God is not on my side, let the first bullets hit me!" Whether or not the Turks believed the bishop had divine support for his campaign against them, it was during Petar's reign (1782–1830) that Sultan Selim III recognized Montenegrin independence.

When Petar died, his corpse was laid out in a winding sheet sent years before by Catherine the Great, a diplomatic nicety that reinforced the traditional alliance between Russia and Montenegro and, so it seems, helped to preserve his remains; their incorruptibility was proof of his holiness, according to Uroš. But Njegoš said he canonized his uncle for political reasons (transforming a theocracy into a sovereignty was no small thing), a detail which did not figure into my guide's narrative. He was busy showing me Petar's casket, then a piece of the True Cross and then, in a gold box, the right hand of John the Baptist.

A study should be made of the role that relics played in the Yugoslav wars of succession. After Tito died, Lazar's bones were dug up and paraded around the country. On 28 June 1989, the relics were displayed in Gračanica monastery in Kosovo, where the Serbian Patriarch, flanked by priests in scarlet robes, marked the 600th anniversary of the defeat at Kosovo; a million Serbs, armed with pictures of Lazar, flocked to nearby Kosovo Polje to hear Milošević say, "The Kosovo heroism does not allow us to forget that at one time we were brave and dignified and one of the few who went into battle undefeated." The veneration of Lazar's relics, then, helped Milošević to focus Serbian discontent into rage against the descendants of the Turkish renegades, and prepare them for war.

"We know Petar's remains are incorruptible," Uroš was telling me, "because his vestments are changed every fifty years, the same time period between war in the Balkans." He let out a sinister laugh. "We have a thousand years of culture, but every fifty years we destroy everything."

The monk grabbed my arm. "The West doesn't see the East," he cried. "It sees only the East infected by the West. The real East remains unknown. But when it is revealed the West will be ashamed, for deep in the heart of every man lies the wish for theocracy, at least in Montenegro." The differences between East and West were thus insurmountable. "Western man pleases himself. Eastern man pleases God. We have a zeal for truth. Faith demands effort, and in the West there's no effort. God must be at the center of the service. You cannot serve God by killing people, like the Vatican."

Enough already.

"So the Serbs are innocent of any atrocities in this war," I said, pulling away.

"A man who is new to his faith, like Bosnia's Muslims, will kill his family because they know he's living a lie," he said, recovering his composure.

"But Islam took root there more than five hundred years ago," I protested.

Uroš shook his head. "Muslims are Serbs," he insisted.

Bosnia's religious history is complex in the extreme, notwithstanding the claims of Croatian and Serbian nationalists who have argued since the nineteenth century that the Muslims are Croats or Serbs who converted to Islam under Ottoman occupation. "May their Serb milk be tainted with the plague," Njegoš wrote of the Montenegrin converts, thousands of whose descendants live in the neighboring Sandžak region. But it would be closer to the truth to say that with the Turkish invasion many Bosnians (and some Montenegrins) embraced Islam as a welcome alternative to Croatian and Serbian persecution. Indeed there were three churches in medieval Bosnia: the Catholic Church, the Orthodox Church, and the Bosnian Church, about which little is known. Myths flourish in the absence of hard facts, and the myth of the Bosnian Church's ties to the Bogomil heresy continues down to the present. Bogomil, a tenth-century Bulgarian saint, founded a dualistic church often accused of subscribing to Manichaeism, the belief that there are two principles of good and evil and that the earth, as the embodiment of evil, must be rejected. It was the Bosnian Bogomils, so the story goes, who converted to Islam, though there is little evidence that the Bosnians *were* Bogomils. "Conversion is a complex process," writes Michael A. Sells. And it is worth noting that the Islamization of Bosnia was 150 years in the making, during which time both Catholic and Orthodox Christians practiced their faith and even gained some converts. The two Montenegrins contemplating joining this monastic order were not unlike those in medieval times who changed creeds to advance their careers.

Or of the monk himself.

"What were you doing while the monastery was closed?" I said.

"Teaching art history at Belgrade University," he replied matter-of-factly.

I was still trying to digest this piece of information when we went outside. Uroš demanded to know why I was traveling alone. When I said my wife had her own career, he became quite agitated. "Then you cannot be truly married," he said. A look of hatred swept across his face, the prelude to a diatribe on why women, who are short-headed, must serve men, who are made in the image of God. His invective brought to mind the Balkan legend, repeated in ballads, poems, and novels, of women immured in walls and bridges in order to keep the gods of destruction at bay. This monk was the sort of man, I suspected furiously, who would think nothing of burying a woman alive.

A Soviet-made MIG fighter jet screamed past. When Uroš asked again why I was traveling without my wife, I pointed at the jet's vapor trail. "That's one very good reason," I said.

"When I don't hear them, that's when I start worrying," he said. "Come back in peace so you can see our new set of ruins." And with that he ushered the prospective monks into the church.

There is no historical basis for the massacre at the heart of *The Mountain Wreath*. It was Njegoš's genius to fashion out of whole cloth this tale of ethnic cleansing, the extermination of the so-called Turkish renegades. Here was an answer to the epic cycles occasioned by the Serbs' defeat at Kosovo, a poetic drama which in the fullness of time became a blueprint for the slaughter of Bosnian Muslims. "Destroy the seed in the bride," wrote Montenegro's last episcopal prince. And there was abundant evidence that JNA troops and Serbian paramilitaries were now taking him at his word, systematically raping thousands of Muslim women, confirming Djilas's assertion in his study of Njegoš that "Art is action." That is why Czesław Miłosz concludes, in "*Ars Poetica?,*" that poems must be written "only with the hope/ that good spirits, not evil ones, choose us for their instrument."

Miroslav did not succeed in finding a beautiful woman to translate General Đamjanović's remarks. And the two men he did line up had even less English than he did. They were waiting for me, along with Miroslav, a liaison officer, and the general, in an ornate private room of the JNA officer's club. The general, a sturdy, slack-jawed man in a suit and tie, placed two typewritten pages down on the coffee table and poured *šljivovica* for everyone. Where would you like to begin? he said, as if from a great distance. It hardly mattered: the answers, which he was anxious to check off with a pencil, bore no relation to the questions I had sent him. It is necessary to explain the causes of this dirty war, he intoned. On January 7th 1941 the Ustaša started killing Serbs. On January 7th of this year the Pope called for intervention against Serbia. History repeats itself, et cetera, et cetera. To hear him read from a prepared text made his answers sound truly bizarre—to wit:

> Serbia is one of the most democratic nations anywhere.
>
> There are many Germans in the Bush administration. The data says the increasing tension comes from the CIA. Muslim countries gave too much money to the Bush election.
>
> I read your papers. You have many problems with your educational system. You must be more like Japanese companies.
>
> If the UN intervenes, war may start everywhere, maybe even in the United States.
>
> We hope Serbs will be the winners in both moral and military senses.

Of course the general refused to discuss Montenegrin atrocities in Dalmatia, though he claimed with a straight face to have captured black Muslims near Dubrovnik. And he said that peace was at hand, because Serbs and Montenegrins wanted to live by negotiations.

"But the peace talks are breaking down," I argued, "because Karadžić won't negotiate."

No one said a word. Miroslav, the translators, the liaison officer, the general, all stared at the typewritten sheets, as if for guidance. I decided to break up the monotony of this performance.

"What kind of MIGs were you flying over Cetinje this afternoon?" I said.

The general's face reddened with anger even before Miroslav had finished translating my question. He stabbed his pencil into the top sheet and resumed reading.

"What kind of MIGs were they?" I said again.

Now everyone at the table was glaring at me.

"It can't be much of a military secret," I continued. "They were flying very low."

The general had had enough. "We have many kinds of jets," he snapped. "Many kinds. What you saw today were not armed."

Stefan Tvrtko, Bosnia's most powerful medieval ruler, built the town of Herceg-Novi, in 1382, to gain access to the sea and freedom from dependence on Dubrovnik ships and merchants. On the bus ride from Podgorica to Herceg-Novi, a half-day's journey through snow-swept mountains, I read up on Trvtko's reign (1353–1391). The story refuted the arguments of some Western politicians who, taking their cues from Milošević and Tuđman, said Bosnia had never been independent. In fact, when Tvrtko was crowned king of Bosnia and Serbia in 1377 (he later added Croatia and Dalmatia to his title), he governed the largest independent realm in the western Balkans. Alas, he was forced to accommodate the Dubrovnik merchants, and Herceg-Novi remained an envious little sister to the Princess of the Adriatic. Nevertheless it became a literary haven: Njegoš went to school here, Andrić had a house by the water. The bus descended along a narrow, twisting road cut into the rocky slopes, where only scrub oaks and wild pomegranates grew, and dropped steeply down to the open plains at the edge of Kotor Bay. It took only minutes to travel from winter to spring, from snow to vines and fig trees and men working in their gardens. There were even Dutch and German tourists among the guests at the Plaza Hotel where I had booked a room, though most came from Belgrade.

Djilas suggested that "the man on the coast could exist without the Montenegrin, but without the man on the coast the Montenegrin would be reduced to savagery," an idea that Mira Matijašević subscribed to. She was a singer-songwriter idled by the closing of the nightclubs, and as we strolled along the stone walkway around the waterfront she recited her love poems (in rhyming English), which left me with no interest in seeing her drawing of Stevie Wonder. Up a hill, past mimosas and oleanders, palms and cypresses, we came to Savina monastery, a complex of Gothic, Baroque, and Renaissance buildings overlooking the bay. A monk with a long stride hurried across the lawn to greet us, lavishing praise on Mira's miniskirt. The less you wear, he told her, the closer you are to God!

The cracks in the walls of the two churches by the monastery were only partially cemented over, the sanctions having stalled the city's recovery from the devastating 1979 earthquake. Inside the larger Church of the Assumption, though, the undamaged iconostasis gave me a measure of the aesthetic distance between the hinterland and the sea. The bitter fervor of the monk in Cetinje, in his unheated cell, was a world away from this screen of icons and gold filigree sealing off the sanctuary. The eighteenth-century painter, Simeon Lazović, working in the ancient style of the Mount Athos iconographers, created nearly eighty portraits of Christ, the Virgin Mary, and the saints, obeying the Byzantine theological injunction "to reflect the image of God" long after the fall of Constantinople. So much for artistic progress. I was reluctant to leave work infused with such spirituality, but there were fifteenth- and sixteenth-century frescoes to see in the smaller church. Newly discovered, they were still emerging from behind layers of paint, which gave them an unfinished air. White patches surrounded the saints and scenes from the New Testament; the restorers had run out of money.

Outside, the sunlight was blinding. An elderly monk called to us from a window on the second floor of the monastery, tossed me an orange, and disappeared. Mira was not hungry. The orange and lemon trees in her parents' courtyard had yielded more than two hundred pounds of fruit last year, some of which her mother brought to Trebinje to feed to refugees, and she had had plenty, *hvala*. I bit into the orange, which was so tart I had to spit it out. All at once there was gunfire throughout the town, the start of the Orthodox New Year's celebration.

"There's such tension here between beauty and life," Mira sighed. "No one wants to work, so they sit in cafés all day and envy people. It's very poisonous."

The Dalmatian coast ends in Prevlaka peninsula, a finger curling around the entrance to Kotor Bay. The disposition of this narrow rock-and-pine covered stretch of land preoccupied Montenegrins and Croats; for the distance to the mainland across the strait is only one kilometer, and whoever controls Prevlaka controls the bay, a deep natural harbor where the Yugoslav Navy kept its entire fleet. If Croatia, citing historical claims dating back six hundred years, positioned guns on the peninsula, Yugoslavia would lose its naval base: a question impossible to resolve.

The bay was awarded to Nikola at the Congress of Berlin—the monarch's shining moment. Like his uncle Njegoš, Nikola staked his soul on the myth of Serbdom, even offering early in his reign to abdicate in favor of the Serbian dynasty. But Montenegro's recognition changed the balance of power in the Balkans, for Nikola more than quadrupled the size of his territory. Then his problems began, not because he was, in Rebecca West's words, "a conscious buffoon" outfitted in the national costume, but because he had expanded into lands where, as historian Ivo Banac notes, "the tradition of Montenegrin specialness and statehood did not obtain." The Montenegrin belief in pan-Serbianism weakened as Nikola moved into Old Herzegovina and around Lake Skadar, incorporating the highland tribes and thousands of Muslims, including Albanians. Then, during the Balkan Wars, he captured southern Sandžak, Kosovo, and more shoreline along Lake Skadar; occupied Shköder, which he gave up only at the behest of the Great Powers; and set his sights on all of Herzegovina.

Even more than land Nikola coveted acceptance in the courts of Europe. He married off five daughters to royal houses, earning the sobriquet of "the father-in-law of Europe." But the Angel of History regarded his matchmaking with a mischievous eye: two daughters, married to Russian princes, introduced Rasputin to Empress Alexandra; and his oldest daughter's marriage to Petar Karadjordjević, Black George's grandson, cost Nikola his crown. Petar seemed a safe choice at the time of the marriage in 1883. He had no use for the Dual Monarchy, unlike the Obrenovići, who held the Serbian throne with backing from Vienna, and he had little chance of becoming king. Nikola could thus adopt a moral stance as guardian of pan-Serbianism without practical consequences: a useful strategy for an autocrat with a poetic bent (and none of his uncle's talent) condemned to govern in a land without economic prospects. Then King Alexander was assassinated in a military coup in 1903. But Petar's unexpected ascent to the throne did not improve Nikola's standing at the court in Belgrade, for the king had designs on Montenegro. Before long Nikola looked like an impediment to Montenegrin unification with Serbia, especially among Montenegrins educated in Belgrade and abroad. The monarchs' competing versions of pan-Serbianism had violent repercussions. Serbian agents fomented revolt against Nikola's authoritarianism and allegiance to things Montenegrin. The Black Hand, a secret society dedicated to Serbian expansion, planned uprisings among the tribes, smuggled bombs to Cetinje, and, in the words of a student leader, insured that "Montenegro became the stage of bloody conflicts, rebellions, protests, bombs, executions, chains, persecutions, of explicit collision between Serbdom and Montenegritude, between love of freedom and reaction."

Things came to a head when the Central Powers conquered Serbia in World War One. Petar's government, fearful that Nikola would take advantage of the situation, convinced him to sue for peace; and though his army was undefeated he left Montenegro in January 1916, never to return. His soldiers, believing he had betrayed them, surrendered to the occupying Austro-Hungarian forces. Then Serbian diplomats outmaneuvered his government-in-exile to let Belgrade dictate the terms of Montenegro's post-war destiny. Finally he was deposed. Petar was crowned King of the Serbs, Croats, and Slovenes, and Montenegrins were henceforth considered Serbs.

But the story does not end there. The Serbian-scripted election, in November 1918, in which Montenegrins voted to unify with Serbia, ending a millennium of independence, created deep divisions in this frontier society. The Whites—their candidates were listed on white paper—favored unification; the Greens did not. It was the ancient split between townspeople and villagers, the educated and the devout, Podgorica and Cetinje. The Greens had more popular support but less organization than the Whites, whose victory at the polls was costly. The Greens laid siege to several towns, where they were routed by White militia and the Serbian army, as well as to Cetinje, where they held on until Christmas Eve. Then life imitated art: as the Turkish renegades in *The Mountain Wreath* were slaughtered on Christmas Eve, so Montenegrins that night fought one another in the flesh. By morning the Christmas Rebellion was over, some Greens surrendering, others fleeing. The Whites retaliated, looting and burning Green villages, raping women, executing prisoners, even flaying two men alive. And pitched battles between Greens and Whites continued long after Nikola's death in 1921, on the Côte d'Azur. Slavko Perović and the Greens' liberal descendants had reversed the terms of the argument. What was once a force of reaction, Montenegritude, was now a movement for freedom. The Whites, or Serbian nationalists, held power, but perhaps not for long.

All the more reason for them to hold on to Prevlaka.

The Great Powers were preoccupied with Saddam Hussein the day I called on two EC military monitors. We sat on the deck of their house facing Prevlaka. While U.S. forces bombed Iraq in retaliation for a foiled assassination attempt against George Bush, in the Balkans the talk was of NATO air strikes against the Bosnian Serbs, who were about to reject the Geneva Peace Accord. Bush might leave his presidency on a decisive note. Or would Bill Clinton enter office with a flourish, marshaling forces to bring the Third Balkan War to a close? But the monitors believed that intervention, which might have worked a year ago, was no longer feasible.

"It's not our place to judge how this should turn out," said one, "since atrocities have been committed on both sides. I say, Let them fight it out, draw up new borders, and learn to live with that. America can't be the world's policeman: there are too many problems." He leaned back in his chair. "This is the most beautiful country in the world—and the most fucked up. And do you know why I'm staying on longer than usual? Because I think both sides are wrong."

A Yugoslav battleship steamed toward the strait. When it rounded the tip of the peninsula and headed out to sea the monitor said, "We haven't seen them do that before. Maybe they're testing the Allies because of the bombing in Iraq. Telephones must be ringing in every Western capital."

One of the two armored vehicles outside military headquarters had a flat tire. Inside, scores of uniformed officers had nothing to do. Accustomed to interviews prefaced by lengthy histories of national grievances, I was surprised when Goran Žugić, assistant police chief in Herceg-Novi, a hulking man who made a habit of tapping his money pouch against the edge of his desk, began by quoting Ivo

Andrić: "Wise people write, fools try to remember." But what followed was a fool's history. The Krajina was ethnically cleansed because of Ustaša atrocities, Vukovar was leveled because of the expulsion of Serbs, Dubrovnik was shelled because if it had artillery it could fire on Trebinje, the village of Čilipi was razed because of the nearby airport, and so on.

When I steered the conversation toward smuggling, Goran brightened. The black market he blamed on the sanctions; the speed boats leaving early each morning for Italy and returning under cover of darkness, laden with goods, he called "winter tourists." Montenegrins needed food, clothing, and medicines, and who was he to keep enterprising men from supplying them?

"Some criminal practices will stop of themselves after the war," he assured me. "And some will have to be stopped by force. But there will be no peace until we settle the question of Prevlaka. We don't necessarily think Yugoslavia has to have it, but never Croatia."

Huh?

"I love the Serbs," Gordana announced, a cigarette in one hand, a beer in the other. She was a tall woman with frizzy black hair and large brown eyes. Her voice was deep and husky, her blouse and tights were black, her violet miniskirt was the gaudiest thing in the bar. "The whole world is against us, and what do we say? Get lost. We're a proud people. The West can't tell us how to act."

It seemed she was an authority on America, having traveled there as a guest of the Rotary Club. "Do you know what I told them at their meetings? 'You know nothing about democracy.' Never have I seen so many fat, self-satisfied people. And the food—it's not like it is here!"

Gordana was an architect from Dubrovnik, and she did not regret the damage done by the JNA to the walls and palaces of her home town. As a matter of fact, she was positively gleeful at the thought of all the work facing her former colleagues at the Institute for Reconstructing Dubrovnik. "I saw the Croats arming themselves," she said. "I knew what was coming. My mother didn't believe me. You know what they did to you in 1941, I told her. They'll do it again." So they had taken refuge in Belgrade. "Serbs are so stupid. We know we're right, but we don't think we have to explain ourselves. Then the world turns on us, and we don't know why."

I told her I had heard Croats and Slovenians speak just as passionately against Serbs.

"But there's one difference," she said with a laugh. "They're wrong. We're right. On Kosovo, too. Do you know what those people do? They have fifteen, twenty babies! Look at me: I have one." She glanced around the room. "And where the hell is he anyway?"

Gordana ran out of the bar to find her son. It was some time before she returned with a little boy who gazed at her as if unsure who she was. "Do you know why the Albanians have their own schools and doctors? Because Serbs give and give, and no one's ever grateful for what they do. They gave to Slovenia, Croatia, Bosnia, and look what happened. Serbs started the last two world wars, and if you don't watch out they'll start World War Three!"

I asked her if she was designing houses in Belgrade.

"I have a new business," she said with a malicious smile. "Smuggling. I import everything. Look at these clothes—straight from Paris! The sanctions will never stop us."

Dinner at the Writers' Club, once the home of Ivo Andrić. Three tables, a space heater, seascapes on the walls. The only other patron, an old man with shaking hands, was not a writer, and the tiny bookstore next to the dining room, where I picked up a paperback edition of John Cheever's stories, displayed none of Andrić's work. The Nobel laureate, born in Travnik and raised in Sarajevo, called himself a Serbian writer despite his Croatian surname. As a student in Vienna and member of the Young Bosnia movement, Andrić fought for the unification of South Slavs. For his revolutionary zeal he was imprisoned by the Austrians and rewarded with diplomatic postings by the Yugoslavs. The villains in his universe were easily identified. In his dissertation, *The Development of Spiritual Life in Bosnia under the Influence of Turkish Rule,* Andrić deplored the contagion that Njegoš feared: "The effect of Turkish rule in Bosnia was absolutely negative. The Turks could not even bring a cultural content or a higher economic mission to those southern Slavs who converted to Islam; for their Christian subjects their rule meant a coarsening of customs and a step back in all respects." Andrić described Njegoš as "the personification of the struggle at Kosovo, the defeat and the indomitable hope." He made himself another "Kosovo warrior" out to redeem that loss. Andrić's bête noir was *devşitme,* the Ottoman practice of taking boys, Muslim and Christian, for training in Istanbul. But as with so much else in the Balkans, *devşitme* was both a curse and a blessing. Some parents mourned the loss of a son, others were grateful for the career opportunities that *devşitme* offered in the military or at the court. Indeed sometimes their sons returned to Bosnia as high officials, like Mehmed Paša Sokolović, who commissioned the building of the bridge in Višegrad, the subject of Andrić's most popular novel, *The Bridge on the Drina.* Few readers will forget the scene where a Serbian rebel is impaled on the bridge, each detail of his suffering vividly rendered. And that bridge is the only significant Ottoman structure still intact in Višegrad. Serbian paramilitaries used it as a killing site last spring when they cleansed the city of Muslims. Its literary history is not finished.

The old Croat refused to tell me his name. A nervous, heavyset man, dressed in a thick overcoat, scarf, and hat even in mild weather, he could not walk more than a few steps along the waterfront without stopping to catch his breath. He would clutch my arm, looking this way and that. He had good reason to be afraid of his neighbors. It was not, as he claimed, his membership in Amnesty International that upset them, nor that he had spent his working years in California, as a checker at Safeway. (His Social Security checks were his only income now that he could not travel to Dubrovnik to withdraw money from his bank account.) It was not even, or at least not entirely, that he was Croatian; at least 6,000 Croats lived in the area, not all of whom received the threatening phone calls and letters he did. Besides, with a bad back and high blood pressure, he could hardly be considered dangerous. But some things are not forgotten. The old man had Ustaša written all over him.

"Lies," he hissed when I asked him about World War Two. "The economy rules in war. Why would you take someone from Dubrovnik to kill them in a concentration camp when you could do it right in their home? This is a poor country. There weren't enough trucks or trains to take many people to Jasenovac. No more than sixty thousand people died there, mostly of disease, and they were Croats, Gypsies, Jews. Not seven hundred thousand. And not all Serbs!"

Small wonder that a band of Montenegrin irregulars had burned out his ancestral home, leaving a cow's carcass to rot in his living room.

Today I am broken, Gordana tells me one afternoon in the bar. I was dancing until daybreak. One week—that was all I was going to stay, but I am having too much fun. She and her friend, Nina, must explain to me why it is in America's interest to have a divided Europe and a weak Yugoslavia. You can't take the competition, says Gordana. You'll take Kosovo and Vojvodina, separate Serbia and Montenegro, then use us as colonies, says Nina, a surly refugee from Dubrovnik. She never tires of saying, That's why you journalists don't tell the truth about this conflict. But Serbia will survive, Gordana insists. We have the best writers, artists, football team. We *expect* to be the best. We have nothing to prove. We created the history, we won the wars, we had our kings and kings and kings. Croatia and Slovenia never had anything. That's why they have inferiority complexes. Nina orders a Coke and asks, Why does the West think we are so bad? Well, I say in exasperation, raping and murdering twenty thousand Muslim women is not very good public relations, is it? She throws up her hands. Croatian soldiers massacre women and children. Our soldiers would never do such things. Whatever raping there was, Gordana adds, was only revenge. Crimes cannot be forgiven in our religion. A crime is a crime until you die, and in the other world you will be judged. Our soldiers could not do what the Western press says, Nina declares. Serbian men would never rape

Muslim women, says Gordana. Why? I say. She glares at me before saying, in a caustic voice, Because they smell. And do you know why UNPROFOR is like a man's balls? They're all observers. We do not like Milošević either, he's too soft. But Šešelj is coming! Then everything will be cleansed.

"They came to our homes and took what they wanted. If anyone opposed them they were killed. They were shooting all around our village. They took away the men and raped women and children." So a young Muslim woman from a village near Prijedor told Helsinki Watch representatives the week before I arrived in Herceg-Novi. Fifteen men had raped her neighbor. "We had to carry her from the village. She didn't know where she was, didn't even know her own name. She was a virgin. I didn't see her being raped. Her mother came into our house crying, screaming that she couldn't do anything. She was raped in the basement of her house. We carried her into her house. She was badly bruised. Her mother watched from the window. There was another group too busy looting to join in. She was very beautiful, but they didn't care. As long as you are Muslim. They scream and yell; are not nice. Our houses are very near. I could hear her screaming and crying and begging: 'I'm a virgin.' 'Well, this is why we chose you,' they said. She spent two months in a hospital with the psychiatric cases. I don't know if she was pregnant, but if she was, she would not want the baby." Needless to say, stories like this will be repeated thousands of times, for journalists and human rights workers and war crimes investigators, before the wars of Yugoslav succession are over.

The phone call from the military monitors woke me from a sound sleep. My bus was leaving early in the morning, and I was in no mood to get out of bed. But at their insistence I dressed and went down to the bar. They wanted to hear about my last encounter with Gordana.

"The hatred here is much worse than anything I saw in Beirut or Cyprus," one monitor said when I had finished. "There they just hate you. Here it's with their whole body."

The other monitor was reminded of a Croat he used to visit in Herzegovina. "Every Sunday he would serve me ham, cheese, bread, wine, a whole feast. One day I had a new translator, a young woman. When she went out to the car to get something, the man said, 'She's Serb, isn't she?' 'Yes,' I said. 'Then fuck her to death,' he said. And right there at the table he showed me and his whole family how to fuck a woman to death. What an example to set for his children. When the translator came back the tension was thicker than you can believe. I never returned to that house."

The first monitor leaned toward me. "There's something you should know," he said in a conspiratorial murmur. NATO was on high alert. The last time this had happened, in December, the monitors had burned their files only to discover that a U.S.-led multinational force was invading Somalia, to the surprise and disappointment of many in the Balkan theater. But perhaps this time the international community would act differently. The monitors were taking no chances: they had lined up a speedboat, for five hundred deutsche marks, to whisk them to Bari.

And what will you do? they wanted to know.

"I'm on my way to Macedonia," I said. "I guess I'll keep my fingers crossed."

12

Macedonia

The inversion was in its second week. Snow had piled up on the streets and sidewalks and the Vardar River was iced over. Rumor had it, managers of the state-run industries were taking advantage of the low cloud ceiling to release dangerous pollutants from their smokestacks. Visibility was less than twenty-five meters. Not that it mattered. Leveled by an earthquake in 1963, Skopje was cluttered with unfinished construction projects; the inversion masked a skyline jagged with idle cranes, beyond which lay the mountains, surrounding the city, that trapped the smog in the river valley. The air had a sickly taste—a fine time for the health workers to go on strike. I longed for the south, not least because I was running short on cash again. Checking into an unheated hotel at six in the morning, I learned from the clerk that no one in the capital would take credit cards.

"But you're not under the embargo," I said.

"Macedonia is a small country," he replied. "No one knows how big it will be. Maybe only as big as Skopje!"

Macedonia *was* a small country, with a population of two million, inside a region nearly twice its size, of the same name, once the heart of Alexander the Great's empire. The former Yugoslav republic—Vardar Macedonia, in geographical terms—was one part of the Macedonian puzzle, which extended into Bulgaria and encompassed northern Greece (except Thrace), or Pirin and Aegean Macedonia, respectively. Albanians, Arvanites, Bulgarians, Greeks, Gypsies, Jews, Pomaks, Pontians, Serbs, Slavic Macedonians, Turks, and Vlachs all lived here; a French chef had even named a mixed fruit salad *macédoine* in honor of the region's ethnic and religious diversity. The Republic of Macedonia, which declared its independence in September 1991, was a mosaic of different peoples, like Bosnia; but while it had thus far escaped bloodshed in the Yugoslav wars of succession, Macedonia was the only former republic *not* to win international recognition.

It could hardly have been otherwise for the Balkan people with the murkiest sense of identity. "Macedonia defies definition," according to one prominent scholar, and sorting out the history of this troubled crossroads is by definition doomed to offend every one of its peoples. So here goes: the Slavs swept into the region in the sixth century, nearly a thousand years after the conquests of Philip of Macedon and Alexander the Great. Then came the Bulgarian and Serbian empires, the Bulgarian centered for a time in Ohrid, now a resort by the lake bordering Albania; the greatest leader of the Serbian empire, Stefan Dušan, proclaimed himself "Emperor of the Serbs and Greeks" in Skopje in 1346. But the Turks had the longest run in Macedonia: half a millennium. Not until the Treaty of San Stefano (1878) allowed Bulgaria to reclaim its ancient lands did Macedonians begin to acquire a national—i.e., Bulgarian—identity. Greater Bulgaria lasted only months: at the Congress of Berlin Bismarck negotiated the reinstatement of Turkish rule in Vardar Macedonia. But the wheels of Balkan nationalism were spinning now. The Macedonian Question—what to do with the land coveted by Bulgaria, Greece, and Serbia?—began to dominate international diplomacy. Some Macedonian students formed a revolutionary movement with two wings, one agitating for autonomy, the other for incorporation into Bulgaria. The cultural approach was eclipsed, in the public imagination, by terrorist tactics. Greek and Serbian clergy, Turkish civil servants, and two Bulgarian prime ministers were murdered, leaving Macedonia in chaos. The stage was set for a century of conflagration.

The First Balkan War began here in 1912 when Bulgaria, Greece, Montenegro, and Serbia joined forces to drive the Turks out of Macedonia only to turn against one another in a second fratricidal war. The Carnegie Endowment report on the Balkan Wars noted that when the Balkan peoples united they miraculously cut the Gordian knot of Ottoman occupation; disunited, they paved the way to continuous war. Meantime, "the accumulated hatreds, the inherited revenges of centuries" were unleashed, and every peasant with a grudge against his landlord or neighbor took vengeance. "The burning of villages and the exodus of the defeated population is a normal and traditional incident of all Balkan wars and insurrections," the commissioners wrote. "It is the habit of all these peoples. What they have suffered themselves, they inflict in turn upon others." The atrocities documented in their report, against soldiers and civilians alike, resembled the war crimes now being committed in Bosnia. Indeed modern ethnic cleansing was born here. Albanians, Bulgarians, Greeks, Serbs, and Turks were all flushed from Macedonia, which became southern Serbia after Serbia defeated Bulgaria in the Second Balkan War. That was when the Greek, Slavic, and Bulgarian parts of Macedonia were formally partitioned. Macedonians, like Montenegrins, were considered Serbs in the Kingdom of Serbs, Croats, and Slovenes—an affront destined to have grave political consequences. The Ustaša plot to kill King Alexander was carried out by a Macedonian.

Bulgaria occupied Macedonia during World War Two, and Tito's recognition of the Macedonians' separate life and language earned him their undying support, first as Partisans, then in Yugoslavia, where they took on all the trappings of a distinct nationality. The language was codified, and the writers set to work singing the Macedonian nation into being, which irritated Greece endlessly. No love was lost between these Orthodox countries after the Greek Civil War, when Slavic Macedonians sided with the Communists. Now the Greeks accused the Slavs of usurping their name, history, and symbols—on both sides of the border signs read, "Welcome to Macedonia"—and of irredentism on Aegean Macedonia. The first was true, the second irrelevant. Though Tito had once harbored designs on Thessaloníki, the Slavic Macedonians had no real army, their first elected president, Kiro Gligorov, having adroitly negotiated the JNA's peaceful withdrawal. Macedonia, already suffering from the UN sanctions, since Serbia was its chief trading partner, as well as from an unofficial Greek embargo (within the year it would be official), could not obtain loans from international financial institutions. The least prosperous republic of the former Yugoslavia was in grave economic distress, with inflation running at 200 percent per month and one in five Macedonians unemployed.

But Greece was only part of the problem. Serbia still considered these lands southern Serbia; Bulgaria said Macedonians were Bulgars (though it recognized the new country); and Albanians in the western districts (a quarter of Macedonia's population of two million) had boycotted the first elections, determined to unite with Albanians in Kosovo and Albania proper. Thus in December the UN had authorized peacekeeping troops for Macedonia, including five hundred from the United States—the beginning of a major American diplomatic effort in the southern Balkans. NATO, of course, had not intervened in Bosnia. And now there was a real possibility that Macedonia would shrink to the size of Skopje (the Greek name for the republic). The capital felt like a transit stop.

Which it was. Because Macedonia lies at the juncture of the only roads through the Balkan Mountains—a north-south Orthodox route (Belgrade-Thessaloníki) and an east-west Muslim route (Istanbul-Durrës)—traders and travelers from many different groups settled in what became a kind of miniature Yugoslavia. Not for nothing was Skopje the Gypsy capital of Europe. And all the minorities complained of the same kinds of discrimination that Slavic Macedonians suffered in Bulgaria and Greece. "Why should I be a minority in your state when you can be a minority in mine?" asked Vladimir Gligorov, the president's son, the unanswerable question at the heart of the wars of Yugoslav succession. The EC had addressed the subject at Maastricht, where the member nations agreed to create a common foreign policy, whose first test was the former Yugoslavia. Any republic meeting certain conditions—democracy, respect for minority rights, guaranteeing existing borders—would be recognized by the EC. But while Macedonia and Slovenia satisfied the requirements, the EC recognized Croatia and Slovenia instead.

On my first morning in Skopje I went to the Writers' Union to hear a speech on Macedonian recognition. I do not remember ever being so cold. After a long unheated bus ride through Kosovo's snowy mountains I had checked into my freezing hotel room only to discover there was no hot water. Under a thin quilt and a pile of clothes I shook like a leaf until it was time to go to the Writers' Union, where President Gligorov asked the writers to tone down their rhetoric. France was taking up the Macedonian Question this week in the Security Council. Now was not the time to stir up trouble.

To save money, I accepted Savo Cvetanovski's invitation to stay in his unfinished apartment on the outskirts of the city. But the attic room I slept in was even colder than the hotel: the central heating lines stopped two hundred meters short of Savo's building, thanks to "the situation," his favorite topic of conversation. He was short and round, a nervous man with bushy eyebrows and thick grey hair. His wife did not let him smoke inside, so he would come upstairs to discuss "the situation," all the while puffing on a cigarette without removing it from his mouth until it had burned down to the filter. Hobbled by an old soccer injury, he limped around stacks of construction materials, talking to me through a cloud of smoke. Always he returned to the same question: what would America do?

Savo had advanced degrees from American universities. He had directed the Yugoslav press and cultural center in New York, and anthologized American short fiction. Once a thoroughgoing Yugoslav (he had put together a composite biography of Tito) and now an ardent nationalist, he had maneuvered himself into the chair of the comparative literature department at Cyril & Methodius University and become vice president of the Reform Forces Party. Independence had cost him his savings in a Belgrade bank, but his apartment was paid off (though it might never be completed), and he owned a house in Ohrid. Savo and his wife would go there to escape the inversion, if not for the threat of war. How real that threat was depended, in his view, entirely on decisions taken in Washington; decisions, he said, that would not jeopardize Washington's interests in Albania, Bulgaria, Greece, and Turkey. Savo just hoped that Macedonia was not swallowed up in the process.

"We're afraid of becoming another Palestine," he said one day. "Believe me, there are angry elements here who could start planting bombs in Thessaloníki again."

I had the distinct impression he knew some of those people well.

Macedonia's finest poet, Mateja Matevski, liked to say he worked simultaneously in the youngest and oldest Slavic literature. Macedonian was indeed the first Slavic language written down (in the form of Church Slavonic, preserved in Orthodox monasteries), but literary works dated only from World War Two. If, as Danilo Kiš writes, "Language is destiny," then Macedonians had the most uncertain future of all. Bulgarians and Greeks called Macedonian a Bulgarian dialect, though linguists recognized a separate language, rooted in a vast oral tradition (distinct from Bulgarian and Serbo-Croatian sources), in which a whole contemporary literature had been created. Matevski, for example, who once described his poetic voice as having "come from far away" (and not because he was born in Istanbul), had married the natural imagery and folklore of western Macedonia, where he was raised, to sophisticated urban themes. A professor of drama, when we met at the Academy of Arts and Sciences he was wearing a blue felt suit and talked, as he said, expansively. Yet the monologue he delivered on Macedonian history was in the main surprisingly correct, given the vehemence of his denunciation of the Greeks, whose ancient plays he still taught. Matevski argued that Macedonian writers had devised a destiny—a language, culture, and tradition—which was theirs alone.

"How can we change our name?" he said. "Today you're Christopher. If tomorrow I tell you your name is Jordan how would you feel? You can't change people to suit your pleasure."

Though he believed Macedonian culture would survive, the fate of his country was another matter. Before the breakup of Yugoslavia, he had published a poem titled "We Will Not Have a Trojan War," which reinforced his conviction that poets have presentiments of catastrophe.

"I am by nature a lyrical, philosophical poet who doesn't write about politics," said Matevski. "But the Greeks are coming. I speak of Troy, which was besieged for ten years by Agamemnon then destroyed, because Macedonia is now surrounded by war."

The noontime crowd was silent. On the low central arch of the Stone Bridge—an Ottoman legacy, from the fifteenth century, and the only structure to survive the earthquake intact—a procession of Orthodox priests came to a halt. Where church workers had broken up the ice around the pillars, the archbishop threw a small cross into the Vardar and several men, including some visiting Russians, dove in to retrieve it. Then the priests marched through the crowd, sprinkling holy water. It was Epiphany on the Julian calendar. Another Balkan irony: the Macedonian Church, autocephalic under Tito, was recognized only by the Vatican; the Ecumenical Patriarch of Constantinople, first among equals in the Eastern Orthodox Church, dismissed it, though it was founded in 893. Indeed in the Macedonian churches and monasteries were the same icons and frescoes as in Athens and Belgrade and Sofia. Strangely, the one Slavic church in which the liturgical language approximated the vernacular was the only one unrecognized by Constantinople. Nonetheless the faithful in Skopje leaned forward to be blessed by the passing priests.

I crossed the bridge and went to the Old Market, where the crowds were larger yet. Along the uneven stone walkway old women huddled over displays of white linens and shoes, moneychangers studied the soldiers patrolling the cobblestone street, dark-skinned boys hawked Marlboros and stolen watches; in the back of the bazaar were arms dealers with an array of weapons for sale—Kalashnikovs, hand and rocket grenades, even surface-to-air missiles. The smell of smoking meats hung in the air. Through the smog the sun resembled a dull coin.

"We're so passive," said the woman who sold me a history of the maps of Macedonia. "Not like the Bosnians. If the Serbs come, we won't fight. It's happened to us so many times before."

The presidents of the six Yugoslav republics met for the last time in Sarajevo, in June 1991. The main item of business was the "Four-plus-two" plan put forth by Kiro Gligorov and Alija Izetbegović, which proposed that Croatia and Slovenia maintain confederal links to a federation made up of Bosnia, Macedonia, Montenegro, and Serbia: a shotgun wedding of sorts. Gligorov and Izetbegović, whose republics had the most to lose from Yugoslavia's impending dissolution, desperately sought a political solution to the crisis. Alas, the Serbs rejected their plan.

Gligorov was regarded as the most reasonable of the Balkan leaders—Izetbegović called him a true democrat—and in my meeting with him on the day that France asked the Security Council to postpone discussion of the Macedonian Question I admired his steady bearing. Caught between radical Macedonian nationalists and separatist Albanians, to say nothing of irredentists in Albania, Bulgaria, Greece, and Serbia openly talking of erasing Macedonia from the map, the former law professor and economist had to chart his course with extraordinary care. Regional peace and stability, he believed, depended upon Macedonian recognition. Without it the war might spread to Kosovo. A general Balkan conflagration, sparked by Greek intransigence, was not difficult to imagine: so many wars had started on Macedonian territory; if this one spread beyond the Balkans, the European diplomatic community would have to accept some of the blame.

"What is the limit to EC solidarity?" he asked. "That a nation must perish? If a family member commits murder, does family solidarity extend to covering up the crime? How can you explain why the only Yugoslav country that hasn't gone to war remains outside the UN?"

The danger was that in this diplomatic limbo relations between Macedonians and Albanians were deteriorating. Albanians in the western counties had proclaimed a proto-state named *Ilirida* with its capital in Tetovo; Albanian students did not want to hear university lectures in Macedonian; and after a government crackdown on smuggling in Skopje's Old Market, in which several Albanians were arrested and one died in police custody, calls for Albanian independence grew stronger yet. At the same time, Milošević might use the "safety" of Macedonia's Serbian minority as a pretext for an invasion: the strategy he had employed in Croatia and Bosnia and might resort to in Kosovo. Solving the problem of nationalities—honoring minority rights—was thus more crucial than ever.

"It is absurd to make people think they can live in ethnically clean areas," Gligorov declared. "We are mixed, we have to live together. Otherwise we'll have war for decades."

As a parting gesture the president gave me a signed copy of an anthology of Macedonian poetry. Over dinner that evening in the Hotel Kontinental I was brought up short by the opening lines of Vlada Urošević's "Forbidden Zone," an allegorical poem with a prophetic tinge: "We're going where it is forbidden to go./ Terrible punishments threaten us." The most prominent victim of this dark future would be Gligorov himself: in October 1995 he would be seriously injured in a car bombing in Skopje. The terrorists would never be identified or caught.

The old man had walked five kilometers to meet me.

"I am a hunter," Tomé Momirovski announced in a loud voice. "I have condition."

He was also a novelist, from Ohrid, who traced his literary origins to his Partisan days. In the First Macedonian Brigade, fighting Italians and Albanians in the western mountains, Momirovski came to know Koća Racin, the father of modern Macedonian poetry. Racin's bravery (he was killed in 1943), his realistic stance, his decision to compose lyrical elegies in the dialect that became Macedonia's official language, all set an example for young writers like Momirovski.

"It is important for our identity, our language, our literature, and our hopes for recognition that we write in Macedonian; and Racin started that for us," he said with a touch of awe.

Momirovski himself was a tall, hardy man in a blue work shirt and boots. His passionate way of speaking caught the attention of the humanitarians and peacekeepers in the lobby of the Grand Hotel. He was a forceful apologist for Macedonia. If it was Tito's brilliant stroke to link Macedonian identity to the Partisan cause, insuring the people's loyalty in the battle for the future of the South Slavs, it was up to the writers, he argued, to tap into their deepest reserves of feeling, the oral tradition in which they had forged and then maintained their identity.

"I counted four hundred eighty metaphors for love in our folklore!" he said. "Imagine. In fifty years we have created a literary tradition. We have a whole range of forms. We are not tired, because we have so many ideas and so much history yet to explore. We are fresh, unlike older languages."

When Momirovski noticed two peacekeepers studying him, he arched his back and spoke with even greater vigor. "My inspiration comes from deep in the earth—our sun, our colors."

"Not very much sunlight these days," I said to change the tone of the conversation.

The novelist shrugged. "Yesterday I was skiing," he said. "It was sunny and beautiful. When I came back I was so unhappy. This weather makes me sick. I am an urban man, but I have an atavistic legacy, which sends me to the mountains. That is my dialectic. If you go to the mountains two days a week you are fresh for five. I can talk with nature. I know all the trees, flowers, animals. Some think we hunters want to kill animals. No. Hunting is my way to return to my roots."

All at once he slumped into the chair, like a popped balloon. Independence had cost him his livelihood: the advance for his last novel had been fifty dollars, and he would receive nothing for his next book. The question of recognition kept him awake at night.

"I think, why literature?" he said. "I may not be able to publish anymore, but I will write for my daughter, my wife, the future. Maybe in your country you will hear that I am dead. I am not afraid for Tomé Momirovski. I am afraid for Macedonia, for what will happen after the Third Balkan War." His voice began to rise again. "When I was young, I joined the resistance with abstract ideas about revolting against the Germans—and I had perspective. Now I have no perspective. We had a socialist system. Now we have a civil society. What more can we do?"

The soldiers were staring at the novelist, who jumped to his feet and started to pace.

"Who's writing this screenplay?" he cried. "I don't recognize this script! Soon we'll be like animals eating little bits of food, because our culture will be gone." He fixed his eyes on mine. "I can't speak about this anymore," he said, and marched out of the hotel.

The inauguration party at the American Center. In the main room stood a large screen for the CNN broadcast of the ceremony, and the air was thick with smoke and laughter. Everyone who was anyone was here. Nebojša Cvetanovski, Savo's son, cornered me by the bar. Though he called himself an architect, engineer, and trade consultant, he was so desperate to get rich he tried to entice me into becoming his partner in an import-export firm. I told him I was a writer, not a businessman. No use.

"Your country wants clean, organically grown fresh fruits and vegetables, and we can provide them," Nebojsha assured me.

"How's that?" I said.

"There are special ships that can freeze them and bring them over," he said, perhaps forgetting that Macedonia was landlocked.

"They won't be very fresh," I said.

"What to do?" he said, and ordered another drink.

The room fell silent when the oath of office was administered to Bill Clinton. By chance I stood next to Kiro Gligorov. The president of one of the newest and most embattled democracies intently studied the chief executive officer of the oldest and most secure. The silence in the American Center held until Maya Angelou stepped up to the podium to read her inaugural poem. Gligorov and I leaned forward to hear her, but she had not read many lines before a look of bewilderment came over him. "She's not our best poet," I offered. "It's a political statement." Gligorov smiled broadly.

When the party broke up, I joined several humanitarians in accepting an American diplomat's dinner invitation. After two months of traveling alone, I found it disconcerting to sit in a restaurant at a table full of my countrymen. The conversation, predictably enough, turned to job opportunities. One humanitarian was thinking about applying to the Foreign Service Institute, another wanted to move to New York City. The diplomat, a beefy, middle-aged man, signaled for the bill. "Anyone in refugee work," he concluded, "will have no trouble getting work for the next twenty or thirty years."

Bogomil Gjuzel, the director of the Skopje Playhouse, began translating *The Tempest* at the start of the war. Even before setting out for an island near Dubrovnik he had dreamed of Macedonia being divided up among its neighbors. With a sense of doom he took lodgings in a nearly deserted monastery and went to work on Shakespeare's final play. His version was well received in Skopje.

"Prospero should forgive Caliban," he told me one morning at his theater. "That's how we played it here."

The cold weather and lack of heat in the building aggravated Bogomil's sciatica; he looked like an old man though he was not much more than fifty. He struggled to sit comfortably, deciding in the end to shuffle about his office. Think of the other versions of Prospero in Macedonia, he was saying, as if trying out different ways to play a part. In the post-Communist variant he corrupts the democrats for his own Communist purposes. Or he embarks on an island, gives Miranda to Caliban, and now Caliban has his own Balkan democracy. He doesn't need Prospero anymore, because his children are everywhere. On every island in the Mediterranean is a Greek who hears that Macedonia is threatening him, and though he has never heard of Macedonia he sends his guests back to their ships. If they capture Ohrid, that will be the end of our civilization.

The Tempest was to be his last translation of Shakespeare.

"Because, you see, everything's coming to an end here," he said, "as it does in Prospero's farewell to his audience and his art. The factories are releasing their toxins into the air, and the state is using this last opportunity to get rid of its poisons. Soon there may be nothing left of Macedonia."

—ʍ—

It was a strange briefing. Every question I posed to the three men from the Macedonian Press Agency provoked an argument among them in Greek. During the lulls, the youngest, an international lawyer who spoke English with a British accent, would say, "I should think"—then the argument would resume. Finally the lawyer would tell me with a shrug, "It is a political question we cannot answer." Or: "It is a legal matter." Or: "You can go to the university for that." The mandate of this newly renamed agency was to provide historically accurate information on the Macedonian Question, but their "facts" were highly suspect. They asserted, for example, that the UN was founded by the ancestors of Greece, and after an hour I had less than a page of notes:

> How could we rename the Greeks of Northern Macedonia? How could we rename the whole region, including our press agency? Six months ago, they founded a press agency with our name. Absurd! Letters from universities around the world go to the wrong place. We do not want to give up a history of 3,000 years. It would be like renaming a family—it cannot happen.

"Now let us ask you a question," said the lawyer. "Why do you think the Western press has turned against us?"

"All I know," I said, "is that this is the most bizarre briefing I've ever had. Everything I ask you is either a political or a legal matter or I can look it up at the university."

Puzzled expressions crossed their faces—then they began to argue again. Eventually the lawyer gave me some printed propaganda and pleaded with me to write the truth. The local journalist who had arranged the briefing led me to the door. He had remained quiet during their arguments, but once we were out of earshot he could not contain himself.

"Do you see why Greece will always be a democracy?" he cried, ignoring its history in chains: four hundred years under the Ottomans, the Nazi occupation, the recent military dictatorship.

I shook my head.

"We have ten million crazy Greeks, each with his own idea about everything," he said in exasperation. "We can't even agree on what to tell you about Macedonia!"

These were the halcyon days in Thessaloníki. The sunlight was brilliant, the winds had died down, and beyond the harbor, where ships were riding at anchor, the sea glittered. I was staying at the Hotel Vergina, named for the sun or star claimed as a national symbol by Greeks and Slavs alike: the sixteen gold rays were emblazoned on the Macedonian flag; on Greek postage stamps and the 100-drachma coin; on jewelry, T-shirts, and letterheads. It was a newly minted symbol: in 1977, in a village near Thessaloníki, a Greek archaeologist discovered in the royal tombs of

Vergina—one of the century's greatest finds—a gold box containing the bones of Philip of Macedon, decorated with a sunburst destined to bedazzle Greek and Slavic nationalists. And the rest, as they say, is history: whether or not Alexander the Great and the ancient Macedonians (descendants, perhaps, of the Illyrians) were Greeks. The truth, as usual in the Balkans, was not easy to uncover.

Alexander, the most gifted military strategist of all time, was indeed tutored by Aristotle, spoke Greek and, despite his disdain for Greek city-states, Hellenized the lands he conquered. But ancient Greeks did not consider Macedonians Greeks. Herodotus and Thucydides called them foreigners; Demosthenes and Isocrates said they were barbarians. The Macedonian elites were fluent in Greek, yet it seems likely that most of the population spoke a different language, which has since become extinct. Greek claims of historical continuity with the ancient Macedonians were thus as misguided as Slavic attempts to link the first Macedonian state to Vardar Macedonia. Which did not, of course, reduce the bellicosity on either side.

And Thessaloníki was the center of opposition to recognition not only of Macedonia but also of the Slavic Macedonians living in northern Greece. At the turn of the century, Slavs made up the majority in Aegean Macedonia, but thousands fled to Bulgaria after the Balkan wars, followed by tens of thousands more after World War One. Before World War Two, the Greek government embarked on a campaign to Hellenize northern Greece, attempting to assimilate the 'Slavophone Greeks,' as the Slavic Macedonians came to be known. Macedonian place and personal names were changed to Greek, Church Slavonic was banned from the liturgy, and the use of the Macedonian language was outlawed. More Slavs left after the Greek Civil War (1946–1949), and, because many had sided with the defeated Communists, the government stripped them of their citizenship and confiscated their property; they were not even allowed to visit Greece, though Greek political refugees from the same era had in the last ten years had their rights and property restored to them. There was no Slav minority in Greece, said the government, notwithstanding the Slavic Macedonians (estimates ranged from 10,000 to 50,000 people) living north of Thessaloníki. Greece, according to this fiction, was ethnically homogeneous. And it was true that many Slavs in Greece no longer spoke Macedonian. But "agents of Skopje" were everywhere, politicians and pundits warned. No wonder the outdoor cafés were crowded with young men and women wearing identical black leather jackets.

Waiting for my appointment at the Ministry of Macedonia and Thrace, I leafed through the English supplement of *Macedonian Life.* The cover story about the Patriarch's journey from Constantinople to his Ecumenical Throne in Greece did not gloss over its political nature. Though he visited Mount Athos, fulfilling a vow to the Virgin Mary of *Axion Esti* (literally, Worthy It Is), the famous icon kept on the holy mountain, his true mission was "to eulogize the flock of the Christian Orthodox church in Northern Greece"—the non-existent Slavic Macedonians, that is. I turned the page to find an open letter addressed to European heads of state concerning Macedonia. The six Greek dignitaries who signed it, including Odysseas Elytis, the Nobel Prize-winning poet, denounced, in language bordering on the hysterical, their northern neighbors' irredentism, warning that EC recognition of Macedonia by that name was tantamount to declaring war against Greece. "Our name, dear Sir, is the immediate jewel of our soul," they concluded. It was sad to see Elytis's name attached to such a document. In *Axion Esti*, his spiritual autobiography, he wrote, "Intelligible sun of justice and you, glorifying myrtle,/ do not, I implore you, do not forget my country!" But in that same poem he cried, "My only care my language on Homer's shores." Now his care in describing the human condition seemed to have given way to bombast. Reasonable discourse on the Macedonian Question was all but impossible to imagine with even Elytis signing nationalist screeds.

And what the director of political affairs told me at the ministry was anything but reasonable. Thrasyvoulos Stamatopoulos's argument went like this: the Macedonian nation and language do not exist. They are a Communist creation in nationalist disguise, a vehicle with which to expand to the warm waters of the Mediterranean. And that was the point of our civil war, wasn't it? To prevent this southward expansion. We were forced to keep the issue of the name low-key when Tito broke with Stalin, and look at what happened. Greeks see themselves as the only losers of the Cold War. The entire population of Macedonia is Greek. There are no minorities. Yes, some people speak a Slavic-sounding idiom, but this is not a language. Slavicized Greek, if you like.

When I raised an eyebrow, he challenged me to challenge him, then cut off my reply. "I call myself Macedonian," he announced, "but not in the sense they do."

"But I thought *they* didn't exist," I said.

"The problem is semantic," he said, "but if it goes any farther it's a different matter, especially if it involves foreign powers. Recognition will not preserve stability."

The White Tower was a good place to reflect on Cyril and Methodius, the ninth-century Greek saints remembered as the fathers of Slavic literature. The sons of a Byzantine official in Thessaloníki, they were not Slavs, as Macedonian nationalists claim, but with their gift for languages the missionaries devised a phonetic alphabet based on the Slavic tongue spoken around them—on Macedonian, that is—and a script that evolved into Cyrillic. Once they had worked out the language they realized its lexicon was too small to address sophisticated theological concepts, so they added Greek words and grammatical forms to create a new language,

Old Church Slavonic—Hellenized Macedonian, if you like—into which they translated the Gospels. Cyril and Methodius thus became the Apostles to the Slavs, countering Rome's missionary work in the region, insuring that the fault line between the eastern and western branches of Christianity would run through the lands of the South Slavs. And Old Church Slavonic, for several centuries the literary language of the Serbs, remains the liturgical language of the Bulgarian, Macedonian, Russian, and Serbian Orthodox Churches.

Which is to say: language is destiny, even when almost no one understands it anymore.

The White Tower dates from the Venetian occupation in the early fifteenth century. Built along the waterfront to stave off the Turks, whose blockade of the city nevertheless precipitated its fall in 1430, under the Ottomans the watchtower was used as a prison. Hundreds of revolutionaries were locked up in it during the War of Independence. And it was known as the Tower of Blood until the Turks painted it white in 1869 to placate the Thessalonians, who remained under Ottoman control until 1881. Now it was a museum—the walls were lined with the most beautiful icons—overlooking the boulevard down which a million people had recently marched in protest of Macedonian recognition. A language, a literature, a culture—any human creation may follow the same course, from lookout to prison to storehouse. A Western diplomat came over to tell me that for ten years the Greeks had predicted the breakup of Yugoslavia. No one had listened.

"They don't want to be proved right again," he said.

The theme of the inaugural session at Art Village was the Trojan Horse. This experiment in artistic collaboration was the inspiration of Sotos Zahariadis, a young painter and art therapist determined to counter the lack of direction in contemporary art. Painters, sculptors, potters, writers, composers, photographers, and architects from all over Europe had accepted Sotos's invitation to spend the month of September working together. What caught their attention, when they gathered at a hillside retreat some distance from Thessaloníki, was a Byzantine tower overlooking the sea, so their first project was to carve a window in a boulder down on the beach, through which to look at the tower—a window onto the past. A German artist built a wooden horse on wheels to accompany the group everywhere, and then they set to work, some making paintings and sculptures of the Trojan Horse, others drawn to the myth of Sisyphus, destroying what they had created because, as Sotos told me one night in his gallery, civilizations rise and fall. Thus a ceramic pot was dissolved in boiling water, a sculptor took a sledge hammer to a finished piece, and an Irishman, who had brought two heavy cubes, one to place in a different setting each day, the other to drag in a line twenty-two meters long, succeeded in "magnetizing" Art Village and digging a furrow in which grass began to grow. Sotos himself painted a room to resemble a church sanctuary, covering the walls with miniature paintings of horses, balls, branches of lightning, mathematical equations, and maxims like *Hair teaches us that harmony is curved and flexible.* I started asking questions: Where is the modern Troy? Macedonia? Sarajevo? And what will artists smuggle into that mythical city?

"In 1942 I was a prisoner of the Bulgarians, but I don't want to fight with them. We sent them food when they were starving to death. No more now, not since Skopje started using our name."

"Bulgarians?" I said.

"Yes, Bulgarians," said John Milarakis.

Macedonians, in other words.

The journalist paced his living room, explaining Greece's side of the conflict with its northern neighbor. A cheerful man with a grey mustache and bags under his eyes, he had built his house with his own hands, and I coveted the round wooden table he had engraved with the figures of two female singers and a man strumming a lute-like *buzuki*. He hoped to sell it to an American collector so he could send his daughter to Juilliard. What luck to find myself in a house given over to books and paintings and music! The coffee was strong. Sunlight streamed into the room. John thought the EC should adopt Greek as its official language: it would make people act better.

"Do you know why our language is so difficult?" he said.

I shook my head.

"Because it has such a long tradition," he declared. "When you say 'mother' in English, you mean the woman who gave birth to a child. But when a Greek says 'mother,' he means 3,000 years of tradition. Do you know what the Spartan mother told her son when he went off to fight? Either you will return dead or victorious. That's the same spirit of Greek mothers in 1940. I remember being taught as a small boy to hide knives in a blanket to throw under the wheels of a tank. So the Greek mother tells her son to defend Greece and its traditions."

He was on a roll. "Take a month to read our mythology," he said, "and you'll understand why we still think the way we do after so long."

I was pondering his assignment when he said, "How can you produce something new without the raw materials of our philosophy? If Frenchmen and Germans learned Greek, that could help the modern world. To me, Americans lost when they didn't make Greek their national language."

My protest that the American pilgrims knew no Greek fell on deaf ears.

"What did the Greeks give us?" John asked. "The idea of freedom." Then he returned to his original theme. "We were sleeping when Gligorov said, 'We are

Macedonians.' It's absurd to base a country on a falsification of history. That's how wars begin."

At the Archaeological Museum, busts of Hercules, Philip, and Alexander. The tip of an arrow bearing Philip's name. Gold coins minted twenty-five centuries ago. A cracked relief of a young girl holding a dove. For once I was grateful not to be able to read the inscriptions: I did not want to know if they had been contaminated by the politics of the Macedonian Question. Yet the pleasure I took from these ancient objects was short-lived. **Macedonia is Greek: 3,000 Years of History,** read the banner above the information desk. The English slogan, ubiquitous in Thessaloníki, irritated me. At last I understood why some younger Balkan writers dreamed of a literary space free of politics. An Orthodox priest once told me the Balkans have too much history, America too little. But as I marched out of the museum I decided there was something to be said for leaving the past behind. The southern Balkans were choking on the legacy of the Hellenistic world carved out by Alexander the Great, whose likeness was etched on the other side of the 100-drachma coin, and the memory of the Greek War for Independence, which inspired Lord Byron to sail to the island of Cephalonia with medical supplies and arms. In one of his last poems he wrote:

> The Sword—the Banner—and the Field
>   Glory and Greece around us see!
> The Spartan borne upon his shield
>   Was not more free!

But Byron died of rheumatic fever before ever seeing battle. "The Greece of the classical heritage and of the romantic philhellene has gone," according to translator and writer Philip Sherrard, "and anyhow has always been irrelevant to the Greek situation."

Tell that to the Greeks.

Crazy—that was how John Milarakis described his driving, weaving from one side of the road to the other as we climbed the steep hill to the fortress overlooking the city. Everything had a story for the journalist, the most vivid from World War Two. Here was the abandoned bakery, for example, where he would get bread for his mother until the Nazis surrounded the building one day and lobbed a grenade inside, which landed by his feet.

"What do I do?" the boy cried.

"Throw it back to them," commanded the baker.

John obeyed him, killing one soldier and wounding three.

And here was the prison where he brought bread for his uncle, who was condemned to death by the Nazis. Certainly the German occupation of Thessaloníki was grim. A large part of the city was destroyed; more than 50,000 Jews were sent to death camps; only one of the thirty-six synagogues survived. For five hundred years Thessaloníki had provided a haven for Sephardic and Ashkenazi Jews fleeing persecution, and by the nineteenth century, when Jews made up two-thirds of the local population, Ladino was the language of commerce. But all signs of what must have been a vibrant Jewish culture were gone, along with a rich tradition of tolerance. Hard to imagine Bulgarians, Castillians, Jews, Turks, Greeks, and Slavic Macedonians all living together.

Harder still to recall this was the birthplace of the Young Turks' revolution, which toppled the Ottoman Empire, as well as of Kemal Atatürk, the father of the modern secular Turkish state—two facts conspicuously absent from Greek textbooks. We came to the white Byzantine walls studded with chips of marble taken from churches razed by the Turks, even bigger bogeymen in John's mind than the Macedonians. For his family was among the 160,000 Greek refugees from the Turkish mainland who resettled here in 1923, a population transfer which changed the character of Macedonia and set the tone for decades of forced migration. The literature professor from Constantinople who lived inside these walls for twenty-five years was for John an emblem of the age. And he said the Turks were preparing for war with Greece: the war of words over the divided island of Cyprus was escalating, each country's air space was routinely violated by the other's air force, and so on.

Our last stop was Vlatades, the only intact Byzantine monastery in the city, built, legend has it, on the site of Paul's mission to the Thessalonians, twenty years after the crucifixion of Jesus. A. N. Wilson suggests the apostle broke new ground in the provincial capital, "plant[ing] the divine seed in virgin soil." And it was to the Thessalonian converts that Paul wrote the first letter collected in the New Testament: "See that none of you repays evil for evil, but always seek to do good to one another and to all," he beseeched them. Easy advice to follow on such a warm, clear afternoon. A wedding was underway in the chapel and guests who had arrived late milled outside. Couples sat on the wall gazing at the harbor. Peacocks, which in the Byzantine era guarded the monastery, strutted in the pen next to the chapel. I was thinking how pleasant it would be to stay here when John showed me into what he said was Saint Paul's cell, its walls blackened with soot; in the frescoes were holes gouged out by the tips of Turkish swords.

"What can you do with a people like that?" John asked. "That is why the West must never trust the Turks. They want to destroy everything civilized."

The joke had lost some of its sting with Communism's passing. Two fishermen, an Albanian and a Macedonian, cast their lines from boats on either side of the border running down the middle of Lake Ohrid. The Albanian has no luck, but the Macedonian reels in fish after fish. "What's your secret?" asks the Albanian. "Our fish aren't afraid to open their mouths," replies the Macedonian.

It is a rare species of trout the fishermen catch, a survivor from prehistoric times found in only two other lakes: Prespa, thirty kilometers to the south, and Baikal, in Siberia. Lake Ohrid trout was considered such a delicacy during the Ottoman era that special couriers brought it to the Sublime Porte in Constantinople. And the trout I ate one evening at a restaurant in the town of Ohrid, grilled with onions, tomatoes, and paprika, was indeed sublime.

The inversion in Skopje had not lifted when I returned from Thessaloníki, so I caught a bus through the mountains to Ohrid, the heart of Macedonian civilization. The black-kerchiefed Albanian women carrying live chickens in cardboard boxes went to Tetovo, a hopeless sprawl of unfinished houses and minarets, idle factories and old men in white skullcaps leading horse-drawn buggies through melting snow and mud. Only one bloodied man remained on board, the victim of a vicious beating delivered by a Macedonian at the bus station; after he got out at Struga, I had the bus to myself. And I was the only guest at the Grand Palace in Ohrid. My room had a view of the lake, but there was no heat, the linens were dirty, and one day the leaking shower became a torrent I could not turn off. When I told the manager my room was in danger of flooding, he just shrugged.

Any resort in the off-season is depressing, but Ohrid's desolation was unnerving. A light snow was falling when I strolled along the stone walkway through the Old Market, much of it empty. The churches were locked, and the archaeological museum, which charged me three times the native rate, closed before I had finished my tour. Then I headed up the hill (the name Ohrid comes from *vo hrid,* on a cliff) toward the monastery in which the first Slavonic university was established by Clement and Naum, disciples of Cyril and Methodius. These monks turned Ohrid into a Slavonic literary and cultural center, which in the tenth century became the seat of the Bulgarian patriarchate and crown of an empire stretching from the Adriatic to the Black Sea. The Ohrid literary school, painters, and musicians flourished under the Bulgars, and I had a mind to tour Saint Clement's Church, which houses a valuable collection of frescoes and icons, but it too was closed. After circling the mosque the Turks built around the ruins of Clement's monastery (the minaret and dome had long since disappeared) I made straight for the lake, which was roiling with whitecaps.

Past rushes and cattails I went, gazing at the snowbound mountains of Albania, a place I dearly wished to visit. There was no bus service from Ohrid, though, and the only car rental agency would not let me take a car across the border. Too many thieves, said the clerk, betraying a general fear. Some Balkan observers thought the Albanians—nine million strong, divided among Albania, Macedonia, and Kosovo—were becoming the next power in the region. In a recent newspaper column a Macedonian writer lamented his failure to befriend any Albanians, now that relations between the two peoples were so strained. I understood that sentiment, my host in Skopje having scuttled the one interview I had set up with an Albanian. So on to Kosovo.

13

Kosovo

I did not know Arkan was living in the Grand Hotel. Not until after I had checked in one evening (the choice was between a room with heat and one with a phone) did I learn that the paramilitary leader had turned the seedy hotel into his base of operations. Kosovo's newly elected parliament deputy was already making his mark on Prištinë. He had opened a gas station to sell his smuggled petroleum, he was looking into banking opportunities, he was said to be preparing his next campaign of terror, this time against Albanians, who had boycotted the election. To his specialties of murder and mayhem Arkan was adding politics, and now the president of the Serbian Unity Party was ensconced at the Grand—where I had chosen the heated room, which was not only unheated but also had no hot water. But it was too late to move. Besides, I was hungry.

At this hour the dining room was nearly empty, unlike the bar, which was popular with the JNA and the secret police. The Grand was a good place for them to unwind from the responsibilities of maintaining a police state, since the target of their oppression avoided it. Indeed what was once an Albanian establishment was now the choice of those plotting the Albanians' persecution. It is worth recalling how Milošević instituted his form of apartheid in Kosovo: first, he revoked the province's constitutional autonomy (twenty-two Albanians and two policemen were killed during two days of protests against the new constitution); then he purged Albanians from every position of influence—political, educational, cultural, journalistic; and then he imposed martial law. Note the logical fallacy: Milošević was waging wars in Croatia and Bosnia on behalf of ethnic rights, yet he based his claim to Kosovo on historical rights. What was once Serbian, et cetera, et cetera.

A curious history, which for Serbian nationalists begins and ends on Kosovo Polje, Blackbird's Field. The decline of the medieval Serbian kingdom may be traced to the sudden death in 1355 of Tsar Stefan Dušan, perhaps a victim of poisoning. At once a patricide and a skilled politician, he had expanded his empire until it stretched from the Sava to the Gulf of Corinth and from the Bulgarian border to the Adriatic. He subdued the Bulgarians by marrying the tsar's sister, fended off the Hungarians, conquered Macedonia, then proclaimed himself Tsar of the Serbs, Greeks, Bulgarians, and Albanians. His son, Uroš, however, was no match for the forces arrayed against him—Serbian noblemen contending for the throne, Ottomans marching on the border—and when he died without an heir in 1371 it was too late for the claimants to the throne to form an effective front against the Turks. True, at the Battle of Kosovo, Lazar teamed up with his son-in-law, Vuk Branković, but they could not withstand the fighting machine of Sultan Murad, who only the month before had taken the fortress at Niš. Both sides suffered heavy losses: Murad died at the hands of the Serbian knight Miloš Obilić, Lazar was captured and beheaded, most of the Serbian noblemen were killed or went into exile. But the conquered Serbs fashioned poetry from the bloodshed. In some versions of the Kosovo epic the defeat is blamed on Branković's supposed treachery, in other versions the crucial event is Lazar's decision to sacrifice earthly glory for the heavenly kingdom. In any case, as Vasko Popa wrote, "The blackbird dries his blood-drenched wings/ At the fire of red peonies," still a common flower on Blackbird's Field. And this "image of disaster of the Battle of Kosovo," according to the critic Svetozar Koljević, "has lived for centuries in Serbian literary and oral traditions with the elusive vividness of a hallucination."

It was in the name of this history that the local Serbian authorities were stripping Albanians of their jobs and attempting to suppress their language and culture. Before leaving office, George Bush had drawn a "red line" at Kosovo, warning Milošević that the international community would not tolerate "ethnic cleansing" of the Albanian population, a hollow threat in the eyes of the strongman's minions. They looked askance at the foreign press, too. A Frenchwoman stopped by my table to say that journalists' notebooks had a habit of disappearing from their rooms in the Grand Hotel. I called for my bill and hurried upstairs, remembering an EC military monitor's warning.

"Kosovo," he said, "is a massacre waiting to happen."

To find the headquarters of LDK, the Democratic League of Kosova (the Albanian spelling), I was advised to write out its initials on a piece of paper and show them to a cab driver. An Albanian would happily drive me there; a Serb might deliver me to the police. As it happened, my cab driver was the same Serb who had taken me from the bus station to the Grand. The good news was that I had tipped him well (he was suffering from emphysema, I felt sorry for him); the bad news, that he could read only Cyrillic. Another driver, offering to help, spat on the ground once he had deciphered my handwriting. He muttered something to my driver, who gave me an ugly look, and off we went. It was impossible to know if he preferred hard currency to escorting me to jail. All he did was wheeze as we bounced and skidded over a stretch of mud behind the soccer stadium—the road had disappeared—before abruptly switching off the engine in front of a small Quonset hut, which housed the Albanian PEN Center and the Kosova Writers Association. Next door was Prištinë's central police station.

Rexhep Ismajli called the main room in the hut the only free fifty square meters in Kosova. Among the copiers and fax machines were writers drinking coffee and human-rights workers documenting Serbian abuses. Ismajli, a social linguist and translator of French at Prištinë University until the Serbian authorities had dismissed him and more than eight hundred Albanian colleagues, was vice president of LDK. He made a point of showing me around the Albanians' one public building. The tour took less than a minute. First, a bookcase of Albanian works, then portraits of three tutelary spirits: a sculpture of Naïm Frashëri, the Muslim apostle-poet of Albanian nationalism; a painting of Gjergj Fishta, the Franciscan priest and Albania's national poet; and a photograph of F. S. Noli, the Harvard-educated Orthodox bishop, translator of Cervantes and Shakespeare, and first freely elected president in the Balkans, who ruled for only six months—because Serbs and Greeks were afraid of democracy, said Ismajli, neglecting to add that Noli was overthrown by his own countryman, King Zog, albeit with Yugoslav support. Ismajli pointed at the ceiling.

"Of course it's bugged," he said.

But he did not mince words. The reunification of Albanians in Kosova, western Macedonia, and Albania was inevitable, he said, because Greater Albania was a valid political idea as well as a cultural fact, notwithstanding eighty years of division. War in Kosova would thus be international in scope (Greece and Bulgaria might also join in), and any Serbian provocation—there were countless daily instances—could lead to war. It was up to Clinton to stop Milošević. No telling how long the LDK leadership could counsel patience to its membership. What choice did they have? There were hundreds of Albanian villages the JNA could overrun without fear of harming Serbs; but convincing Albanians to hold out for peaceful change grew more difficult by the day, since they were now required to hand over their guns to the authorities (some villagers were buying hunting rifles to give to the police in order to avoid arrest) even as the JNA was arming the Serbs.

Then came the obligatory history lesson. What Ismajli emphasized, though, was a tradition of tolerance instead of grievance: four religions living side by side, Sephardic Jews having migrated into the region after their expulsion from Spain. Albanians were the oldest Balkan people, and the religious diversity of this ancient crossroads, between East and West, Byzantium and Rome, had of necessity fostered tolerance. Like Sarajevo, with this difference: the war in Bosnia was an internal matter. If fighting broke out in Kosova the tragedy would be much greater.

"It's not a classic war," said the beautiful young architect. "Just every day there's a beating, a killing."

She was the prettiest of my Albanian acquaintances to give the lie to the ethnic joke that they could not learn languages. The language Albanians refused to learn, I realized in LDK headquarters, as I heard English, French, and German, was Serbian. Albanian is a language unto itself in the Indo-European family, distinct from the Slavic tongues, and Kosovar Albanians had the same cultural and political interests in using their own language as their Slavic countrymen: language *is* identity. The architect, who had never been allowed to work at her profession, looked up.

"Something is happening," she murmured.

The man in the doorway, a dental surgeon on "forced holiday," had in tow three students just released from jail. Their crime was studying with him, in Albanian, the construction of bridges and crowns; two fellow students were still in custody. When they had gone to the surgeon's house to get their fall semester grades, his Serbian neighbor had called the authorities. The police arrested the students and beat them on their hands. But they did not betray their teacher.

"It was my fault," he said. "I shouldn't have had so many come at the same time."

Of course, he should have been able to teach them at the university. Meantime, he worked in the parallel system of schools, hospitals, cultural institutions, and media the Albanians had established just to survive. You had to admire their tenacity. One dental student, for example, who was rubbing his bruised hands, had already fled from Sarajevo. He commuted ninety kilometers a day to attend the surgeon's lectures. No wonder he had done poorly on his exams. Now this.

"He's seen everything," said the surgeon. "He has so much to think about!"

"What has never been can never be," Vuk Karadžić wrote. "One land only but two masters." It was true: power sharing was not in the political lexicon of Albanians or Serbs, thanks in no small measure to Karadžić's work on the Kosovo epic, sung by *guslars*, at the heart of Serbian nationalism. The Serbs had won this war of poetries. "The Downfall of the Kingdom of Serbia," for example, tells the story of Lazar, who on the eve of the Battle of Kosovo has a remarkable dream. A grey falcon flies from Jerusalem, with a swallow in its beak. But the falcon turns into Elijah, and the swallow becomes a letter from the Holy Mother spelling out the choice the prince must make between earthly glory and the heavenly kingdom. The poem's concluding lines:

And Lazar chose heaven,
not the earth,
And tailored there
a church at Kosovo—
O not of stone
but out of silk and velvet—
And he summoned there
the Patriarch of Serbia,
Summoned there
the lordly twelve high bishops:
And he gathered up his forces,
had them
Take with him
the saving bread and wine.
As soon as Lazar
has given out
His orders,
then across the level plain
Of Kosovo
pour all the Turks.

Lazar's church of silk and velvet, a lovely metaphor for Kosovo's Orthodox monasteries and shrines, had given way to a darker figure, the black-and-blue fatigues and flak jackets of the soldiers patrolling Prištinë. The Turks? In Kosovo, they built bridges and mosques (most Albanians converted to Islam); recruited Albanians to serve in the Sultan's army; made Turkish the official language. And they gave Serbian poets a powerful theme: in the Kosovo epic Lazar became a Christ-like figure, betrayed by Vuk Branković, the Serbian Judas Iscariot. Crucified by the Muslims, Lazar, and thus the Serbian nation, would be resurrected only when these lands were cleared of the Christ killers, the Bosnian Muslims and Albanians whom Serbs viewed as turncoats, terrorists, and Turks.

"Just as we are not and do not want to be Turks, so we shall oppose with all our might anyone who would like to turn us into Slavs or Austrians or Greeks, we want to be Albanians." This memorandum from the Albanian League in Prizren to the British delegation to the 1878 Congress of Berlin garnered little support in diplomatic circles. Bismarck for one did not believe that Albanians were a distinct nationality, and while the Great Powers could no longer ignore a people their own historians and poets regarded as direct descendants of the Illyrians they did not hesitate to divide up most of the Albanian lands among Serbia, Montenegro, and Bulgaria. Five hundred years of Ottoman rule were ending, even as the Porte's last Muslim subjects were developing a national consciousness. They codified their language, created a poetry, and in the First Balkan War declared independence—only to be occupied by Serbia. One month later, when the Great Powers granted Albania autonomy (ceding vast tracts of land to Serbia, Montenegro, and Greece), Montenegro laid siege to Shkodër, et cetera, et cetera. In short, occupation and partition are the watchwords of Albanian history. Yet Greater Albania is easier to envision, at least from a cartographer's perspective, than Greater Serbia: only borders separate Kosovar Albanians from their brethren in Albania and western Macedonia.

Ah, but borders are drawn in blood.

Serbs, Montenegrins, Greeks, Bulgarians, Italians, Austrians all fought over Albanian lands in the Great War. And no one was happy with the decision at Versailles to restore Albania's 1913 borders. The Serbian dynasty enlisted none other than Ivo Andrić to sort out its Albanian policy. In a secret memorandum, dated 1939 and finally published in Croatia in 1977, the Nobel laureate advocated partitioning Albania, assimilating Catholic and Orthodox Albanians, and deporting the Muslims to Turkey—a hateful document that reinforces Charles Simic's observation that it is time to dismantle "the myth of the critical independence of the intellectuals."

"Ali Podrimja is not from here," the beautiful young architect told me when Kosova's greatest poet arrived at LDK headquarters. "He's from the cosmos."

As it happened, the poet was on his way to the Vienna Human Rights Conference. One circle was closing for me: it was Podrimja who had read at the publishing party my Slovenian friends had taken me to in Carinthia. He spoke warmly of the poet Tomaž Šalamun, recalling an Albanian poetry night in Ljubljana, in 1988, disrupted by bomb threats called in by Serbs wary of Slovenian-Albanian ties. Kosova was indeed Slovenia's dark twin. Slovenian nationalists had used the Albanian cause in their own independence drive; Kosova had lost its autonomy. Slovenia's population was declining; Kosova had the highest birth rate in Europe. Slovenian poets had international reputations; Kosovar Albanians were almost unknown; and so on. Twice that night in Ljubljana the poets had to move to a new location. A two-hour event stretched into six. Šalamun was inspired by the danger.

"I would love to have my reading blown up," the Slovenian poet announced.

Podrimja was accustomed to such threats. At this year's Frankfurt Book Fair, during a symposium on the Balkan War, he had just begun to read when police swarmed around him. Don't be afraid, said one policeman. Unruffled, Podrimja asked him to go have a coffee with him. The policeman's reply astonished the poet: I want to hear your poem about the Berlin Wall. And when Podrimja, stopping in Ljubljana on his way home to Prištinë, told this story to Šalamun, the Slovenian poet said, You're a lucky man. You will have a long life. No one can hurt you.

Decidedly less certain of his destiny, Podrimja put his faith instead in Albanian poetry. "It takes the side of a people who are not conquerors," he said. "It is against violence, because its source is not only in our struggle for Albanian existence but also for universal existence."

Then he was gone, but not before leaving me with a copy of the poem the German policeman admired, which Rexhep Ismajli and I immediately translated:

Fates

I tiptoe through the Berlin Wall
with a rose in my hand

I'm afraid of hurting
the fallen souls

When I wanted to go through the Albanian Wall
my feet my head were soaked in blood

Over hills and fields
a woman dressed in black
looks for my grave

And my body
wakes up every day among you
civilization's terrible torso

Doesn't it bother you
the way my torso
accuses

J. J.'s hand, when I shook it, was as soft as a sponge. He was unshaven, his eyes were bloodshot, and he had shuffled into LDK headquarters with the stiff bearing of an old man, though he was not much more than thirty. The doctor with him explained that J. J. had spent the last twenty-four hours in police custody—for selling videocassettes of Albanian folk singers. His family company was licensed to sell them, and he had come to Prishtinë yesterday from his home in a nearby village to check on his business. But when he arrived at his office his brother was out and the only customer was an undercover policeman waiting to beat him. Two uniformed policemen arrived to take him to a Serbian café. There, in the water closet, one policemen (both were drunk) put a bullet in his revolver and said, Do you see? He cocked the revolver. Now you will learn who the Serbian police are, he said, shoving the gun down J. J.'s throat. Then I will kill you. The other policeman started to hit him, then the first one used his gun to strike him on his head, his back, his legs—everywhere. They tortured him for four hours before adopting a new strategy: threatening to shoot him if he did not lead them to his brother. But his brother was nowhere to be found, so they went to the police station, where the beating resumed until a civil inspector arrived. Did the police beat you? the civil inspector asked. Until they were tired, J. J. said. What shall we do with him? the civil inspector asked the policemen. You know what you can do, they said. What they did was take his money and release him, promising to kill him in ten days. That was why J. J. would tell me only his initials. And your brother? I said. In hiding, he said. Will you leave the country? I said. Who will give me a visa? he said. Besides, he was the sole provider for his parents, his brothers, his wife, and son. Do you want to see what they did to him? said the doctor. I nodded. J. J. slowly pulled his shirt over his head. His upper back was a solid bruise the color of the sky at sunset, shot through with lines of black and blue; more bruises and welts covered his calves, thighs, and buttocks. And his hand, said the doctor, lifting J. J.'s swollen right hand, it's probably broken. We haven't been able to X-ray it yet. Why, I wondered, did he shake my hand?

We're lucky in one way, said the former tour guide. Our president is a nice, calm writer who has kept our people calm. And we're lucky in a special way: whatever the Serbs do is wrong. That's why we're patient. It's unnatural to keep doing something wrong. If someone tells you you're drunk, you'd better believe it. And the whole world is telling them they're drunk, but they don't realize it yet. They call this the cradle of their civilization. I say, Let's turn to the encyclopedia. It's not the churches they want, it's the minerals—the lead, copper, gold, silver. But the mines are closed down. What I want is to drive to Albania one day, have a swim in the Adriatic, then a nice dinner, a cognac, and drive back. We're the same people as them. We just haven't seen them for fifty years.

What George Orwell called "the evil atmosphere of war" pervaded Prištinë, a dingy, unfinished city. MIG fighter jets buzzed the drab apartment buildings. Soldiers and policemen armed with automatic weapons lined the muddy streets. The poet Eqrem Basha and I, in search of a car, walked toward the distant snow-capped mountains, discussing the relative merits of Ivo Andrić and my beloved St.-John Perse. Fate had linked the diplomats, awarding them the Nobel Prize for Literature in successive years. Juxtapose the photograph of Andrić at the signing of the Tripartite Pact, alongside Ribbentrop, with the story of St.-John Perse staring Hitler down at the Munich Conference, and you will have a history of the conflicting political engagements of twentieth-century writers. Sadly, this is a parable of talents in which the muse does not distinguish between the cowardly and the courageous. *The Bridge on the Drina* is a masterpiece, notwithstanding the novelist's political sympathies.

"Only after Andrić won the Nobel Prize did we learn he was a fascist," said Eqrem, motioning me quickly into a battered Fiat. "The Serbian project has always been to purify this area, spiritually and ethnically. But the more they tried to assimilate us, the stronger we became." This was because the Albanians had a revolutionary tradition to inspire them, he said, and in the linguist-president Ibrahim Rugova they had a leader with the moral authority of Václav Havel to guide them.

It was up to the writers to articulate democratic principles, relying on metaphors to evade the censors, as in Eqrem's poem, "Urban Planning," which we had translated before setting out:

> First place the city where it will catch
> The sun morning and evening
>
> Then plan for the sewers
> To remove the remains
> Of assassinations on dark nights
>
> Set the monument to the unknown hero
> In the center where he can breathe
> Freely when the seasons change
>
> And
> Don't put skyscrapers
> Where their shadows will fall on people
>
> When you design the streets
> They should lead out from the city's heart
> So blood will flow
> To every limb
>
> If there is no river
> Pour the tears of the despairing
> If there are no parks
> Plant a forest with hair that stands on end
> And
> Before everything else
> Leave room for solar panels
>
> Because the city
> From the very first will catch the sun

And the sun was shining when we drove to a neighboring village to visit Eqrem's best friend, Agim Çavdarbasha, Kosova's foremost sculptor. Every day the sculptor and his students—he too had been dismissed from the university—walked two kilometers to this studio, which he heated with scraps of leftover wood; he worked in marble and bronze only during the summer. The sculptor had no money, no commissions, and his galleries had been shut down, yet he seemed remarkably content.

"*Il travaille toujours*," marveled Eqrem.

Outside, large abstract pieces, which owed a debt to Henry Moore, took up most of the available space. But it was a series of smaller works that caught my eye: "Cages," haunting forms trapped in bars. They were Eqrem's favorites, too.

"The very first sentence of our history books," he sighed, "says we are descendants of Slavs."

Hydajet Hyseni had spent ten years, almost his entire adult life, in prison for "counterrevolutionary activities"—writing and publishing poetry. But the dictates of the recording angel were what he now obeyed, documenting Serbian abuses for the Council on Human Rights and Freedom. The volunteers in this unheated office had no shortage of work.

Hydajet, a thin, dark-haired man, apologized for not speaking English: a prison legacy, said his elderly translator. The translator was a former banker, the last person you might expect to find working for an "enemy of the people," as the poet was called. Hydajet's problems, the translator explained, dated from his university days when a friend published some of his poems, without his knowledge. Soon after, the secret police kidnapped him and took him to a house in the woods; at the end of a lengthy interrogation he was ordered to choose between informing on other students or going to prison. He managed to escape, hiding out for the next four years, in Kosova and abroad, with his friend, Kadri Zeka, until the secret police assassinated Zeka in Stuttgart. Then Hydajet joined the student movement, writing poems and articles for their publications.

"Propaganda," he said in English.

March 1981. The first demonstration in Prištinë was spontaneous, a food fight at the university, which spiraled out of control. Student demands for better food, housing, and working conditions gave way to mass protests in towns and villages across the province—and sharper political focus: the province should be granted republican status. The Serbs' worst nightmare thus became the Albanians' rallying cry. Riot police moved in, and for the first time since World War Two Yugoslavs fired on their own countrymen, killing at least a dozen Albanians and wounding hundreds. Martial law was established; half of the adult Albanian population was arrested or reprimanded; some died in detention. It was not long before Hydajet was taken into custody and interrogated for four months.

"The police tortured me in every way imaginable," he said, opening an album of photographs.

I cannot describe the horror I felt looking at these pictures of beaten, murdered, and mutilated Albanians—teenagers, old men, priests, a woman with her ear cut off—which formed the backdrop to Hydajet's recollections of his year in solitary confinement in a Serbian prison. Then he was placed in a cell with murderers and rapists who were ordered to abuse the Albanians. Yet the Serbian prisoners came to see their folly, even banding together with the Albanians to fight the guards. We found a common language, said the poet, whose hero was Adem Demaçi, the Albanians' Nelson Mandela. After twenty-eight years of imprisonment, Demaçi still believed the Albanians' enemy was the Serbian government, not the Serbian people. I would rather kill my own son than someone else, said Demaçi. What Hydajet learned in prison was that the Serbs and Albanians were in the same bind.

"It's all a prison here," he said, closing the album. "You just don't see the bars. I'm convinced the Serbs in Kosova will understand the truth one day. My only fear is that will be too late, and the government will repeat the tragedy of Bosnia. As we say, Arkan didn't come here for fun."

More aggrieved men and women filed into the room. Foreign human rights workers were recording stories, in French and German—of a woman beaten in front of her children, of a man shot for selling cigarettes. The poet gave me copies of several lists of missing and murdered Albanians.

"We are determined to find a peaceful solution," he said, "but the people are beginning to think freedom must be fought for. They say that if we don't fight we'll be the only ones who don't get their land. You must beware of a patient man when he finally becomes angry."

My hands were shaking as I wrote, and I was having trouble catching my breath.

"I hope the next time we meet we can talk about poetry," said Hydajet before we parted.

The banker-translator escorted me back to my hotel, bidding me farewell once we could see the building. On your next visit, he said, please stay at my house. It's not much, but at least it's not as dangerous as the Grand. And then he vanished into the crowd.

Shoeshiners were doing a brisk business. A line had formed at a kiosk named McDonald's. On every corner policemen wearing flak jackets were pointing their automatic weapons at passers-by. I went to the National Library, a modernist building enclosed in ornamental bars, like a giant cage. It was surrounded by mud, the front door was locked, several windows were broken.

I circled the city, peering into empty stores and closed museums, churches and a Muslim cemetery. At dusk I came to a plaza, in the middle of which stood a tall, white tri-pronged monument. All at once the sky was full of blackbirds, thousands upon thousands of blackbirds circling the white spires, cawing hysterically. This ominous scene brought to mind a distant Sunday morning in Seattle, when I had awakened from a troubling dream whose effects I could not shake off. The sky on that spring day had the greenish tinge of an impending hurricane, though the forecast was for clear weather, and from my front porch I saw nothing to suggest a storm brewing—except for the flock of barn swallows flying furiously up and down the street, zigzagging around the apple trees that lined the parking strip. I was staring at the birds, wondering what was going on, when my neighbor opened his door. Did you hear the news? he cried. Mount St. Helens had erupted.

The cackling blackbirds (reincarnated Serbian warriors, according to local legend) settled on the roof of a nearby apartment building. I did not want to spend another minute here—or in Kosovo, for that matter—but before going I wanted to learn the name of the monument. Off to one side of the plaza was a young couple, and it was not until I had introduced myself to them that I noticed the man's beard—a Četnik fashion statement in these parts; sewn into his leather jacket was a red patch emblazoned with Greater Serbia insignia. Though I was tempted to beg their pardon and leave, nevertheless I asked them what the monument was called. The man answered pleasantly enough.

"What does that mean?" I said.

"Togetherness," he replied.

"The three cultures?" I said. "Muslim, Catholic, and Orthodox?"

With a nod of his head he offered to write its name down for me. I opened my notebook, and several loose papers fell out—the lists of missing and murdered Albanians that Hydajet Hyseni had given me. My heart was pounding as I skipped over my Prištinë interviews in search of a blank page. When he had written out the name in Cyrillic (which I could not read), I thanked him profusely.

"Another time," he said, meaning, I presume, "Any time."

The blackbirds flew up into the sky again and just as quickly resettled on the roofs. Over a mosque's loudspeaker came the chanting of a muezzin calling the faithful to prayer. MIG fighter jets buzzed the city. Under the circumstances the Serb's slip of the tongue seemed appropriate.

14

Flight

"*Americano!*" the driver called.

I stumbled through the dark to the front of the crowded bus.

"Sofia," said the driver, staring straight ahead.

"Where?" I said.

"Sofia," he repeated, and pointed to an empty tram station.

I stepped out into the cold. It was not until the bus roared on toward Istanbul that I realized it was snowing. I dragged my duffel bag to the tram station and considered what to do.

Two a.m. I had been traveling on buses for eighteen hours. At the Kosovo-Macedonia border, when the Serbian policeman examining my passport asked me where I was coming from, I searched my memory for the name of the city I had been so happy to leave. An anxious minute passed before the man next to me said, "Prištinë." The policeman shook his head. It was some time before he returned my passport—and longer yet before we continued on to Skopje, where I boarded a chartered bus filled with Turks. The clerk who had directed me to it spoke no English. I hoped it was going to Sofia.

Icy, twisting roads. Squalid villages. Then a scene from hell: four hours at the Bulgarian border, in a smoke-filled, overheated bus. *The Untouchables,* dubbed in German, was playing on the video monitor. An old man argued at the top of his lungs with the driver. The couple in front of me gnawed on greasy chunks of meat, the toothless woman behind me coughed and burped. A short, loud woman walked down the aisle, demanding food and cigarettes. She stuffed pieces of meat into her mouth, lit a cigarette, and leered at me. My last meal had been in Prištinë, and my back was in a spasm. It took all my self-control to keep from shouting at the border guard who searched my duffel bag three separate times. Smugglers coming from Bulgaria were waved through.

Now I was somewhere on the outskirts of Sofia (which, as it happens, means wisdom), without a hotel reservation, stamping my feet in the snow. Trucks rumbled by. The wind picked up. An hour passed before I hailed a cab—the buses and trams had stopped running at midnight.

"Hotel," I told the cab driver.

"Sheraton?" he said.

"The Sheraton's fine," I said. A half hour later I was at the front desk of the first hotel in post-Communist Sofia to be refurbished.

I did not think to ask the clerk about the room rate until I had already checked in.

"Two hundred and fifty dollars," he smiled.

The four hours of fitful sleep I caught before being awakened by a misdirected telephone call were the most expensive in my life.

The Orthodox churches were being restored, and in the open air market stolen icons were for sale; even monks were hawking religious souvenirs. After I changed $100 with a black marketeer, a half-dozen money-changers descended on me, begging for my business. A retired economics professor from Sofia University was certain I would buy her entire collection of lace doilies and tablecloths.

"God watches out for me," she explained.

Another woman pointed at the economist's head. "She's touched," she said. Then she drew her finger across her own throat and laughed.

When I told the economist I had no money to spend, she gave me one of her doilies.

"I can't take this," I said.

"Of course you can," she said. "You're an American."

"Bulgarians are only free when they're enslaved," said the Western diplomat.

English was the language of choice in the lobby bar of the Sheraton. Foreign businessmen were exploring investment opportunities, and what they found, as elsewhere in the former Soviet Bloc, were inefficient state industries, scant interest in privatization, and no secondary markets. Bulgarians, said the diplomat, are a dishonest, inward people feeling their way forward.

"They're still trying to sort out what democracy means, politically and economically. They're historically pessimistic, and Communism destroyed any social consciousness. But they have hidden resources: a peasant mentality and peasant cunning, even in Sophia. Because this is still an agricultural country. They know how to hide what they have from each other—and from us."

Even so, the diplomat thought Bulgaria was acting responsibly during this Balkan crisis, despite irredentist claims on Macedonia from hardline nationalists, the kinds of claims that fueled the first Balkan wars. Indeed the government was pressing its neighbors to recognize Macedonia (though it refused to acknowledge any ethnic distinction between Macedonians and Bulgars). And its failure to prevent smugglers from using the Danube to deliver gasoline and other goods to Serbia was understandable: the lost income from the sanctions only magnified Bulgaria's economic crisis; and there was enough popular sentiment for the Serbian cause to make politicians wary of succumbing to international pressure to take action against the smugglers.

"Look," said the diplomat, "Bulgarians are a cautious, passive people. They know they'll have to live with Serbia after the UN leaves."

The demonstration was half-hearted. A small crowd marched up the street in a festive spirit. The peaceful fall of Bulgaria's Communist government in 1989 stood in marked contrast to the violent overthrow of the Ceauşescu regime in neighboring Romania, but since reform initiatives had come to a halt and the economy was in a steep decline, these demonstrators were blaming all their problems on their first non-Communist president. The policemen watching from the sidewalk smoked and laughed, others drifted through the crowd. I remembered something the diplomat had said: "Bulgaria has never had an extended period of happiness."

It turned out the Bulgarian king, Tsar Simeon II, was still alive.

He was named after Bulgaria's first tsar and most colorful ruler. During Simeon's reign (A.D. 893 to 927), the first Bulgarian literary works appeared (an account of the creation, lives of the saints, translations of legal texts, and liturgical works), and the first Bulgarian empire expanded to include Macedonia and most of Serbia. Trained for a religious career, Simeon detested knavery, wine, and sensual pleasure. What he liked was waging war, particularly on Byzantium. He occupied Adrianople and repeatedly threatened Constantinople, drove the Magyars into Hungary, and suffered disastrous defeat in Croatia. "God has enlightened the Slavs," wrote Hrabr, a Bulgarian monk of Simeon's time. The tsar devoted his life to enlarging that enlightened circle.

Simeon I met his death during an expedition to Constantinople. Legend has it, an astrologer in the Byzantine capital discovered in the forum a marble statue looking west: Simeon's double. If you strike the statue on the head, the astrologer told the emperor Romanus, Simeon will die. So the statue was smashed, the tsar's heart gave out within the hour, and within a hundred years Basil the Bulgar Slayer had annexed the Bulgarian empire, capturing and blinding the larger part of its army. One man in a hundred was spared an eye to escort home the 14,000 blinded prisoners. Samuel, the Bulgarian king, fainted "at this grisly sight," as one writer puts it, and two days later he was dead.

The second Bulgarian empire (1186–1396), which extended over the entire Balkan Peninsula (except Greece), fell victim first to Serbia, which conquered Macedonian Bulgaria, and then, after the Battle of Kosovo, to the Ottomans, who tried in vain to destroy Simeon's greatest legacies, the Bulgarian language and the Bulgarian Church. Not until the early nineteenth century were Bulgarian schools reopened, and by the time an autocephalous church was reestablished, in 1870, Bulgarian nationalism was in full force. With Russian help, the Bulgars were granted autonomy in the Ottoman Empire in 1876; in 1908, during the Young Turks' revolution, Bulgaria declared independence. In the First Balkan War Bulgaria joined with Serbia, Montenegro, and Greece to defeat the Turks, then turned on its allies, hoping to restore Greater Bulgaria, and was roundly defeated. Boris III, who came to power after World War One, established a dictatorship in 1935. Nevertheless when his wife produced an heir to the crown in 1937 Boris pardoned tax delinquents, released thousands of prisoners, and raised by one point the grades of all schoolchildren. The euphoria was short-lived. Boris sided with Hitler, hoping to recover lands in Romania and Macedonia, but his reign ended in 1943 when, after refusing to deliver Bulgaria's 50,000 Jews to Nazi concentration camps, he died under mysterious circumstances. Simeon II succeeded him, under a regency, which was continued the next year when Russian forces occupied Bulgaria. When the monarchy was abolished two years later in a Communist-led plebiscite, Simeon went into exile.

A life of waiting. For almost a half century Simeon lived in Spain, plotting his return. He married a Spanish heiress, gave his children Bulgarian names, and went into business. "I believe in *time*," he said. And some Bulgarians believed his time was at hand. A movement to restore the monarchy was gaining momentum, spurred by the government's inability to improve the standard of living and the growing sentiment that democracy did not work. What Bulgaria lacked was Simeon's authority, the monarchists argued. As he himself said, people were "a good nice *herd*, in the best sense, and normally they get to be quite naughty, and it's a herd of sheep which start biting or kicking at a given moment, which is also normal. But if you have somebody to lead them, they follow."

It was said the tsar was waiting for the right moment to return.

The cab driver could not find the address. Though he had precise directions, in Bulgarian, we drove along one street after another, waving down dozens of cars and pedestrians. No one could help us. One hour passed, and then another. The meter was still running. Finally, the cab driver stopped in the middle of the road and pointed to a tall building. I fished out the money I had changed on the black market, peeled off the first banknote, and discovered that my wad of leva was blank paper. I threw it down in disgust. The cab driver shrugged. The ride, he gave me to understand, was free.

"No one knows where anything is anymore," my host said after I had explained why I was so late. "Since the fall of Communism we've changed the names of all the streets and squares, even the bus numbers." He looked at me. "Never do anything on the black market in Bulgaria."

Luban, a third-year law student, was studying human rights law with a visiting American legal scholar. A remarkable development in a country still suspicious of its former enemy, and still suffering from the effects of "Bulgarization," a 1984 assimilation campaign in which Turkish language broadcasts and publications were abolished, mosques were closed and destroyed, and 800,000 ethnic Turks were forced to adopt Bulgarian names. The beleaguered minority responded with hunger strikes and violent demonstrations, which led the authorities to imprison activists and expel Turks—some 300,000 by the fall of 1989, when the government of Todor Zhikov was toppled. Indeed it was international pressure to stem the Turkish exodus that precipitated the Communists' demise.

"We're afraid of the Turks," Luban admitted. "They have such a different mentality than ours. We can't trust them, because we don't understand them."

Yet he was one of the few in his country who believed that protecting minority rights was the only way to insure peace in the Balkans. During this transition period Bulgarian nationalism, coupled with increasing economic hardship, would, he thought, probably lead to more violence against the Turks and Gypsies. What Bulgaria needed was a good constitution, said Luban, not mindless efforts to wipe out all traces of its Communist past. Someday there would be a functioning legal system, and he wanted to be one of its codifiers, though he knew nothing would change until the mafiosi and robber barons had made their fortunes. That was why he called himself a "tempered idealist."

"Bulgarians have been hurt so often it's hard to be otherwise."

After dinner, Luban and I took a bus back into the center of the city. Though a light rain was falling, which would soon turn to snow, blanketing Sofia by morning, it was pleasant to walk along the deserted streets. I had moved into a much cheaper room in the Grand Hotel, and I was in no hurry to return to it. So we went to see the Russian Orthodox Church, which was under renovation. Scaffolding surrounded the gold-plated onion dome.

"God is too high and freedom too far for us," Luban said before he disappeared into the dark.

Graffiti on the wall of the Orthodox seminary: *Some things are truly, and they never changes.*

Rousted from bed, naked, and all my belongings are gone—glasses, clothes, wallet, passport. I have no idea where I am, and whoever barged into my room to yank the blankets and sheets from my bed—whoever stands now at attention in the dark—will not speak. Just before I am to be taken away to some nameless destination I wake in a cold sweat. The room spins. Am I awake, or is this just another dream? Only when I switch on the light and see my glasses and duffel bag do I begin to calm down . . . And I dreamed this dream in every Balkan city I visited.

It was a modest house overlooking a dirt road—the Institute for Slavic Studies. In its climate-controlled basement were more than 600 ancient manuscripts and scrolls, in various stages of restoration and repair. An elderly woman opened an illuminated text from the tenth century, a Byzantine artifact perhaps secured by Tsar Simeon I himself. The elaborate script was figured on parchment and bound in leather; the religious paintings were done in gold. The scribes had used the blank pages in the back as a journal in which to describe a series of earthquakes and wars. The elderly woman urged me to touch the manuscript.

"Rub the page," she said. "It will bring you happiness."

Bears? Impossible. But when I looked again there were four honey-colored bears chained to the rocks below the sea wall. The largest, padding up and down, strained at its leash; another rocked back and forth, like a keening woman; two slept by the empty rowboats moored to the rocks. Freighters were riding at anchor in the Bosporus. A light snow was falling. One of the sleeping bears, rousing, stretched like a cat. I decided I had lost my mind.

Surprise, surprise.

The twelve-hour bus ride from Sofia—another smoky, unheated affair, complicated by severe stomach cramps and lengthy searches of my luggage at the Turkish border—had done me in. Arriving in Istanbul at daybreak, shaking with cold, I checked into a cheap hotel in the Old City and tried to sleep. But by nine, still queasy, I was out on a run, unable to contain my excitement over leaving the Slavic world. I made straight for the Bosporus and hopped onto the sea wall. A fresh wind off the Black Sea was at my back, and on my right were the remains of the land walls, dating from the fifth century, from which Byzantine defenders turned back more than twenty invasions. Persians, Arabs, Avars, Bulgars—no one could conquer the city that Constantine had made the capital of the Roman Empire in 330. Indeed the walls were not breached until 1204, during the Fourth Crusade, and then only by treachery. The Crusaders, taking advantage of Byzantine hospitality to scale the low walls in the harbor of the Golden Horn, sacked the city. After three days of killing, pillaging, and book-burning, they carried off the largest haul of booty since Creation, according to one chronicler, and dealt a deathblow to Byzantium. The Greek kingdom was restored in 1261, by which time Ottoman

Turks had begun to chip away at the empire; by the spring of 1453 only Constantinople remained in the sights of Mehmet II. Introducing the brass cannon to warfare, and with a vast naval force at his command, the Sultan besieged the New Rome from every side. He even circumvented the iron chain protecting the mouth of the Golden Horn: teams of oxen pulled a fleet of his ships up a steep hill and down into the harbor. Yet the city held out for nearly two months, and when it finally fell the Turks went on their own three-day rampage; its scale and cruelty dwarfed the ravages of the Crusaders. Women and children were raped, churches were plundered and desecrated. Blood stained the waters of the harbor. Historian Andrew Wheatcroft notes that a Venetian doctor who lived through the siege was reminded of the rotten melons that sometimes clogged the canals of his native city when he saw hundreds of severed heads bobbing in the water just offshore.

With disbelief I stared for several minutes at the bears before deciding there was nothing to do but continue running toward the Sea of Marmara and pray that on my return they were gone.

There is a short story by Tommaso Landolfi, in which a woman leaves Parisian society to live in a provincial castle, only to discover that in her new surroundings everyone hibernates. The villagers, her domestic help, a prospective suitor, all sleep through the winter in goatskin bags hung from beams in the ceiling. The woman cannot cook, and the horses—her only means of escape, unless they, too, have gone to sleep—frighten her, and so she falls into despair. That was how I felt about my winter among the Slavs. From the distance of this teeming metropolis, which once reined over the Balkans, the entire Slavic world seemed to be caught up in a nightmare, which in this season took the form of suspended animation. Everyone was waiting for something to happen.

I picked up my pace. Cold as it was, the longer I ran, the better I felt. It was fine to gaze at the Bosporus, a sheet of gunmetal grey stippled with freighters and fishing boats. Dividing Europe from Asia, the narrow strait takes its name from Greek mythology, Ford of the Cow. Zeus was in love with Io, and to protect her from his jealous wife, Hera, he changed her into a white heifer. Hera was not fooled. She sent a gadfly to torment the heifer, and Io fled through Europe, crossed this strait, and settled in Egypt. This old story of seeing through a rival's disguise and driving him or her out repeated itself with endless variations in the long history of palace intrigues that shaped this city. The Ottomans' bloody siege brought to a close 1,100 years of Byzantine rule, one glorious cultural tradition and unremitting chronicle of assassinations, battles, and massacres making way for another. The emperors were always going mad or meeting violent deaths. And it was little better for the Sultans, who paid for their excesses with their wits and necks.

But outsiders rarely saw through Ottoman disguises, which generally took three forms in the Western imagination: the Terrible Turk who carved out a vast empire and nearly conquered Vienna; the Lustful Turk trading the discipline of military life for the pleasures of the harem; and, as the empire declined, the Sick Man of Europe unable to stand up to his rebellious subjects. What sealed the Ottomans' doom, in fact, was an uprising in Bosnia in 1875. Serbia and Montenegro joined the Bosnian cause, and then Russia, which for three centuries had fought the Turks for control of the Balkans, the Caucasus, the Black Sea, and, most important, the Straits of the Dardanelles and the Bosporus, waged its most decisive military campaign in the region. The Ottomans lost not only Bosnia but also Bulgaria, Serbia, and Montenegro. And the losses mounted in the Balkan Wars and World War One until the empire was reduced to a small shell containing the modern Turkish state.

Expansion and contraction—one way to think about the vicissitudes of Turkish history and the character of a people who presented such contradictory faces to the world, now advancing on the Continent and Persia and Arabia, now turning inward. Another was to invoke the figure of Janus, the Roman god of beginnings, who looks both ways at once, in this case, to East and West. (It gave me a thrill to see Asia just across the water.) And the Bosporus itself, which connects an amazing variety of peoples and cultures, offered a third metaphor, since its currents run in opposite directions: the one on the surface flowing from the Black Sea, the other, forty meters down, bringing the saltier waters of the Sea of Marmara to the Black. Which current was stronger—what you saw or what you did not see, the fragile institutions and trappings of a secular state or the memory of Ottoman glory? This was a difficult moment in Turkish history, with Islamic fundamentalism on the rise, Kurds fighting for their own state, and relations with Greece falling apart. Meantime, the recent death of President Turgut Özal had plunged the country into the sort of crisis once dear to the Czars and now welcomed by Turkey's opportunistic Arab neighbors, Iran, Iraq, and Syria.

The run back along the sea wall, with the wind in my face, was slow going, and it was some time before I learned that my prayer had been answered—the bears were gone—which, of course, only increased my agitation. Perhaps I *had* lost my mind. But climbing the hill to my hotel I saw them again, in the middle of the street, only now there were ten bears, each led on a chain by a trainer.

What a marvelous world! These were the dancing bears of Istanbul.

This is still the center of the universe, said the Western diplomat, whether you call it Constantinople or Istanbul. Look at all the third- and fourth-generation Bosnians living here. They came when the empire collapsed, and now they're coming again. The Turks don't care about the Serb-Croat conflict. Bosnia's what they think about. It's another log to add to the fundamentalists' fire. He glanced up the street, where a crowd had gathered to watch the filming of *Indiana Jones, Part III.*

In the '30s, he said, this was a Greek part of the city, but with each flare-up over Cyprus more Greeks leave. That's why so many empty buildings are falling to pieces. The Greek owners either can't or won't return to fix them up or sell them. The Turks think the Greeks should realize that since 1453 this has been a Turkish city. The Greeks say that in the history of this region that's nothing. So the Turks make provocative flights into Greek air space, and Greek jets scramble and crash. It's not war and it's not peace. But the Greeks keep leaving. And old Istanbulans are afraid of the refugees coming from Anatolia, because they vote with the mosque. But the Kurds represent the greatest threat to the fabric of Turkish society; even Westernized Turks are getting sick of them. At least the Ottomans were decent about minorities, if not human rights, but now that the empire's down to Anatolia they're trying to figure out what a Turk is. There's no future for the Kurds.

The recent murder of newspaper columnist Uğur Mumcu had galvanized the intelligentsia. When the Islamic Jihad claimed responsibility for blowing him up in his car, secular Turks took to the streets. Four days ago, in Ankara, more than 200,000 people marched past the Iranian Embassy, proclaiming their devotion to the secular ideals enshrined by Mustafa Kemal Atatürk (1881–1938), the founder of the modern Turkish state. It was Atatürk (literally, father of the Turks), a daring military leader and nationalist, who organized resistance to the Treaty of Sèvres (1920), by which Ottoman holdings were reduced to northern Anatolia and the Allied-occupied Zone of the Straits. First, he reclaimed territory awarded to Armenia, then he routed the Greeks in southern Anatolia. The Treaty of Lausanne (1923), which gave an imprimatur to ethnic cleansing (800,000 Turks expelled from Bulgaria and Greece, 1.5 million Greeks from Anatolia), defined the borders of present-day Turkey (making no mention of the state the Kurds had finally won for themselves), which Atatürk rebuilt along Western lines. He replaced the Arabic script with the Latin alphabet and drove Islam out of public life—closing down religious orders, outlawing polygamy, forbidding men to wear the traditional fez. Though some restrictions on religious practice were lifted after Atatürk's death, three times since 1960 the military had staged coups to defend the secular system. Perhaps another was in the offing, given the corruption and fragmentation of the secular parties, the growing strength of the fundamentalists, and the post-Cold War pressure on Turkey to return to the Islamic sphere of influence. Still, a majority of Turks believed in Kemalism. *Long live the constitution*, Mumcu's mourners chanted in Ankara.

Turkey was not a safe place for journalists. Last year's figures on the war against the fourth estate included dozens killed and imprisoned and hundreds of newspapers and periodicals confiscated. Mumcu's murder ruined any possibility of a truce with the government, which was doing little to find the assassins, because most of the victims wrote for left-wing or pro-Kurdish journals; the Kurdish separatist insurgency's terrorist campaign provided the government with an excuse to clamp down on the media; security forces were implicated in some journalists' deaths. And there were stories of Kurdish activists being tortured and dying under mysterious circumstances while in custody, some of which Mumcu had reported on in his column in *Cumhuriyet,* a mainstream newspaper. But he had made enemies across the political spectrum, criticizing Islamic fundamentalism as well as government corruption, drug trafficking, and terrorism. Though two unknown Islamic groups also claimed responsibility for his death, most observers blamed Iranian intelligence agents.

One journalist trying to sort out the truth was Haluk Şahin, a close friend of Mumcu's. Şahin himself had received death threats, and on the rainy day I visited him he was in a pensive mood. *It's Not Easy Being a Turk*, the title of his first book, might have made a good caption for a photograph of this small wiry man who kept rising from his desk to call in his assistant. Şahin, who had traded the print medium to produce *Arena*, Turkey's top-rated investigative news program, was preoccupied with organizing a press conference to determine who was behind Mumcu's murder.

"It's fine to die for something you believe in," he said. "But for targets like my journalist friends and me it's frustrating not to know what the plot is. Who's using whom? Who's the master? You're just a piece they play at will."

The way he saw it, with terrorists to the south and the Balkans to the north, Turkey was in a critical position. The conditions were right for an Islamic revolution or a coup, and Şahin thought Iran, Syria, Greece, and the United States all had an interest in destabilizing the regime. If democracy failed in Turkey, the Great Powers could divide it up again, as they had at the end of the Ottoman era. And Turkey's fate was tied to Bosnia's. The West's indifference to Serbian atrocities bewildered secular Turks like Şahin.

"For us it's family," he said. "Almost everyone here has some Bosnian connection. They're central to our culture—you see many blond, green-eyed Turks—and they're being persecuted for religious reasons. Islam was our gift to them. We feel a little guilty about leaving them behind."

Western hypocrisy compounded the problem. During the Gulf War, Turkey had complied with Allied requests to shut down the pipeline carrying Iraqi oil to the Mediterranean and provide bases for American warplanes. Saddam Hussein was no hero to most Turks, but they believed he had been unduly punished for what seemed like a minor infraction when compared to the systematic rape of Bosnian Muslims. You did not have to go far to see how Islamic groups exploited the discrepancy. The perfidy of the West, a common theme in the mosques, boosted the sales of cassettes, compact disks, and videos glorifying Ottoman efforts to preserve Islam. And the longer the war dragged on, the larger the audience for the hundreds of publishing houses and journals, unlicensed radio and television stations, that had sprung up since the end of the Cold War to propagate Islamic theology.

"The Balkans remain in our cultural sphere," Şahin said. "Our outlook, our food, our music are Balkan. Yes, we've looked inward since the creation of the modern Turkish state. We'll protect every inch of our land without coveting anyone else's. But a change came with the oppression of the Turkish minority in Bulgaria. Before that we were afraid of Russia—the Big Bear. Then we started to speak out, and now we're sending weapons and troops to Bosnia."

His assistant was in the doorway with a question about the press conference. "I just hope to discover who wants to murder us," Şahin said in despair. But the press conference was destined to be canceled. No one would ever learn the identities of Uğur Mumcu's killers.

"Who, after all, speaks today of the annihilation of the Armenians?" Hitler asked when he sent his Death's Head units into Poland, with orders to kill every Jewish man, woman, and child. It was indeed the international community's indifference to the Armenian genocide in the Great War that emboldened the Nazi leader to carry out the Final Solution. Even now the full horror of the century's first exercise in ethnic cleansing remains unknown. What is known is that in 1915, when the Turks suffered disastrous losses at the hands of the Russians, they turned on their own subjects, the Armenians caught between the two armies. True, some were agitating for autonomy, others for a Russian victory, but most were loyal to Istanbul. Armenian soldiers fighting on the Ottoman side were nevertheless ordered to remove their uniforms and dig their own graves; the intelligentsia was hunted down; at least two million Armenians were killed or expelled from eastern Anatolia. Long columns of deportees set out for Syria, and for six months, in the words of one American witness, "practically all the highways in Asia Minor were crowded with these unearthly bands of exiles. They could be seen winding in and out of every valley and climbing up the sides of nearly every mountain—moving on and on, they scarcely knew whither, except that every road led to death." To save gunpowder, Turkish and Kurdish mobs attacked the deportees with axes, scythes, and saws; thousands died of exposure, starvation, and disease. The roads were lined with corpses, and suicides floated down the Euphrates. Yet no Turkish government had ever owned up to this massacre—one reason why History had perhaps already singled out an infant, asleep in his crib, destined to turn to his generals forty years hence and ask, Who, after all, speaks today of the annihilation of the Bosnians?

Everything was for sale. And the salesmen would not take no for an answer. A fast-talking opium dealer even followed me down the street until I disappeared into the labyrinth of the Covered Bazaar, where I wandered among the rugs and silks, leather jackets and luggage. There was so much *stuff* it made me dizzy. And once I found my way back to the street a boy lugging shoeshining equipment ran alongside me for two blocks, begging for my business. I waved him off until he leaned down, in full stride, and smeared red shoe polish across one of my brown shoes. Now you'll have to get them shined, he said with a smirk. Yes, I said, but not by you.

I'll have my day, Pinar Soyupak assured me at lunch. I had no reason to doubt her. She was a shrewd businesswoman, a Georgetown graduate who analyzed the Turkish economy for Citibank—no small task in a country with a 70 percent inflation rate, a thriving black market, and corruption at every level. The restaurant was emptying, and our conversation, ranging from the consequences of the bank's cost-cutting measures to the impossibility of collecting revenues from the open-air markets to the history of empires, had turned to the changing role of women in Turkish society. All the pretty young women you saw in my office, said Pinar, are *definitely* not secretaries.

Not so long ago the most important women belonged to the Sultan's harem. The concubines, inhabiting their mysterious world, wielded considerable influence. The harem (which means forbidden: few men ever visited it) was guarded by black eunuchs, who also acquired great power, though not nearly as much as the concubines (slave girls, often from the Caucasus) who delivered heirs. Once one became the Valida Sultana, the Sultan's mother, control over state matters was hers for the taking, as long as the sultan was young or incapacitated; his drunkenness and debauchery were thus encouraged; hence one commentator's description of the harem as "a cancer within the Ottoman system." The story of Ibrahim the Debauched (d. 1648) is emblematic. Condemned to death by his dying brother, when Ibrahim was spared—thanks to his mother's intervention—he was obliged, as the only Ottoman descendant, to produce an heir. He designed a special robe, sable on both sides, for sleeping with each of his twenty-four slaves—on the same day. Naturally, he went mad. He decided his pleasure must be proportional to the woman's size. Messengers were sent out to find the fattest women in the empire; and when a huge Armenian became the Sultan's favorite his mother had her strangled. The Valida Sultana then obtained a *fatwa* from the Islamic judges—The madman is ruining his empire!—with which to depose her son. She replaced him with her eight-year-old grandson, Mehmet IV, best remembered for his ill-fated siege of Vienna in 1683, which marked the end of Ottoman expansion. And now as another era came to a close women from the farther reaches of the former empire were again wreaking havoc in Turkish society.

"All the university-educated women," said Pinar, "coming from Romania to sell their bodies—they don't think of themselves as prostitutes. They're just trying to make enough money to go back and buy houses. But look what they're doing to family life!"

She ran her fingers through her hair. "The women in the villages have five children before they're thirty, and they don't care about their appearances. But I heard of one who learned her husband was visiting these new women. So she got her hair done, bought some nice clothes, and met him at the door when he came from work. What do you think happened? He got mad at her. 'Why do you want to look like a Romanian woman?' he yelled. 'Get rid of that makeup!' But the women who used to feed their husband and children then eat by themselves in the kitchen are now sitting at the same table. And now we're about to elect our first female prime minister."

Pinar herself had her sights set on high office, though she had passed up the traditional route into Turkish politics, the Foreign Service, because, as she explained, diplomats were always in danger of having their heads chopped off.

Figuratively speaking, she said with a grin.

The stereotype had lost none of its force on the Continent. This year the EC was to decide on Turkey's suitability for membership, but European politicians were finding reasons—human rights, the economy, Cyprus—not to act on its application. "The real reason," according to the late President Özal, "is that we are Muslim, and they are Christian." Indeed the memory of Turks at the gates of Vienna burned brighter for many Europeans than that of Napoleon's military campaigns or of the two world wars. Though the Ottomans were more enlightened in their treatment of their subjects than their European contemporaries, in Continental mythology they remained barbarians. It would take a collective leap of imagination for Europe to accept Turkey into its common house.

Turkey had thus turned to its near abroad—Azerbaijan and the Turkic-speaking Central Asian republics of the former Soviet Union: Uzbekistan, Turkmenistan, Kazakhstan, and Kyrgyzstan—to forge new political and economic ties. Low-interest loans, humanitarian relief, joint business ventures were Turkey's weapons in the war for the hearts and souls of its eastern neighbors, along with a satellite television station, telecommunications equipment, and scholarships to Turkish universities. Businessmen, diplomats, and military officers from the Caucasus and Central Asia were traveling to Ankara instead of Moscow for training, and teachers of Turkish were being sent abroad. A pipeline to carry oil from Azerbaijan and the Central Asian republics to the Mediterranean was in the works. Empire-building, in short, in the modern guise of cultural affinity.

"Do you know that Turkish is the only language spoken all the way to Japan?" Pinar said with a gleam in her eye. "A Turk can travel from here to the Pacific without an interpreter. Imagine what that will mean in the New World Order. Politics is economics now."

The Greek poet George Seferis called Hagia Sofia "a spaciousness of lines that breathe," a good description not only of its architecture but also of its multilayered history. Built by Constantius in 360, Hagia Sofia (Divine Wisdom) embodies the principle of revision: the Byzantines kept restoring and rededicating the domed basilica; the Ottomans added minarets and turned it into a mosque; Atatürk converted it into a museum. And in the interior of the vast edifice, the spiritual center for Byzantium as well as for the Ottomans, are lines of column and arch and window, of Christian motif and Islamic decoration, of light filtering in at every angle, all quickened by the sense of history on the march. How much was consecrated within these walls! First the images—mosaics of the archangel Gabriel with a missing wing, of Christ flanked by the Virgin Mary and John the Baptist, of emperors and empresses—then, in green panels, the words—the Holy Names of Allah and Mohammed, of the Caliphs and the Prophet's grandchildren. The poet's words about Hagia Sofia applied to every large undertaking, religious, political, and cultural. Istanbul, for example. On this overcast afternoon, when police cars and ambulances jammed the streets and the sidewalks were overflowing with peddlers and men in blue skullcaps and women wearing silk scarves, I saw the city as a series of breathing lines laid atop seven hills—lines of thought and desire, one of which I followed to the nearby Topkapi Palace.

Mehmet II built this enormous (700,000 square meters) structure, which housed the sultans and harem until the nineteenth century. I went from the courtyards where the entertainment used to consist of parades and public executions to the rooms in which great and terrible decisions were taken and examined the collections of porcelain and silverware, sacred relics and weapons, clocks and calligraphy. I lingered in the harem, thinking: what a glorious waste of power. And: a people who did this once may do it again. By the end of my tour I was very tired.

From the palace I strolled along the Bosporus and crossed a bridge over the Golden Horn. I wanted a good look at the Black Sea, though it was pleasant just to wander under the cypresses and umbrella pines lining the road. A glance at the magnificent gardens of the Dolmabahçe Palace, where Atatürk had died, convinced me to save the tour of the sultans' last home for another day; and when I regained the sidewalk a beautiful woman was striding by. I snapped to attention. Into the marina I followed her and onto a ferryboat, which in minutes brought me to the Asian quarter of Üsküdar, once known as the City of Gold, where all the roads of Asia Minor used to end—and pilgrimages to Mecca began. Next to the dock were old women hawking flowers and fishermen displaying their silvery catch. A freighter steamed toward the Black Sea. A stiff wind blew into shore, and as I pulled up the collar of my overcoat the woman who had caught my fancy disappeared into the crowd.

How easily borders dissolve.

Yes, I agreed. Very beautiful.

The rug, woven by an old Kurdish woman, was patterned after a palace floor plan—three red-and-green rectangles, with quadrangles jutting out of either end, connected by short passageways. Animals of every color and shape surrounded the geometrical figures, an imaginary bestiary on a white background. The border was made up of crosses.

And the price is right, said Omar. You know that, of course.

Of course, I said.

To escape the rain, I came to his shop every afternoon to drink tea. There were three things he liked to talk about: rugs, women, and the war. Omar would unroll his rugs, each with its story—I bought this from an Azerbaijani woman. She traveled here forty hours on a bus. You can hang it from a door. And I would say, The Serbs just blew up a dam. Croatia's going back to war. Extraordinary, really, but I can't afford it. Undeterred, Omar would pour me another cup of tea and show me rugs until they were piled up to my knees. My favorite, the one with the design of the palace floor plan, inevitably landed on top. Turks are fighting and perhaps dying in Bosnia, Omar would say, clicking his tongue. No one knows what's happened to them.

He came from a Kurdish village in eastern Anatolia, a handsome young man with a wardrobe of expensive Italian clothes—hand-me-downs from his older brother, the first in his family to move to the city to sell rugs. Here was a story Omar loved: when a Canadian businessman cheated his brother out of his savings, he became the gigolo to a vacationing German woman who promised him $200,000 if he would introduce her to his parents. But taking her back to his village was out of the question—his father would kill him—so he rounded up Kurds from the neighborhood, brought them to his shop, and told the German woman they were his family. The ruse worked. With her money he opened a string of shops, including Omar's, and became a millionaire many times over.

An unusual story, no? Omar smiled.

The story of the Kurd who had buttonholed me into taking him to lunch earlier in the day was more typical. He was a gaunt man, with bad teeth and no prospects, who had moved here in search of a better life and instead had joined the ranks of the dispossessed, Kurdish and Turkish, swelling the slums. While poorer Turks found solace in the mosque and the dream of an Islamic republic, this man was one of millions of Kurds praying for the restoration of the state promised them in the Treaty of Sèvres. But I was hardly the first foreigner to hear his diatribe against the government, and he was still eating when I left him, pleading another appointment—with Omar, as it happened.

The Kurdish problem interested the rug merchant only insofar as it drove tourists away. He thought of himself as a Turk, having lived here long enough to ruin his eyes repairing carpets; his spare time was devoted to the baths and belly dancers. For he was one of Cavafy's "champions of pleasure," and of the hundreds of women Omar claimed to have slept with he remembered one most vividly—an American writer, researching a book on devils, whose nose he had punched in a jealous fit. And, yes, he had slept with Romanians, he admitted sheepishly. It was because his country was changing. You could even watch pornography on TV. He just hoped he did not have AIDS.

"Turkey's in the worst place in the world," he said, "but with the Turks in Russia we could be like the Ottomans again. Just think: a hundred and sixty million Turks waiting to join up with us."

Omar rolled up my favorite rug. He was, of course, a skillful salesman. Look, he said, it will fit into your duffel bag. And it did.

Light. I could not get enough of it. Leaving my hotel early in the morning, I walked all over Athens, sun-dazzled, soaking up the light like a plant removed from a darkened room. As I climbed the steep hill to St. George's Church, an entry from Seferis's journal came to mind: "This light, this landscape, these days start to threaten me seriously. I close the shutters so I can work. I must protect myself from beauty, as the English from the rain and the Bedouins from the chamsin. You feel your brain emptying and lightening; the long day absorbs it. Today I understand why Homer was blind; if he had had eyes he wouldn't have written anything. He saw once, for a *limited* period of time, then saw no more." At the church were three Albanian refugees, a sculptor and his students, studying the Acropolis through binoculars. Why risk visiting it in person? The Greek government wanted to send its half million Albanian refugees back to their homeland.

Athens was in its third year of drought, and I was glad my afternoon appointment took me to the suburb of Glyfada, by the sea. The salt air, the palms, the mimosas flowering by the front gate to the handsome three-level house, all made me eager for conversation. And Despina Meimaroglou was hungry for talk. She was the sort of well-to-do artist, schooled in Marx, whose passion for left-wing ideas might have made her ridiculous, if not for the rigor of her thought. Indeed the political cast to her work was rooted in dispossession. Her father was one of the thousands of Greek refugees who fled from Smyrna when it fell to the Turks in 1922 and was destroyed by fire. She herself was an exile from Alexandria, once the center of Hellenic and Jewish culture and now, in her words, just another Arab city.

Her living room was an essay in the cosmopolitan spirit she prized. Persian miniatures hung on her walls alongside her own paintings and pages of calligraphy from the Koran. A display case was filled with pocket watches; in one corner were stacked all manner of gilded frames, a favorite subject. She made paintings of frames suspended in the air; photographed frames and used a photocopier to enlarge the images, which she then transferred onto linocuts; placed frames in different settings to explore the ways we discover and invent meaning. I live and move inside a frame created by society, she said, setting out a tray of coffee and chocolate cake, but inside that frame anything can happen. In her daily life she had the frames of family (a husband, two teenaged sons) and her fellow artists' nationalist fervor to contend with. In her creative hours the media's framing devices preoccupied her.

"All the atrocities we see on television, so close to us—we get it all as an image," Despina said. "Nothing effects us. We watch it, and that's it. We don't choose what we see. Someone else sorts it out for us. I want to comment on that in my work."

Her method was to reconstruct images—of a field in Aix-en-Provence, of a dead terrorist, of refugees from Bosnia—in order to reveal their enduring qualities. Trained in England as a printmaker and painter, when she took a job as a graphic designer for an advertising agency she learned to focus on single details until they assumed dramatic shapes of their own—a process suitable for her creative work, too. After dividing a photographic image into rectangular segments, she would enlarge the segments on a color photocopier and rearrange them, abstracting the original image by fragmenting and blurring certain elements of it. Nearsighted, Despina liked to look at objects without her glasses on; the haze through which she saw them helped her to forge new connections between traditional masterpieces and contemporary events. *Montagne Sainte Victoire,* then, her rendering of a place made famous by Cézanne, began with a snapshot, taken by chance, of a field with a view of the mountain in the distance, and initiated a conversation with Impressionism, conducted with new technology. In *Deposition* she juxtaposed a photograph of a Greek terrorist slumped in a policeman's arms with Caravaggio's *Entombment of Christ*, offending some of her countrymen. What caught my attention was a work in progress: an enlarged photocopy of a picture of a mother and child from Sarajevo, superimposed on the figure of Mary clutching the infant Jesus in Carracci's *Landscape with the Flight into Egypt*. It was a question of layering, one image on top of another, just as the story of Joseph and Mary spiriting the Savior away could be read as a New Testament version of Exodus. Exile was the theme, however short-lived the political component of the image might be.

"The image itself," said Despina, "is timeless."

"Our word for freedom means to walk toward the sacred," said the philosopher Emilios Bouratinos. He searched through his briefcase for a pen, undistracted by the noise in the crowded hotel lobby. In this setting he might have been mistaken for a traveling salesman, if not for the books in his briefcase. "Early on," he said, writing out the word *eleutheria* (from Eleusis, the site of the mysteries), "I realized that a careful study of ancient Greek could teach me everything, because even if you exhaust all the possible meanings of a word you can still find new contexts for it and new meanings. But ever since Cicero translated our philosophers we have studied them through a Latin filter. We Greeks are schizophrenic. We had four centuries of Ottomans, and before that Venetians, Byzantines, Franks, et cetera. Because our modern state was a Western creation, designed to needle the Ottomans, our first intellectuals, in the nineteenth century, learned about ancient Greece not from the sources but from Western scholarship." The philosopher had thus devoted his life to reinterpreting ancient Greek thought. In his seminars on the science of consciousness he revived a custom practiced in the Agora twenty-five centuries before, when citizens gathered to discuss anything they wanted—politics, religion, economics, health, metaphysics. It was out of the question to think one opinion was more valuable than another; hence a field would develop through the discussions, in which the right answer would emerge, like a bird taking flight, an answer the authorities had to abide by. But democracy failed right from the beginning, he said, when the schools of rhetoricians and sophists destroyed the idea of discussion. "The purpose was no longer to find the truth but to sell your ideas. In Greek, truth means to lift the veil of slumber, not the Latin idea of *veritas*: what can be proved, not what moves man. The Greek idea of truth had no content; its essence was awareness, not what you are aware of. You must walk to the sacred, to freedom; no one can do it for you. But the modern citizen says to the politician, If you'll let me run my life the way I want, I'll let you do whatever you want. When individuals hand over their consciences to others, freedom, democracy, and truth are perverted. And the problems in the Balkans are the price of our inadequate understanding of those very words."

My last stop in Athens was the Hellenic Foundation for Defense and Foreign Policy. I had traveled enough in the Balkans to expect only partisan reflections from my host, Thanos Veremis, a political scientist, so it came as a surprise when he took a sober view of the conflict with Macedonia. Very legitimate, was how he described Macedonia's birth as a nation, which, he went on to say, was no more artificial than any other political entity, including Yugoslavia. Indeed he thought it was a brilliant stroke on Tito's part to encourage a Macedonian ethnic identity, an incentive for the local population to side with the Partisans and thwart Bulgarian claims on the region. After all, Basil the Bulgar Slayer was a Macedonian, not a Macedonian Slayer, Veremis smiled.

He gave me a copy of his collection of essays on Greek nationalism. In the introduction I read, "If all nations are imagined political communities, their constituent elements must be sought in the works of their most imaginative members, the intellectuals." I sat up straight. Benedict Anderson's work of political theory, *Imagined Communities*, had provided one theme for my Balkan travels, and Veremis explored the sense of nationality from the same perspective. The problem, he said, is that intellectuals like to appropriate cultures and histories other than their own. Then all hell breaks loose. Thus a geographical name—Northern Macedonia, Vardar Macedonia—might offer the best solution to the crisis. Anything smacking of irredentism was to be avoided. Better to promote human rights over statehood. Ethnic rights and autonomy led only to disaster.

"We need a universal idea," he insisted, "like the French Revolution or your Constitution. We can't ignore smaller groups anywhere, since nation-states are collapsing. All rights must be accepted, except the idea of creating states, which would lead to another medieval state of affairs."

The Balkans were approaching such a state. The war had caught the West unawares, because it was preoccupied with Russia. Then Germany's recognition of Slovenia and Croatia had condemned Bosnia, and now the Balkans were destined to remain a backwater of Europe, especially with Clinton unable to find his bearings on foreign policy issues. His first presidential statement on the war—the Vance-Owen peace plan might "work to the immediate and long-term disadvantage of the Bosnian Muslims"—frightened Veremis. Any Western encouragement of the Bosnians threatened Serbia, and the destruction of Serbia would leave a power vacuum in the Balkans. Imagine planting an Islamic power as the administrative center of this region, he said. Then you're back to Turkish Islam versus Russian Orthodoxy. The prospect of fragmentation beyond control was in his view quite real.

"A Turk came by this morning," said Veremis, "and he was saying, Wouldn't it be wonderful to revive the Balkan Pact? Yes, I told him, but where are the leaders with the vision to do that?"

The Balkan Entente (1934), an alliance of Greece, Turkey, Yugoslavia, and Romania designed to safeguard their borders against Bulgarian revisionism, had dissolved with the outbreak of World War Two. A new entente was taking shape in the Balkans, an Orthodox alliance of Bulgaria, Greece, and Serbia to counter, on the one hand, Slovenian and Croatian integration into Western Europe and, on the other, Turkey's efforts to reassume its traditional role as protector of Albanian and Bosnian Muslims. It was a classic fault line war exacerbated by the poverty of political leadership in the Balkans as well as in the international community. Veremis's feelings for one politician, however, brought me up short. Milošević, he said matter-of-factly, is our best hope for peace.

I could not believe my ears. What he had said until now lulled me into thinking I had finally met an objective Balkan thinker. Then he made another bizarre suggestion: bring in an accountant with no memory of anything before 1989, show him how many have been killed, raped, and maimed, and have him calculate how many more will be injured.

"He can think in concrete terms," said Veremis. "Politicians can't, because they're crusaders. I'm more concerned with atrocities to come than with what's already happened. And Milošević"—he began to twirl his pen—"Milošević is the best hope of appeasement in Serbia."

"But he's a war criminal," I protested.

"I'd rather let war criminals go free than let the war drag on," he replied. "Without a strongman in the Balkans the Big Bear will move in."

"What?" I said, incredulous.

He gave me a significant look. "We need Milošević."

Part III

May 1993–April 1996

15

Sarajevo I

30 May 1993

The shelling begins long before daybreak. I pull my sleeping bag up to my chin, dreaming of a thunderstorm. But when a shell screams past our house on Koševo Hill, landing less than a hundred meters away, I go downstairs to the living room, where three relief workers have rolled out their sleeping bags. A fourth, Pat Reed, is pacing in the hall. She says it is better to have a floor between us and the Serbs. So I lie in the dark, listening to a weird mixture of sounds: artillery and birdsong. First a series of blasts, then silence, then the birds resume singing.

"A good day to donate blood," John McCormick mutters.

It was a BBC report about a group of Bosnians trudging through the mountains, fleeing the Serbs, that convinced John to give up environmental lobbying and come to Sarajevo. One detail from that report haunts him: a woman recovering from childbirth left her newborn in the snow. But his original plan of working for a children's relief organization came to nothing. And his latest idea—to restore the sewage treatment plant—is unfeasible: the plant is in Serbian hands. If John has yet to figure out how to stave off outbreaks of dysentery, cholera, and hepatitis, he has nevertheless picked up some useful information. On my first day here he spread out a map to show me the most dangerous places in the city. With each explosion now his breathing grows heavier.

"They're slaughtering *kids* up there," he sighs.

Machine guns start firing in the hills surrounding the city, where Serbian shells are softening Bosnian front line positions. The shooting sounds like a hailstorm on a tin roof.

"Fucking Četniks," Vinnie Gamberale grumbles. A soft-spoken Australian with a military background, he has no respect for the armies fighting in Bosnia. "They're a drunken lot," he likes to say. "They have no discipline." This morning even Vinnie cannot sleep through all the noise.

"Did you hear the whistle from that shell?" says Vic Tanner, whose most recent assignment was to Somalia. "I hate that."

"When I look out at the ruins," says Pat Reed, "all I can think is, What a senseless war."

"This is the end of Sarajevo," says John.

"This is the end of something," says Vic.

Vinnie climbs out of his sleeping bag.

"Careful," says Pat. "I'm an old woman. My heart can take the shelling, but not the sight of Vinnie in the nude."

At daybreak, when there is a pause in the shelling, I walk out into the haze and sit with John McCormick and Muha, a technician now commanding a squadron of troops on Sarajevo's western front. Muha is off this week, since there are not enough guns to go around. During his leave he studies English. John is teaching him grammar. This morning they are working on prepositions.

"*The book is under the table*," John says to Muha. "Which is the preposition in that phrase?"

There is a detonation nearby. I flinch.

"No problem," says Muha. "That's outgoing."

Since my last journey to the Balkans, the Clinton administration has adopted a feckless policy on Bosnia. The new president has reservations about the Vance-Owen peace plan—and no alternative. One week he fears that "the terrible principle of 'ethnic cleansing' will be validated, that one ethnic group can butcher another if they're strong enough," the next week he urges the Bosnians to negotiate within the framework of Vance-Owen. His campaign promise to lift the arms embargo on the Bosnians and use NATO air strikes on Serbian positions, which during his first weeks in office divided Europeans and gave the Bosnians false hope, is one of many he will break. Tyrants everywhere take note of Clinton's indecisiveness. Warren Zimmerman, our last ambassador to Yugoslavia, is not alone in believing the president has squandered American authority.

"I don't believe we can call ourselves the leader of the free world or the greatest superpower anymore," he said not long ago. "It rings hollow now because we haven't shown the fortitude that goes with that."

The moral dimension of American policy is also missing, as Elie Wiesel publicly reminded Clinton at the April opening of the Holocaust Museum in Washington. "I have been to the former Yugoslavia," cried the Nobel peace laureate, "and Mr. President I cannot not tell you something: we must do something to stop the bloodshed in that country." The president has a moral imperative to end the genocide in Bosnia, Wiesel went on to tell State Department officials. The president has a higher moral obligation, they replied: to maintain the liberal coalition he assembled to win the 1992 election. Bosnia is not worth sacrificing his presidency over.

Nor is Clinton alone in acting feebly. Though NATO planes patrol the skies over Bosnia to enforce the no-fly zone recently mandated by the UN Security Council, Operation Deny Flight, at a cost of $2 million a day in fuel alone, is just an exercise in overflying. Three weeks ago, the UN declared Sarajevo and the besieged cities of Bihać, Goražde, Srebrenica, Tuzla, and Žepa "safe areas," not "safe havens," the stronger designation, because UNPROFOR, the UN peacekeeping force, is not prepared to defend them. Bosnian Serb General Ratko Mladić, who herded tens of thousands of Bosnians into the eastern enclaves of Goražde, Srebrenica, and Žepa then shelled them with impunity, happily agreed to the UN's terms: among other things they required the Bosnians in the "safe areas" to disarm. Mladić knows he can overrun them whenever he likes.

Indeed the peacekeeping mandate will remain weak as long as the United States refuses to send troops to Bosnia, a Bush administration decision continued by Clinton. This only worsens relations with our NATO allies, particularly Britain and France, the countries with the most troops on the ground. Like the arms embargo, UNPROFOR is a fig leaf for the West's inability to act decisively, reinforcing the status quo in Bosnia, i.e., the Serbian military advantage. Milošević's reaction is telling: "I appreciate very much that the United States will not be the world's policeman, to put everything in order in its own view." And the disarray in the international community has prompted Bosnian Croats, supported by Zagreb, to turn against the Bosnian government, overrunning Muslim villages in central Bosnia and besieging the Muslim side of Mostar, the capital of Herzegovina. The plan to divide Bosnia between Serbia and Croatia, hatched by Milošević and Tuđman, is nearing completion.

Meanwhile, despite American ambivalence about the peace plan, Vance and Owen convened a conference in Athens earlier this month to bring the warring parties together. Milošević, playing peacemaker now that Greater Serbia is within his grasp, persuaded Radovan Karadžić to sign on to the plan. The Bosnian Serb leader did so knowing that his assembly in Pale would reject it.

"This is a happy day," Lord Owen announced at the conclusion of the conference, "and let's hope that this does mark the moment of an irreversible peace process for Bosnia-Herzegovina."

He could not have been more wrong.

Bosnia suffered another blow this month when Secretary of State Warren Christopher, traveling to several European capitals, failed to convince anyone to endorse Clinton's plan to lift and strike. Christopher's lackluster performance surprised European diplomats accustomed to receiving their marching orders from Washington, and infuriated the Bosnians. The Serbs hold 70 percent of Bosnia, and with no prospect of military support from the West the Bosnians launched an offensive three days ago to break the siege of the capital, now in its fourteenth month, where there is still no water or electricity. A Serbian counteroffensive is what awakened us on this first anniversary of the Security Council decision to impose sanctions on Serbia and Montenegro.

Just beyond the small apple orchard in our front yard is an elementary school—yesterday I heard the thwack-thwack of two Bosnian soldiers stroking a tennis ball on the playground. Across the valley, 400 meters away, houses climb the side of a hill; a Serbian shell struck one house last evening, killing eleven. At the foot of the hill, the gutted railroad station; at the top, the mangled radio and TV tower the JNA use for target practice, says Muha. From the backyard we can see the Olympic hockey stadium destroyed by the Serbs and the soccer field that has become a cemetery. The park below us was thickly wooded until last fall, when Sarajevans began cutting down the trees to heat their houses and apartments. New grave markers fill some of the empty space. Above the park is Koševo Hospital, where in the last year more than 20,000 wounded have been treated, the unpaid surgeons operating around the clock, often by flashlight. The hospital floors are streaked with blood—no water can be spared to clean them—and the refrigerated space in the morgue is too small to handle all the dead.

"This winter there were so many wounded they just dumped the amputated parts out with the trash," says John. "When the dogs got into the trash heaps they turned wild."

Muha shrugs. He is a wiry young man in blue jeans, with a patch of black hair growing on his Adam's apple. Shells fly overhead in both directions. "Stereo," he says with toothless grin. "The shooting at night is beautiful." He pulls out a knife and waves it at the hills. "Fire," he commands.

A shell lands by the hospital, striking a building (which houses an army logistical center) at the College of Medicine. Smoke rises into the air. "That's life," Muha says in a steady voice. But a moment later he is crying, "Why? Why no America? Why no Europe?" Another college building is hit. "What are you doing, Četniks?" Muha demands. "Sunday is no good for this."

He points at Mount Trebević, the steep face that forms a natural boundary on the southern edge of the city. "Četniks," he says. He points to the southwest, at Mount Igman. "Četniks," he says again. This he solemnly repeats, turning in a circle under the grape leaves, for each hill and mountain around Sarajevo. "Četniks everywhere."

A NATO jet roars overhead.
"Taking pictures," Muha sighs, "like a tourist."
At the pop of an outgoing round I flinch again.
"It's good," says Muha. "It's good."

Sarajevo was known as the Golden Valley before it fell to the Turks in 1429. From its founding in 1263 the frontier town was said to be blessed with wealth. The surrounding mountains yielded no gold, but merchants from Dubrovnik made fortunes mining iron ore. Under Ottoman rule Sarajevo became one of the empire's major cities, rivaled only by Thessaloníki and Edirne. It was far enough from Constantinople to insure some autonomy, and since it was a crossroads—the Bosna River, from which the country takes its name, has its source just outside the city and runs north into the Sava—trade flourished here. Even Sarajevo's name marks the way a variety of people, settling or passing through, gave the city its cosmopolitan air: *serai*, which descends from Turkish and Persian sources for inn, shares a root with the English word, caravansary.

There is a legend that Turkish forces conquered the town by a ruse, setting thousands of fires in a nearby mountain village to frighten the soldiers in the garrison. The Sultan swept into the Golden Valley from the north, and at the sight of the Bosnians he cried, "*Kos! Kos!*"—"Run! Run!" So the part of the city where I am staying is called Koševo. And the village in which the fires were set the Turks named Pale: to light. A circle closed, then, half a millennium later, when Pale, now a ski resort, became the capital of the self-styled Republika Srpska, from which the siege of Sarajevo is being directed, a systematic violation of human rights, according to international legal scholars. In fact, the Security Council has just established an International War Crimes Tribunal at The Hague to investigate atrocities, including the indiscriminate shelling of cities. The Serbian gunners targeting Sarajevo this morning do not seem to be afraid of being charged for their crimes.

I am in the bathroom when a shell strikes the building next door. I rush out into the hall, pulling my pants up, mindful of Erasmus, the Slovenian knight killed, by cannon fire, in a squatting position.

On our way to the basement Pat says, "Whenever I go to the bathroom I wonder if this will be the time they get me."

Next to the garage is a small room with a mattress pushed up against the picture window. An open fire is burning from a gas line dangling in the wood stove, heating water for coffee; on the table are freesias and oleander blossoms in a vase. An old woman who has just come in from the street is holding her hand over her heart; her daughter, rubbing her shoulder, keeps looking at her leg, as if to see if it is still there. Another shell lands nearby.

"If that's how they're are going to retaliate for every little bit of land we get," says Mirna, a young woman with cropped blond hair and bad skin, "let them just take the city."

Mirna is curled up on a cot, clutching her arms and pouting. She and her husband, Muha, moved into the basement when the International Rescue Committee (IRC) rented the house. A pharmacist by training, Mirna helps her father cook for the relief workers; her mother does the cleaning. She takes a kind of perverse pleasure in the knowledge that her father built the munitions factories supplying some of the shells landing all around us.

"Last night we drank a bottle of whiskey," Mirna says. "And when Muha tried to wake me up this morning I couldn't move, not until I heard that old woman screaming."

Muha smiles at her. She looks away. At the sound of a third explosion, this one less than seventy-five meters from the house, Mirna throws up her hands.

"Every night last summer three shells would hit somewhere in the neighborhood. I lost my mind," she says. "Once I tried to run outside, but my aunt held me back. She asked my father for some water so she could give me a pill. But his hands shook so much he spilled it all over her."

"The good thing is, it will probably be quiet for a while," says Pat. "They usually fire in threes. Don't ask me why. Of course I could be proved wrong at any moment, and that would be that."

The old woman and her daughter bow to us and leave.

"This is the kind of day the Četniks fire two thousand grenades, and we fire twenty," says Mirna. "That's how we spent last summer—each family in its cellar, with the music turned up as loud as it would go to drown out the shelling."

Muha switches on the radio. Sting is singing "Message in a Bottle," after which Vatican Radio broadcasts a Mass.

"It's a big day," says Muha. "I may have to go back to the front tonight."

"We have a friend on an anti-tank brigade," says Mirna. "He can see a Četnik get into his tank, fire a grenade, then climb back out and lie in the sun."

"The old part of town is suffering the most today," says Muha. "It's just old people there."

"The ninety-year-old man next door has lived through five wars," says Mirna. "This is the worst, he says, because wars used to be fought between armies, and now it's soldiers against civilians. The Četniks will kill a thousand today if we fight back just to show us how futile it is to do anything."

Then the pop of an outgoing mortar.

"I have a feeling we'll be here for a while," says Pat.

"No tennis today," says Slaven, "only grenades."

He is standing in the doorway to the basement. Bosnia's former top-ranked junior tennis player works as a guard for IRC. Before the war, when injuries cut short his playing career, he made good money as a teaching professional in Germany. Now he is a heavy smoker who writes folk songs in his free time. Slaven hopes to work as a musician after the war.

"It will be difficult," he admits. "In Bosnia there may be only twenty thousand people to buy my records. In Yugoslavia there were ten times as many. So I may have to keep teaching tennis: fifty-five minutes on the court and five minutes to smoke!"

His wife and son are living in Belgrade; his marriage ended before the war, and not because she is Serbian. "Too much money, too much girls," he explains, showing me a photograph of the woman who"ruined" his marriage. "My wife's very angry at me."

He looks at my tape recorder.

"I have one of them," he says. "I put it under the bed when I make love to girls."

"Really," I say.

He tunes the radio to hear the news. "One of our commanders has lost his mind," he says after a minute. "They ordered him to stop firing his mortar, but he won't listen. Unfortunately, he's very close to us." He lights a cigarette. "It's bad," he says. "The news is very bad. The Četniks have retaken three positions and Sting is having troubles with his voice."

Everyone wants to hear about Damir's adventure.

"The other night I was walking home late," IRC's engineer begins. "I came to an intersection, and four cars converged on me, just like in the movies. They looked at my blue card [the identification card distributed to humanitarians and journalists by the UN High Commissioner for Refugees or UNHCR]. 'Why are you working for them?' they said. 'You should be defending your country.' They put me in a car with eight others and drove us up into the hills. In the morning I was digging trenches a hundred meters from the Četniks."

The commanding officer of Damir's work detail has a reputation for never excusing anyone from digging trenches. But by that evening IRC pressure on the government won the engineer his freedom; after all, he is restoring gas and water to the city.

"They took me to see the commander," says Damir. "I thought he would blackmail me or make me smuggle something for him. But he just wanted to know who caused him to spend all day on the telephone, and then he let me go."

Damir's story is exceptional in only one respect: his quick release. The warlords in the army often kidnap men and boys and force them to dig trenches at the front or work on the tunnel under the runway at the airport. Once the tunnel is finished, linking Dobrinja and Butmir, two government-controlled suburbs, arms and supplies can be smuggled into the city.

John McCormick repeats something he said earlier. "They're slaughtering kids up there."

Just before noon a shell lands in the top floor apartment of the next building over, severing the leg of an old woman who lost her husband to a sniper's bullet last month.

Ćiki appears in the doorway, clutching a five-pound piece of shrapnel from the shell, which left a hole in the façade of his in-laws' house. Ćiki is a Serbian architect married to a Muslim; at the start of the war they moved in with her parents, and when she fled the country, he went on living with his in-laws, though they despise him. No one trusts Ćiki, who manages the IRC warehouse: rumor has it he is calling in coordinates for strikes in this neighborhood. He chalks the date on the piece of shrapnel. Muha props it on the table next to the flowers.

"It looks like the devil's tooth," says John McCormick.

Now there are twelve of us in the basement—humanitarians and engineers and soldiers; a housewife, an architect, a tennis coach: Australian and American, British and French, Bosnians of all stripes—Serb, Croat, and Muslim. Ours is an unlikely caravan in a makeshift caravansary. And we have different ways of coping with the terror. John breathes heavily, Pat paces, I record in my notebook every conversation. Vinnie calmly explains military strategy.

"The way the momentum is going now the Bosnians will lose," he is saying. "Arms are getting in, but the Serbs and Croats are getting all of them. I try not to be pessimistic about it."

Mirna goes into the bedroom. "How dark it is," she exclaims—and then there is another explosion.

Pat opens a book called *The Destruction of Yugoslavia* but immediately closes it. "What should you read during an artillery attack?" she says. Not for the last time do I recall the story of the Serbian writer calling his Slovenian friends during the Ten-Day War to ask if they have read his new novel. I clutch my copy of St.-John Perse's *Collected Poems*. Vic has an anthology of erotica in his backpack. His is the better choice, everyone agrees; the ancient link between love and war, Venus and Mars, is proving stronger for the time being than poetry or *The Destruction of Yugoslavia*, which, in any case, we are experiencing firsthand. In fact, we are too scared to read at all. But Mirna has the best answer to Pat's question: "Marx and Lenin," she calls out from the bedroom. "They burn forever."

You cannot talk to John McCormick for long without hearing about John Jordan, the carpenter who came to Sarajevo to fight fires. Last fall, when Jordan learned that Serbian snipers were targeting firemen, he left his home in Rhode Island, vowing to stop them. He knows war as well as fire: the Vietnam veteran is a volunteer firefighter. What he lacks in diplomacy he makes up in passion.

"He's going on raw anger," John McCormick says during a pause in the shelling, "but the UN thinks he's playing with a full deck. In the States, on a handshake, he got three hundred firefighting suits, and now the fire brigade is fitted out with protective gear. That doesn't mean much when it comes to snipers and grenades. They've lost at least a dozen men and thirty wounded."

The shooting begins again. More shells land around us. It is some minutes before the lobbyist can resume his story.

"John Jordan believes this is the first time in history that firemen have been military targets. It's another development in warfare, like targeting schoolchildren or systematic rape. And he wants the UN to provide protection for his men. If they come under attack, UNPROFOR should return fire, then catch the sniper and take him before the world court."

Suddenly water is running from every faucet in the house. The Bosnians jump to their feet and, with shells falling nearby, rush to the garage to collect the dozen plastic fuel containers Vinnie scrounged up for just such an occasion. Soon we are all lined up like a fire brigade, filling and passing containers back out to the garage. A shell has struck a water line in the neighborhood, and there is no time to waste, since there is no telling how long the water might run. We fill the toilets and tubs on every floor. Mirna begins to wash dishes and clothes.

"It's a miracle," she says. "A real miracle."

The longer the shelling goes on, the closer we grow. Vic stayed out after curfew last night, playing a computer game in his office. With no telephone service in the city and his walkie-talkie inexplicably switched off, he could not be contacted. John and Pat were worried about him. But when he strolled in after ten, their questions only upset him. He stormed off to bed. Now he apologizes. Meanwhile, Slaven is offering Ćiki cigarettes, and Vinnie, who regarded me with suspicion from the moment I moved in, seems to have decided that I am not all bad.

Vinnie hates journalists, especially free-lancers. At dinner one night he attacked them at such length it was difficult for me not to take his words personally. "They arrive with fifty dollars in their pockets and expect us to show them around," he complained, perhaps forgetting that he too had come here without money or backing, relying on his wits—and a false ID—to set up IRC's operation. Only after prodding from Pat did Vinnie reveal another source of his anger: returning to Sarajevo from Split last week, he had lost his place on a UNHCR flight to a journalist from Alaska.

"Steffan Patterson?" said Pat.

"What about him?" Vinnie muttered.

"He's no journalist."

Pat and I had met Steffan (if that was his name) one night in the bar at the Hotel Split. He had been drinking, and when we said we could not help him—he was looking for Martin Bell, the famous BBC correspondent—he got a wild look in his eyes.

"Are you journalists?" he said.

"Why do you ask?" said Pat.

"They just fucking shut my operation down," he said.

"They" were UNPROFOR, and his operation was helping Sarajevans to escape from the city. Steffan claimed to have smuggled out hundreds of Bosnians, Croats, Macedonians, Muslims, and Jews. For the last eight months he had also snuck medicines and 6,000 pieces of mail a week into the capital. He had come to Bosnia with $40,000. He had $4,000 left.

"I've never asked for one fucking nickel from anyone," Steffan declared. "Plenty of mercenaries do what I do. But I'm a stand-up guy. And I'm going to pay the price on this."

His specialty was falsifying documents: press credentials, blue cards, medical evacuation orders; his mistake had been to try to pass off three children as journalists. He had managed to board them and five other Bosnians on a plane in Sarajevo (only peacekeepers, diplomats, humanitarians, and journalists were allowed to travel on UN flights), but by the time they landed in Split UNPROFOR had caught on to his ruse. They took away his own doctored blue card and threatened to arrest him. It was not clear why the Bosnians were allowed to stay in Croatia.

"Here's the bottom line," he said. "Strong people have to stand up for the weak. Who are you? I always knew who I was. This is my time. And I'm only talking now because I got caught."

It was no surprise to learn he had studied creative writing in college. An inventive storyteller, Steffan claimed to have earned his money working on the Exxon Valdez cleanup. With dummied-up ship captain's orders, he was made skipper of a boat responsible for clearing dead animals from the beaches in Prince William Sound. If Hemingway provided the myth for his adventure, Oscar Schindler offered him a model. "I would have been the man in World War Two," he insisted.

"But the first thing to know about refugees is they want to leave, then they want to go back," said Pat, pointing at the old Bosnian men scattered around the hotel lobby. "Look at their life. You smuggled out 236 people? That many go back every week."

Steffan was not listening. "There's a need for people like me who can walk between both worlds," he declared. "There's nothing more noble than trying to save people." He drained his beer. "The Bosnians realize their hope is false. They thought Bush might save them at Christmas. In the winter they held on because Clinton said we'll stand up to the Serbs. Now he tells the Bosnians, 'Fuck you.

You're on your own.' And they're tattered. When their soldiers go to the front lines at two in the morning they look like they're walking out of Siberia. They know they'll lose, they're fighting with sticks and stones. Every night they put on a big show with a little gunfire. They have a couple of places around town to launch mortars, but they're defeated.

"The Serbs are brilliant! Like Goebbels. Now they're presenting themselves as peacemakers, and we buy it. But soon, very soon, the Sarajevans will try to break out, and they'll be slaughtered. You should listen to the wind. You can hear them. Do you know what I'm talking about?"

"Who are you working with?" said Pat.

"The Jewish Center," he replied.

"Is that all?" she said.

"I don't remember." He smiled. "But there's a saying, 'When the Jews leave, a city dies.' It hasn't happened yet, but it will. Just listen to the wind. Listen to the outgoing. Listen to the incoming."

I needed a flak jacket to get on the UNHCR flight to Sarajevo. (In Split they were renting for $100 a day.) Steffan was willing to trade his if we agreed to bring a duffel bag of mail into the capital. A fair exchange, said Pat. And my reservations about the veracity of his story vanished in the morning, when I met him at a bus stop under a palm tree. The sky was overcast, and a brisk wind swept in from the sea, where a freighter rode at anchor. The Bosnians Steffan had smuggled out were huddling together. We are very grateful to the American, said a woman holding an infant. What ever made him think he could pass off a baby as a journalist? I wondered. But I said nothing. I needed his flak jacket.

The board members do not look happy. The three older men flew in this morning to determine if IRC should suspend operation. From the moment that John Fawcett, IRC's Sarajevo director, brought them into the basement they have sat stiffly on the cot, unwilling to remove their flak jackets and helmets. John and Vinnie, having failed to get a radio message through to tell them to delay their trip from Zagreb, adopt a new strategy: to convince them that it is not too dangerous to work here. When a shell strikes just outside the house Vinnie calmly explains that it landed 350 meters away.

"That's no problem," he says with a straight face. "The Serbs are trying to get a mortar in a house across the valley." He waves in the general direction of the house the Serbs blew up last night. "I hope you won't take that to mean we should pull out of here. Safety's our first concern."

He asks Ćiki to describe, in his nervous, halting fashion, a new program: IRC is distributing vegetable seeds around the city. Soon gardens will be growing on balconies.

A NATO jet rumbles overhead.

"Our useless military," John McCormick sneers. "The Powell Doctrine is, We do deserts, we don't do mountains, jungles, or people who fight back."

"It's good to have someone careful at the Pentagon," a board member counters.

"The pundits are always making the analogy to Munich," says John Fawcett, "but for the wrong reason." He is leaning against the door, a thin man with a drooping mustache, and I assume he will defend the board member, because I do not yet know how courageous he is. As it happens, he is one of those humanitarians who will stay in Sarajevo until the bitter end. An unmistakable air of tragedy clings to him (a daughter's death, a broken marriage), which may explain the nonchalance with which he sets the board member straight. "In 1938 the British and the French needed time to rearm," he tells him. "We don't have that problem now. We could stop this."

"What do you know about Banja Luka?" said Steffan Paterson. "Do you know about the fertilizer factory? The pig farm? At night the Serbs pulverize the people they've killed and turn them into fertilizer. Once they took thirty people from the hospital, shot them, and fed them to the pigs. What about Sanski Most? Last year the Serbs executed thirty-five hundred on the bridge in Sanski Most and pushed them into the river. Gypsies were brought in as a burial detail. But a few people escaped, and they came here to the mosque. I've verified their stories. You see, the Serbs are learning, just as the Germans did. It's very easy to fight an enemy that no longer exists. And guess what, boys and girls? I'm here to tell you the Germans have a new disguise, but the world isn't listening."

When John Fawcett drives off with the IRC board members, the conversation turns to flak jackets, prize items for the Bosnian warlords.

"My sister wants me to wear one that covers my private parts," Pat says with a laugh. "I tell her I'm not interested in having children."

"I don't know if it's better to be mugged by a Bosnian for wearing one or to walk through the streets without one," says Vic.

"I won't wear mine again," Vinnie declares, "until the Bosnians change their attitudes towards relief workers. They think we're just stuffing the chickens before the slaughter."

John shakes his head. "It's amazing to see a UN armored vehicle going fifty miles an hour down Sniper's Alley, and pedestrians in summer jackets scattering to get away from it. You can imagine what kind of attitude that creates. And of course the driver's wearing a flak jacket."

"I hate the way our jackets keep disappearing from the house," says Vic.

"As long as we have enough for the end," Pat says. "The Serbs are sending reinforcements here from Banja Luka. If they take the airport, it's all over."

"Which means what?" I say.

"We evacuate overland," she explains. "The UN estimates we'll lose 50 percent."

"That's peacekeepers and relief workers," Vinnie adds. "Not journalists."

John repeats a popular saying: "Easy to get into Sarajevo, impossible to leave."

Pat looks at me. "Whatever you do, don't get separated from your passport and blue card. You'll need them to get out."

Slaven tunes his guitar and begins to sing folk songs in Bosnian, English, and German. His nose is stuffed, his voice is raspy, his music is unbearably sad. But while he plays we feel—all of us there in the basement—a kind of peacefulness, until another shell lands just outside the house.

"They haven't had enough bloodshed yet," John says angrily.

Mirna takes Muha's hand. Ćiki lights a cigarette. Slaven sings a little louder.

Late in the afternoon, during a lull in the shelling, I go upstairs and out onto the balcony. Dust hangs in the sunlit air; an acrid smell rises from the ground. Ćiki's father-in-law is sweeping the street with a hand broom; his mother-in-law dusts the window next to the new hole in their façade. Someone is hammering in the distance. A house across the valley, which took a direct hit an hour ago, is burning. A year ago, during a similar assault on the city, an independent Belgrade television station broadcast an intercepted radio transmission of Mladić giving shelling instructions to his unit commanders. "Fire on Velušići," he said, mispronouncing the name of the neighborhood across the valley. "There aren't many Serbs living there." His strategic goal was more precise: "Shell them until they're on the edge of madness," he ordered. Needless to say, I am on that edge.

I lean on the railing, watching the smoke rise toward the radio tower. I am recalling a scene from Czesław Miłosz's *Native Realm*. Nazi-occupied Warsaw. August 1944. Miłosz and his future wife are on their way to visit a friend when all at once machine guns start firing. Pinned down in a potato field, one hundred yards—"a whole journey"—from the safety of their friend's house, Miłosz never lets go of the book in his hand, T. S. Eliot's *Collected Poems*, because he needs it.

A bullet whizzes over my head. Through the doorway I dive, and while I lie on the floor, breathing heavily, the Bosnian officer in our neighborhood fires his mortar. I hurry down to the basement, needing the poems of St.-John Perse, which I left under a cot, in the same way that I need to put another layer of concrete between me and the soldiers firing at us from the hills.

"Do you want to hear another story?" Steffan Paterson said. "Ahmići. Vitez. The massacre. I was there, and I know who did it. I've been drunk with the son of a bitch. HDZ? Do you know about the HDZ? It goes right up to that prick Franjo Tuđman. I know who did it, I walked into the houses, I stepped over the bodies.

There was a woman, a mother. One of her children was shot, and the blood from her body was splattered against the wall, and they pushed the mother down on all fours, with her head pulled back like a dog, and they fucked her up the ass, and then they poured benzene all over her, and they torched her. Her burned body stayed in the same position, like she was still getting fucked. And upstairs there was a boy, about twelve years old. Do you know what a roast pig looks like? It doesn't have any features. He was lying face up, completely burned. His little penis was sticking up in the air. Half his leg was burned off, six inches from his body. That's how hot the fire was. He had a bullet hole in his left side, and fresh blood was on the ground four days after he was shot. You know he lingered. You know he suffered. I will never be the same."

Dusk. Ćiki returns to his in-laws, a new guard replaces Slaven, and Muha leaves the house without saying where he is going. In the hills above Sarajevo the dead and wounded are carted away, trenches are dug, soldiers keep watch; in the city friends and families gather to remember the dead, journalists file their stories, and UN observers write up their reports. I am in bad shape, having wrenched my back diving away from the sniper's bullet; and when Mirna begins to cook dinner I find I am neither hungry nor thirsty. In fact, I have not eaten in two days, not since taking a sip of water from a bottle Mirna's father then warned me not to drink from. The fever I am running I attribute to nerves. The shelling persists into the night: there are tracers everywhere.

Pat says to me: "You know, you don't have to be here."

"I know," I say.

John and Pat put mattresses in the cluttered cubbyhole next to the garage. John fashions a bunk out of cardboard boxes, and Pat rolls out her sleeping bag. From the shelves I take three suitcases, stack them one on top of another, and lie down on the floor, with my legs resting on the suitcases to take the pressure off my back. Mirna climbs up on John's bunk, where he is lighting a candle. Pat switches on her short-wave radio to hear the BBC World Report.

"The good news is, we have earth on one side and two floors above us," says Pat. "The bad news, there are three barrels of fuel outside our door!"

BBC is reporting 24 dead and 170 wounded in Sarajevo. More than 1,000 shells struck the city today—an average day, by Bosnian standards. Upstairs, the new guard is rifling through my duffel bag; in the morning I will be $100 poorer. John lights another candle. The radio plays on.

16

Sarajevo II

October–November 1993

Vanessa Redgrave was furious. Yet on the telephone she patiently explained to the UN why she and her contingent of actors and directors should be allowed to attend the opening of the Sarajevo Film Festival. Earlier in the day, at the airport in Ancona, Italy, where part of the Bosnian airlift originated, Redgrave had tried in vain to convince UNHCR to reverse its directive barring her and what it called "eight famous film stars" from taking a humanitarian flight to Sarajevo. So the actress and her entourage had repaired to a conference room in the nearby Hotel Federico, where for several hours she pleaded her case to a succession of officials at the UN and the British Home Office. I happened to be sitting with the luminaries at a long table, having played a minor part in acquiring films for the festival, and I am fairly certain that by nine o'clock it was clear to everyone but Redgrave that they would not be going to Sarajevo.

"This is what negotiating is all about," said the actress, placing a call to UNPROFOR Civil Affairs. "You just try and get the truth across."

The "truth" she was attempting to convey was that she and her colleagues, including Jeremy Irons and Daniel Day-Lewis, were journalists who deserved places on a UNHCR flight. It *was* true she had convinced the British Film Institute to accredit them, but this was the first that Day-Lewis, for example, had heard of his reporting assignment. When Civil Affairs told Redgrave the stars could not provide "objective accounts" of the festival, Day-Lewis shrugged. He gave the distinct impression that Sarajevo was low on his wish-list of places to visit. As a matter of fact, only Redgrave and Irons seemed determined to get to the film festival.

The whole thing was a farce. Redgrave and her friends, armed with ideals, vitamins, and uncut versions of *The House of the Spirits* and *In the Name of the Father* to screen, were artists, not journalists. Theirs was the latest in a series of celebrity gestures on Bosnia's behalf. In Sarajevo, musicians like Joan Baez performed, Susan Sontag staged a production of *Waiting for Godot,* and a French conductor, suspended in a sling hung from the ceiling of the Holiday Inn, led the Sarajevo Philharmonic in a concert. Sontag likened the war to the Spanish Civil War, and while Western intellectuals rallied around the Bosnians with the same passion that poets and writers once mustered for the Spanish Republican side, few made the dangerous trip to Sarajevo. Hence the film festival organizers were keen to have Redgrave's troupe in attendance.

"Film festival life is entirely to do with the press and reporting," Redgrave said before hanging up in frustration. When she saw me taking notes, she demanded to know what I was doing.

I might have told her I was working, but I decided to try another tack. "Don't you think this is a political problem, not administrative?"

Redgrave fixed her piercing blue eyes on me. A critic's description of her was accurate: even a ravaging close-up did not diminish her beauty. "What do you mean?" she said.

"They've taken a decision in London and Washington not to let you in," I said.

"Perhaps the time has come for us to solve this on our own," she said.

Crestfallen, I went to the bar, where I fell into conversation with an American stringer in love with a Bosnian woman. He wanted to bring her home with him, but he lacked the money, political connections, or fame without which her evacuation from Sarajevo and resettlement in the United States would be very difficult to arrange. Could I help him? I shook my head.

It was not long before we were joined by the film stars—minus Redgrave and Irons—who proceeded to drink us into oblivion. Jim Sheridan, the director of *In the Name of the Father*, seemed particularly relieved to be on his way back to Dublin. "I have a *fillum* to finish," he kept saying, pronouncing the word in the Irish way, with two syllables. By the end of the night *film* on his tongue had three syllables. That was when he and the stringer came up with the idea of doing a live television feed into Sarajevo for the opening festivities. The actors could beam their greetings to the Bosnians. Rock musicians might be enlisted to play. A three-way show was not out of the question.

"I have Bono's number in Vienna," said Sheridan.

"I can call somebody at MTV," the stringer cried.

They ordered more drinks—for a shimmering moment everything seemed possible—and began to work the phones. It was too late, though, and they were too drunk to convince anyone to sign on to such an expensive and complicated telecast. But everyone agreed it was a good idea.

In the morning, at the airport, the bleary-eyed stars generated more excitement. It appeared, briefly, that one member of the troupe, Irish screenwriter Terry George, would be allowed to go to Sarajevo. And here the story takes another bizarre turn: George's press credentials were legitimate—he worked as an editor at *Travel Holiday*—but certain items on his résumé, well known to the British Home Office (membership, in the 1970s, in a splinter group of the Irish Republican Army; a six-year prison term for carrying weapons for the IRA) might have given authorities a genuine excuse for scuttling his travel plans. Not that it mattered: UNHCR soon discovered he shared a screenwriting credit on *In the Name of the Father.* How could he report "objectively" on the festival?

Presently a crew from RAI TV arrived and directed the artists to stand in a half circle to tape a message to the Bosnians. I was surprised to see them perform as well as they did, considering how hungover they were. And I was even more surprised when Redgrave and Irons decided to interview me for the BBC. The Surrealists were right: the world is a marvelous place! We were in a warehouse, among humanitarian supplies, forklifts, spare tires, and airplane engines. The UNHCR sign above us read: **If they survive, they're refugees.** How hard is it for a journalist to fly to Sarajevo, the actors wanted to know. Their colleagues were signaling for them to hurry: the chartered jet which would take them home was ready for boarding. Easy to get into Sarajevo, impossible to leave, I started to say, then thought better of it. No problem, I told them. Redgrave's last words to me were cruelly innocent. Of the Bosnian people she said, "They're saving us"—and then she flew away.

Maybe Airlines was what peacekeepers called the UNHCR airlift. I thought of this when the transport plane I was on made an assault landing at Sarajevo Airport, coming in at a steep angle over Serbian positions, touching down just beyond the front line, and then taxiing quickly to the terminal, which was surrounded by sandbags and barbed wire.

Strangely enough, I was happy to return to the capital, having cut short my first trip. My back injury had turned out to be a blessing in disguise, for no sooner had I reached my hotel in Split than I began to shake uncontrollably. Nerves, I thought. Typhus, said the doctor who took my temperature of 106°. Maybe dysentery. But it was hours before I was admitted to a military hospital. By then I was delirious; and of the next several days what I remember most vividly is the pleasure the nurses seemed to take in letting my intravenous bottles run dry. I was sharing a room with two convalescing Croatian soldiers, who liked to watch my arm swell, or so I imagined, since only they could convince the nurses to bring me a new bottle and they sometimes waited as long as an hour to call for help. My fever broke early one morning. I fled the hospital the next day.

And now I was back for a film festival titled "Beyond the End of the World."

It was a fair description of life in Bosnia. Alija Izetbegović's bitter joke—that choosing between Milošević and Tuđman was like discovering that you have either leukemia or a brain tumor—had taken on new meaning since the spring. Croatian forces were overrunning Muslim villages in central Bosnia, and this summer, in Herzegovina, they had laid siege to the city of Mostar, blocking humanitarian aid convoys for two months to the Muslim side of the city. Ethnic cleansing, summary executions, concentration camps—these staples of the Croatian offensive showed how much they had learned from the Serbs. Meanwhile, the siege of Sarajevo was in its eighteenth month. Ten thousand people had lost their lives (including almost two thousand children), tens of thousands more had been wounded, perhaps billions of dollars worth of damage done. Serbian gunners and snipers continued to target civilians and cut utilities. For food Sarajevans depended upon the black market and a relief operation the scale and size of which had already eclipsed the Berlin airlift.

Of the humanitarians I had stayed with on my previous visit only Pat Reed remained. At the airport she told me Sarajevans had lost all hope: nothing had changed. UN peacekeepers and NATO warplanes monitored the city's destruction; diplomats negotiated with the warring factions; journalists documented more savagery; the world watched in horror. What was to be done?

Pat had two projects. IRC had imported thirty thousand meters of plastic pipe, promising it to any neighborhood willing to dig trenches in which to bury it, then teaching people how to tap into the natural gas line running from Russia. The humanitarians reasoned that the Serbs, who also needed fuel, would not shut off the line. Explosions were inevitable, and in addition to preparing for medical evacuations Pat was helping out at the film festival, a humanitarian effort to open a cultural corridor into the capital and help break the siege. Though "Beyond the World" would receive little media attention, I took it as an opportunity to explore the artist's role in war. "When people challenge me and ask what is the place of art in that sphere of death and horror," Aharon Appelfeld writes of the Holocaust, "I reply: who can redeem the fears, the pains, the tortures, and the hidden beliefs from the darkness? What will bring them out of obscurity and give them a little warmth and respect, if not art? Who will take that great mass which everyone simply calls the 'dreadful horror' and break it up into those tiny, precious particles?" The novelist admits that "art cannot replace faith. Art lacks the power for that task, nor does it pretend to possess such power. Nonetheless, by its very nature, art constantly challenges the process by which the individual person is reduced to anonymity."

Black humor was another line of defense. Pat said one joke was especially popular this fall: what is the difference between Sarajevo and Auschwitz? At least in Auschwitz they had gas.

"The artist's responsibility?" said Ademir Kenović. "It is to define precisely what is happening."

The filmmaker, a tall, gaunt man with sleepy eyes and a beard, was the *éminence grise* of the film festival office, which also housed SAGA (Sarajevo Group of Auteurs), the production company Kenović had founded in the early 1980s with a musician, Ismet Arnautalić. The office was a whirlwind of activity, powered by a diesel generator and the frenetic energy of a dozen high school- and college-age students, some watching or editing videos, others working on computers, and still others vacuuming the carpet or dusting the artwork forged out of pieces of shrapnel.

What they had accomplished! They went out into the streets to shoot hundreds of hours of video footage, from which fifteen directors created more than fifty documentary films of life under siege. Though much of the artistic community had fled the city, many of the remaining painters, writers, and musicians were involved in Kenović's war production house, and young people were replacing the exiled artists much faster than anyone might have expected.

"We had the sad opportunity to be witnesses to this," said Kenović. "We at least had to register what happened, if only to serve long-term research. It's as if you had a camera crew in the Warsaw Ghetto. Unfortunately, there was nothing in Warsaw, but maybe we can contribute a little toward insuring that this horror not become the paradigm."

The artist's other obligation was to state his opinions, according to the filmmaker, raising his voice. If Sarajevo's future is Hell, then Hell will be the world's new paradigm. If the extreme nationalists are not stopped here, they will never be stopped. Politicians have fooled everyone into believing the Balkans are doomed to endless cycles of ethnic conflict, but it is up to the artist to clear the people's blurred vision.

"No one who sees even a single frame of the pictures from Sarajevo will ever rest easily," he said. "People everywhere are disturbed unconsciously by the siege: they know something horrible is happening. If 400,000 people are in concentration camps, with the cooperation of the political and moral zeroes on the world's stage, then it's obvious this cannot end up good for anybody."

And SAGA's films were unsettling. *I Burnt Legs* told the story of the man at the hospital responsible for incinerating amputated limbs. *Traveling Children* showed children steering the wheels of cars burned and blown up in the siege. *Confessions of a Monster*, the most provocative, was an extended interview with Borislav Herak, the Bosnian Serb soldier captured in Sarajevo after making a wrong turn in his car. Herak, who had witnessed or participated in more than two hundred murders, was charged with genocide, mass murder, rape, and looting. And here, in black-and-white, was a haggard man with large ears and a shaved head dropping to his knees to demonstrate how he had cut the throats of Muslim men, a technique learned from his commanding officer.

"He trained us on pigs," the soldier told the filmmakers. "That's how we did it, too."

The same officer ordered his soldiers to rape and murder Muslim women—to raise the troops' morale, said Herak. He betrayed no remorse, though he admitted to seeing in his dreams the faces of those he had killed. But, he added, "My dreams are not connected."

"The growing power of television has led all coup plotters to change their tactics: formerly, they would assault presidential palaces, governmental and parliamentary seats; now they try first and foremost to gain control of the television-station building," writes Ryszard Kapuściński in *Imperium*. "The screenplay of the latest film about a coup d'état: tanks roll out at dawn to capture the television station while the president sleeps peacefully, the Parliament building is dark; there isn't a soul around. The plotters are headed for where the real power lies." The Serbs shot from the same script in Sarajevo, destroying the radio and television tower before they took out the *Oslobođenje* building. No news is better than real news, and nothing is as good as propaganda, reasoned Karadžić and his media henchmen, the ones who explained away the Bread Line Massacre in May 1992 in an ingenious fashion: Bosnian TV cameramen could not have filmed the carnage unless the mortars were fired by their countrymen. How else would they have known to go to that blood-soaked street to get footage of the dead and wounded? Moreover, not only was there no evidence of Serbian involvement in the atrocity but most of the victims were Serbs; their corpses were simply exchanged for Muslims and Croats wounded earlier in the day. A breathtaking performance.

"People have needs beyond food and water," said Dana Rotberg, the Mexican filmmaker and co-organizer of the film festival. I caught up with her in the London Café, which had just opened near the offices of IRC. A petite woman with short brown hair and large glasses, she smoked Marlboros, drained cup after cup of espresso, and talked in a staccato fashion.

"In every city of the world people have the right to watch films," she said, explaining the genesis of the film festival. "Why not Sarajevo?"

Sarajevans were suffering from a kind of boredom peculiar to war. Without electricity, they could not watch TV or videos, and there was no light to read by—if, that is, they had not already burned their books for fuel. How to survive the Serbs' slow strangulation of Sarajevo, in aesthetic and spiritual ways—that was what Rotberg and Haris Pašović, Bosnia's celebrated theater director, asked themselves. "Beyond the End of the World" was their answer. In ten days they would show more than one hundred films; it was usually standing room only in Radnik Theater, despite the efforts of the sniper who liked to make it "hot" by the entrance.

"Snipers are a technical problem for us," Dana said indignantly, "like trying to line up enough diesel to show a film."

These were hardly the only problems she and Pašović faced. The festival got off to a bad start: a nine-hour artillery attack on Sarajevo, which left ten dead and scores wounded. But this did not deter the organizers and their volunteer staff of eighty. They worked around the clock, staying up all night to subtitle videotapes (there was no projection equipment), printing daily schedules (which were subject to constant change), and risking their lives to get to the theater.

"Yet the show goes on," I said.

"A civic attachment to life is a form of resistance," Rotberg replied. "People here try to live normally in this abnormality. They can't defend themselves because of the arms embargo, they're shelled, they're massacred, yet they're determined to live, not just survive. And this festival is a way of breaking the siege. The films are a window onto the world. You see people running through the sniper fire, laughing, hiding behind cars, just to get to the cinema."

Rotberg, the director of *Angel of Fire*, had moved to Sarajevo three months earlier to make a film about the siege. But she could not do that without understanding the texture of people's lives. "Otherwise it's just CNN reportage," she said. "So I thought, if I can't do a film, I'll bring a film festival to town."

Were all the problems worth it? I asked.

"One day we showed *Aladdin*," she said, brightening. "Hundreds of kids who were only two years old when this stupid war started went for the first time to the cinema. If that doesn't justify a film festival, what does? Going to the cinema is a kind of resistance. You can stay in your cellar, or you can go to a film. That's the point of this city: it's not in a cellar. That's why it's still alive."

Everyone knew the story of Ismet "Nuño" Arnautalić's return to Sarajevo.

Nuño was a professor of psychology—he had never practiced his profession—better known for his guitar work in one of Bosnia's best rock groups, *Indexi*; his gift for arranging he put to good use at SAGA and TV Sarajevo. Short and wiry, Nuño had the look of an aging sound man—black beard and ponytail, jeans and leather jacket—accustomed to life on the road. That he was even here was a testament to his improvisational skills and perseverance.

In March, Nuño had led the musical delegation to the annual Eurovision Song Contest, a venue for Bosnia to showcase its cause. The preselection competition was in Ljubljana; and when UNHCR would not let Nuño fly out of the city, rejecting his press credentials, he had to sprint across the airport runway, avoiding Serbian snipers *and* peacekeepers (who made a practice of capturing Bosnians and returning them to their starting places), hike over Mount Igman (in the sights of Serbian gunners), then hitch rides to Croatia and Slovenia. No wonder the singers sought political asylum when Bosnia's entry in the contest, "The World's Pain," went on to win in Dublin.

Not Nuño. He succeeded in getting press accreditation upon his return to Split, but then the airlift was suspended because of shooting. So he and the remaining members of his delegation took a bus to the Bosnian border and hitched a ride over Mount Igman in a military vehicle, arriving at dusk in Butmir, the war-ravaged village at the edge of the airport. Then the real excitement began. On one side of the runway were Serbian snipers equipped with night vision scopes, on the other, peacekeepers patrolling in armored vehicles. The delegation picked its way through the barbed wire, waited for the searchlights to switch off, then dashed across the runway. Two made it to safety. The peacekeepers who caught Nuño, his brother, the pianist, and the producer took their papers and asked their destination. Sarajevo, said Nuño. The peacekeepers thus drove them back to Butmir and pushed them out into the mud. Only Nuño tried to enter the city again that night, and he was caught seventeen more times, often by the same peacekeepers, before giving up at daybreak.

The producer, too sick to continue, returned to Split in the morning, and when darkness fell Nuño's brother and the pianist abandoned their efforts after three attempts. Nuño himself tried ten times that night and eight the next. Then, on the fourth night, after the pianist also went back to Split, Nuño came up with a clever plan. He and his brother ran to the barbed wire and fell to the ground, facing Butmir, as if that were their destination. Snipers were shooting all around the men, who lay beside their duffel bags, waiting for the peacekeepers to arrive. But now it seemed to take forever for them to be caught. At last an armored vehicle pulled up and the men were ordered to get in. Where are you going? said a peacekeeper. Butmir, said Nuño—and soon they were in Sarajevo!

Nuño had another story with a musical twist. His was one of countless families torn apart by the siege. His mother, sister, and Montenegrin brother-in-law were in Belgrade, and his two daughters were in Italy, the oldest having left with her boyfriend at the start of the war to study design in Milan. When Nuño and his wife sent their ten-year-old out on a Children's Embassy convoy, a cameraman filmed their leave-taking, which became the lead story that night on SKY TV. What the world, and Nuño's scattered family, saw was a girl crying at a bus window with her parents in tears on the sidewalk, footage that Phil Collins used in his music video, "Tell Me Why."

A bullet passing within a foot or so of your head sounds more like an explosion than a whine. I learned this on my way to the world premiere of Romain Goupil's *Lettre pour L....* Sarajevans were anxious to see this documentary of Goupil's wartime journey from Vukovar to Belgrade to Sarajevo. What is the world doing to stop the fighting? Goupil wanted to know. What about Vukovar? was his refrain. And he would be here for the screening. Unlike Vanessa Redgrave and her friends, the unknown filmmaker had managed to take a humanitarian flight into the city.

Dusk was falling when I set out for the Radnik Theater with Maria Blacque-Bellaire, a French humanitarian, and Miro Purivatra, the director of Obala, an art gallery/cultural center. The sky was overcast, and it was sharply colder. Đure Đaković, the north-south running street connecting the Olympic ice arena and downtown Sarajevo, was empty. Something is not right, said Miro. Then a yellow flash, an explosion in my ear, a bullet skipping off the sidewalk—a sniper was working the street. We hugged the wrought iron fence in front of the Residency, the home of UNPROFOR's commanding officer, and slipped through the next white gate into a courtyard, where we lined up behind a tree. Shots rang out all around us in the growing darkness.

Miro, a small thin man with dark hair, lit a cigarette and launched into a story about a recent show at Obala. He was remarkably calm, and as he explained how eight artists had created "Witnesses of Existence," a series of conceptual works with titles like "War Trails," "Ghetto Spectacle," and "America likes Sarajevo and Sarajevo likes America," I began to get control of my nerves. Miro said the most stirring piece was a terra cotta sculpture in the shape of a mass grave from which the glassed-in faces of victims of the siege stared out, an answer of sorts to the Tower of Skulls the Turks had erected in Niš. "Our skulls are flowering with laughter," Vasko Popa wrote of the Serbian shrine. "Look at us look your fill at yourself/ We mock you monster." At the opening of the Obala exhibition the artist covered the faces with dirt. Just outside the gallery, a man was wounded by a sniper.

Miro's words had a soothing effect on Maria and me. Strange to think that only earlier this afternoon I had seen her outside the London Café, surrounded by children. They were tugging at her long red hair, shrieking with delight; and now she and I were ducking with each explosion, with each yellow trail that left an acrid smell in the air. Miro did not flinch. We'll have to stay here until it gets dark, he said. Though the sniper also shot at cars racing down the street, at a woman on a bicycle, and at an old man lugging water, we seemed to be the primary target.

There was a story of a Serbian woman who became a sniper after her husband brought shame onto the family by helping Muslims and Croats to escape from

Serb-held parts of town. She turned him in to the local authorities, and he was promptly executed. For all I knew, we were now in her sights. The shooting was personal in Sarajevo—snipers often knew their victims—and when I was not imagining what it would feel like to be shot I wondered why this sniper was interested in us. Sowing terror in the civilian population was one Serbian strategy—hence the number of children killed—and I was certainly terrified. George Orwell's view of his brush with death in the Spanish Civil War came to mind. After being wounded at the front (a sniper's bullet through the neck), he is assured, repeatedly, that he must be the luckiest creature alive to have survived. "I could not help thinking," he remarks, "that it would be even luckier not to be hit at all."

When at last Miro declared it *might* be safe to leave, I decided to go write an article titled, "On Not Attending the World Première of Romain Goupil's *Lettre pour L...*" Miro and Maria dashed across the street, into the bare stretch of ground that until last winter was a heavily wooded park. I ran back up the street. An old women waved me down by the UNPROFOR checkpoint. I pointed at the hill. Sniper, I said. The old woman sighed and shuffled on toward the city.

The IRC office turned out to be a good place to write my article. A driver was describing to his colleagues how a bullet had just shattered his windshield, missing him by inches. Then a Canadian peacekeeper walked in, shaking with anger. The same sniper had fired a shot right between his boots. The peacekeeper vowed to take care of the bastard. I typed feverishly until I came to the last line of my story. But how to finish it? My frenzy gave way to an ordinary bout of writerly anxiety, and I was staring out the window when Maria returned from the theater. Moments later I had my ending.

How was the film? I asked.

"Not worth getting killed over," she said.

No real Serbs still lived in Sarajevo, Radovan Karadžić liked to say. But the poet Goran Simić, director of the Serbian cultural society, was living here with his Muslim wife and two children. In this he was not unusual: of 1.3 million Serbs living in Bosnia before the war, 200,000 stayed in Bosnian territory (50,000 in Sarajevo alone), while 400,000 more fled to Serbia and abroad (many to escape mobilization). Indeed Serbs made up a third of Bosnia's Territorial Defense Forces, and Brigadier General Jovan Divjak, a Serbian deputy commander in the JNA, was now helping to defend Sarajevo. It was only "Alija's Serbs," as the nationalists called them, that kept the so-called "good Serbs," including Goran's father and brothers who went over to the Serbian Republic at the start of the siege, from flattening Sarajevo. The Bosnian government had identified one of

Goran's brothers as a war criminal, but the poet did not want to talk about his family the day we walked down Marshal Tito Boulevard. We had not gone far before a burst of automatic weapons fire sent us ducking into a doorway, from where we watched peacekeepers atop a white armored vehicle take aim at a sniper who might have spent part of the previous afternoon shooting at Maria, Miro, and me.

"Jesus Christ," said Goran.

In my possession was a manuscript of his new book of poems, *Sprinting from the Graveyard*, and just before the gunfire interrupted us I had asked him about his lament for the National Library, a prose poem that seemed to sum up the writer's obligation under siege:

> When the National Library burned for three days in August, the town was choked with snow. Those days I could not find a single pencil in the house, and when I grabbed one it had no lead. Even the erasers left a black trace. Sadly my homeland burned. Liberated from their bodies, heroes of novels wandered around town mixing with passersby and the souls of dead warriors. I saw Werther sitting on the collapsed graveyard wall, and Quasimodo on the minarets; for days Raskolnikov and Meursault were whispering in the cellar, Gavroche walked by dressed in camouflage, and Yossarian was already trading with the enemy. Not to mention young Sawyer, who threw himself into the river from Prince's Bridge for pocket money.
>
> For three days I lived in the spectral town with a terrible suspicion that there were fewer and fewer people alive, and that the shells were falling only on my account. I locked myself in the house and leafed through tourist guides. I went out on the very day the radio announced that people had removed ten tons of coal from the library cellar. And my pencil got its lead back.

Unfortunately, we never finished that conversation, because the UNPROFOR operation was still unfolding when Goran decided to return to his flat. And perhaps Karadžić would have the last laugh. For this day Izetbegović was reshuffling his government; by nightfall his only Serbian and Croatian appointees would be ministers without portfolio. Soon Goran's wife and children would flee to Italy, and before the war was over he would join them in emigrating to Canada.

"It was strange, it was heart-rending," Rebecca West wrote of Sarajevo, "to stray into a world where men are still men and women still women." Women went to great lengths during the siege to keep up their appearances. Beauty parlors thrived, and with little water Sarajevans prided themselves on wearing clean clothes whenever possible. Applying makeup, said Ljiljana Šulentić, an architect, is how we tell the primitives we will not surrender. Ljiljana needed no makeup to stand out, even in Sarajevo. Tall and slender, with shoulder-length auburn hair and deep brown eyes and flushed cheeks, she always looked as if she had just returned from a long walk. Her suits came from Italy—she had studied and worked in Rome—and it was her passion, as the head of a household of women (her mother, her aunt, a younger cousin), to explain why the women of Sarajevo were so attractive.

"Mixing the races is what makes the richest and most beautiful people," she told me one overcast afternoon as we walked to her apartment. "It's the nature of life to mix."

Ljiljana was a fine illustration of her thesis. Her father, who died in the first winter of the war, was Croatian, her mother was Serbian, and in her extended family were Muslim, British, and French relatives. Her best friend, a Montenegrin woman, was killed in Serbian-occupied Grbavica, when she sneaked across the river to bring a packet of food to her parents. It was a Bosnian sniper who shot her. Most of the mourners at her funeral were Muslim.

"It's so stupid, so primitive," said Ljiljana. "But I have decided to survive because I would like to go to concerts and travel again. For this someone has to be alive."

But she was tired, winter was coming, and she had not heard from her friends and colleagues in Belgrade and Zagreb. A thoroughgoing Sarajevan, she kept her distance from what she called the "half-people" destroying her city, the villagers who came here after World War Two and during this war, some to fight on the Serbian side, some to occupy abandoned flats. Designing homes for refugees who would transform the city into a large village in which the arts had no place was the last thing she wanted to do, so she would probably leave Sarajevo when the war was over.

We climbed the stairs to what was left of her family's corner apartment. Ljiljana had prized its panoramic view of the mountains until it was struck by a pair of rocket-launched grenades. She thought she was prepared for the devastation, having already toured some 5,000 destroyed flats in her new job of estimating architectural damages. But this was the first set of ruins she visited in which she could not tell what had been there before the shells struck. Only her father's books and record albums survived; everything else was smashed under the falling walls.

"My friends think I have an American flat," she said with a bitter laugh, "because the walls are gone. But I don't think American flats have rain and snow coming into them."

A cold wind was blowing through the sheets of plastic covering the window frames; plaster and dust swirled on the floor. Ljiljana, who came from a family of engineers and urban planners, was now the architect in charge of demolishing Sarajevo's damaged buildings. Before we joined her mother and aunt for a cup of coffee she caught my arm.

"Silly job, silly people, silly country," she said.

In the market eggs were selling for five deutsche marks apiece; a liter of cooking oil was forty-five. The prices were roughly ten times what they were abroad, and the average Sarajevan took home only ten marks a month. But since there was money to be made on the black market soldiers traded with their counterparts at the front, civilians carried bags of food across the runway, peacekeepers used the airlift to sneak in supplies. Ingenuity was the key: two men smuggled a slaughtered cow across the runway on a gurney, with a sheet pulled over its head and an IV drip hooked up to one leg; when UNPROFOR stopped the men, they said they were taking a wounded man to the hospital, and the peacekeepers let them go! I remembered what a Bosnian friend had said: It is impertinent to be fat in Sarajevo. And this: It is dangerous to have a wooden leg here because we need the wood. And this: When the last deutsche mark leaves Sarajevo, the war will be over.

No films were shown the day the government cracked down on a pair of local warlords. Sarajevans were told to stay inside, and the heavy fog hanging over the city (ordinarily a welcome sight, since snipers could not see their targets) deepened the mystery surrounding the raids on the warlords' headquarters. Rumors swirled through the city: of civilians taken hostage, policemen killed, and the arrests of hundreds of soldiers. For a better picture of what was happening I called on Haris Pinjo, a slight young man who served as an assistant to Haris Silajdžić, the new prime minister—the official responsible for reining in the rogue elements of the Bosnian Army, the enemy within.

Ramiz Delalić, nicknamed Čelo (literally, bald), and Mušan Topalović—a.k.a. Caco—were two of the criminals who helped to organize Sarajevo's defenses early in the siege. In addition to guarding key positions on Mount Trebović their gangs "requisitioned" private vehicles; kidnapped men to dig trenches at the front; murdered, raped, and robbed with impunity. When six children were killed by a Serbian shell in June, for example, Caco's paramilitaries killed ten local Serbs in retaliation and dumped their bodies in a pit above the city. They assaulted foreign journalists, stole from aid agencies, hijacked APCs from UNPROFOR, and monopolized the black market. The renegade army commanders grew so rich and arrogant that the news of Caco's "mysterious" death in police custody was greeted with jubilation. Now the government could concentrate on breaking the siege that had turned the city into a vast black market. The gangsters' operations were the most violent manifestations of a societal meltdown that forced Sarajevans to break the law to survive, selling cigarettes, exchanging coffee for food, trading humanitarian supplies for other necessities on the black market. This infuriated Haris Pinjo. He was accustomed to buying his clothes in Italy and keeping up with the latest music, film, and literature. He proudly called himself Bosnian—i.e., someone with the open way of thinking that led the Serbs to besiege Sarajevo, fearful of its spread.

"Karadžić has the tradition of eating two meals from the same plate," he said, "because he's a peasant, not a citizen. If you cut the tail of a snake, it's still alive. But if you smash its head it will die. That's what's happening here. They want you to believe this is an ethnic war, but it's not. It's the war of people who want to make their own rules, and they're killing Sarajevo."

It was a corner apartment, and during a late lunch the humanitarians were playing a game: what was the most revolting meal they had eaten in their line of work? A rat, one offered. Cockroaches, said another. A fruit bat, said a third.

"That's nothing," said the psychiatric nurse who once explained to me the memorable phrase Sarajevans used to describe certain relief workers and journalists. They were "going on safari" to see a new species of animal: men, women, and children living under siege. Some humanitarians, the nurse admitted, had altruistic motives, just as some journalists were devoted to the truth, but others came in search of a level of chaos corresponding to what was going on in their minds. A safe, peaceful life led them to such anxiety that only in dangerous situations did they feel calm. War is where you can act in an uncivilized manner, she said, because you think it's an arena in which there are no rules. Only rarely do you discover the rules that determine whether you live or die. A peaceful life was what she wanted, having spent her career in war zones and disaster areas. Soon she would marry a Bosnian tank commander who said that when he went into battle he carried 150 bullets: "149 for the Četniks, and one for me. I won't let myself be taken prisoner. I know what the Četniks do to their prisoners."

"That's nothing," the psychiatric nurse was saying. "Once I ate the fetus of something."

"The fetus of what?" someone asked her.

"Something," she repeated with a look of wonder.

Filming was almost complete for the first televised program of *The Surrealist Hit Parade,* the hit radio comedy. The setting for this modern fairy tale is a place forgotten by the rest of the world, a city reached after traveling for days through the mountains and over rivers. Subdila, the heroine of "Cinderella in Sarajevo," must attend to her three older brothers and father hiding in the basement to escape the draft. Every day she has to stand in line for water, fetch wood, cook, clean, and do the laundry. Never satisfied, the men make her life miserable. One day she comes upon a peacekeeper in a stalled APC, which she fixes then pushes up the hill. The peacekeeper drives off with one of her boots and cannot find her to return it, compounding her problems. The abuse heaped on her by her family is more than she can bear. One night she falls asleep on the staircase and dreams she is Cinderella. A Good Witch appears in a flak jacket waving the keys to the radio station instead of a wand, promising to grant her any wish. But no one can help me, cries Subdila. Let me show you my magic, says the Good Witch. All at once water gushes from the public taps. The lights go on in the city. The Prince—the peacekeeper—arrives, in a blue flak jacket and white silk shirt, and takes her to a restaurant where a band is playing. He tells her to eat and drink as much as she likes, and then they dance through the night. Before they part the peacekeeper asks her to the ball. What a glorious time! *The ball, the ball*, Subdila is saying in her sleep when her father shakes her awake, demanding an explanation for such crazy talk. But this is her secret, and although she will never see the peacekeeper again she can wait in line now for bread and water in a much better mood.

Rumor had it, a contingent of Afghan *Mujahedeen* lived in the elementary school behind the IRC house. Like the Russian and Greek mercenaries fighting on the Serbian—i.e., Orthodox—side, the Islamic warriors had come to the defense of the Bosnians. CIA-trained, fresh from fighting the Soviets, they brought arms (including Stinger surface-to-air missiles), experience, and religious zeal, which Bosnians welcomed more on the battlefield than in society. In central Bosnia, Pat remarked one morning, *Mujahedeen* were dragging women out of offices to keep them from working and stopping them in the street to make them put on more clothes.

"It's the logical consequence of not defending the Bosnians," she said. "They don't want the *Mujahedeen,* but what choice do they have? They have to take what help they can, which will drive out the educated, secular Muslims, Serbs, and Croats who won't want to live under the Mullahs."

A classic Balkan reversal. Izetbegović was a pious realist, yet some Western leaders believed the Serbian and Croatian nationalists who branded him a fundamentalist. In *The Islamic Declaration* (1973), he addressed, in theoretical terms, the idea of an Islamic republic, but he did not imagine he could foist such a system onto Bosnia's Serbs and Croats. The European fear of Islam persisted nonetheless. One symptom was the arms embargo, which led the Bosnians to turn to the Muslim world for help. Faith was on the list of weapons they received.

I remembered the trio of Iranian "journalists" I had met at the airport in June. They wore thin white flak jackets, more cotton than Kelvar, and as one bowed his head in prayer the others urged me to visit their country. At least it is ours now, they said defiantly. A young Slovenian cameraman offered them brandy, which they angrily dismissed. Nor were they interested in following him onto the tarmac to film a wounded peacekeeper's evacuation. Indeed when I asked them what kinds of stories they had filed, they shrugged. They were journalists of the same ilk as Vanessa Redgrave and her friends, yet the intelligence they provided to the Bosnians was perhaps more immediately useful than the messages of solidarity the celebrities beamed into Sarajevo. The Iranians' interest perked up when a Danish peacekeeper chased down the cameraman and confiscated his film. "If you say another word before you get on the plane you won't be allowed back into Sarajevo," said the peacekeeper. "Then you'll just have to stay here till the war ends," I told the Slovenian, who looked aghast. But it was the peacekeeper who should have been worrying: he had come to Bosnia for adventure, he said, and within the week he would be dead of a sniper's bullet. The Iranians were praying.

Ferida Duraković was the striking young poet featured on an ABC *Nightline* special about Sarajevo. I recognized her at a meeting of the Writers' Association, and we had just begun to talk when she said, "They're all old men here!" Off we went to a café, where she announced that the literal meaning of Ferida was *unique*, of Duraković, *foolish* or *durable*. All three applied. She was a kind of holy fool who spoke outlandishly—e.g., "Happiness is the main task in this war"—and soon she was pressing on me a small painting by an artist friend. On one side of a shard of glass from a broken window was a hollow face on a green stem; on the other, a black figure, with two blank eyes, curling around the base of the stem and looming behind the head, like a three-cornered hat. My friend is painting the city, said Ferida. I can't accept this, I said. There's much more where that came from, she laughed, and led me outside. A firefight stopped us in our tracks.

"What was that about?" I said when it was over.

"I don't know," she said. "I don't know anything anymore."

Homeless for more than a year, somehow she retained her *joie de vivre*. Serbian shells had struck the bookstore she managed and burned it to the ground. Then her parents' flat was destroyed, and her personal library went up in flames. Usually we are just happy to be alive, she said when we walked on, sidestepping the so-called "Sarajevo roses," the marks left by mortar shells. Never have I been so happy to write, to live, to meet people, to eat. I don't have anything complicated in my life, just life and death. I have to choose, and I choose to write. She paused at a crosswalk we would have to sprint across to avoid a sniper. This is a war against civilians, she said, glancing at the houses on the hill, and calmly told me that 85 percent of the casualties in World War One were military, 15 percent civilian: figures inverted in this war. The fate of the poet Georg Trakl was never far from her thoughts. After the battle of Grodek, in August 1914, Trakl, a pharmacist in the Austrian Army, was put in charge of a barnful of seriously wounded men for whom he could do nothing to alleviate their pain. Outside in the trees were the convulsing bodies of hanged deserters. Inside, the poet, who saw one of the wounded shoot himself in the head, was also preparing to commit suicide, though not before writing for these men "the wild lament/ of their broken mouths," the inspiration for Ferida's newest poem:

Georg Trakl on the Battlefield Again, 1993

Our dear Lord dwells above the planes, in the highest Heaven.
His golden eyes settle on the dark, on blackened Sarajevo.
Blossoms and shells are falling outside my window.
Madness and me. We are alone, we are alone, alone.

Ferida looked at me. "Ready?" she said.

I dined with Haris Pašović, the acclaimed director, on the last weekend of the film festival. Blankets covered the windows of the IRC house to black out the candlelight. The night sky was lit up with tracer rounds. But Pašović was in a buoyant mood. He believed the festival's success augured well not only for Sarajevans but also for the world. Here was proof that the free and democratic future prefigured in *2001: A Space Odyssey* was at hand. Of course he was extravagant, but he backed up his words with achievements. A slight grey-bearded man of thirty who looked fifty, in the last year he had put together a musical collage based on the writings of Cavafy, Zbigniew Herbert, and Sylvia Plath; enlisted Susan Sontag to direct *Waiting for Godot;* and staged Euripides' *Alcestis*. His next project was equally ambitious: to organize an exhibition of architectural drawings for the new city of Sarajevo, with work solicited from architects around the world. The forces of barbarism cannot destroy the core of what has been achieved this century in the arts and sciences, he announced. Then he told a new joke. Why was the gas cut to Sarajevo? Because there are no Jews.

But five hundred Sephardic Jews remained in the city (out of a prewar population of 2,000), where their families had lived for a half millennium. All but obliterated by the Ustaše, Sarajevo's Jews (some still spoke Ladino) were, at least in theory, not under attack by the Serbs, having suffered alongside them in World War Two. In practice this meant that hundreds were allowed to leave when the siege began, and theirs were the only humanitarian convoys that consistently reached the city. Their soup kitchen, pharmacy, and infirmary were open to all; their cemetery was a main confrontation line. Suffice it to say the Jews suffered the same indignities, privations, and dangers as any Sarajevan.

There was a crowd at the top of the circular staircase, above the lobby in the Holiday Inn. It was the morning before the Day of the Dead, and Ferida Duraković was taking me to the War Congress.

"Don't look now," she said when the president of the Writers' Association approached us, "but you're about to be invited to give your greetings to the congress."

"What do you mean?" I asked.

"You can't refuse," she replied.

Nedžad Ibrišimović, a tall, dark man with a full beard, had traded thirty years of drunkenness for faith in Islam, the fighting in his part of town being particularly savage. He punctuated each phrase by waving his arms over his head. Ferida, translating, could not keep a straight face. "In the tradition of your countryman, Ernest Hemingway," Ibrišimović said in a deep voice, "who gave his greetings to the War Congress of the Spanish Civil War, I invite you to do the same here in Sarajevo."

I nodded uncertainly.

Into the meeting room filed sixty poets, writers, government officials, and local journalists. Modeled after the international literary congress of the Spanish Civil War, this was a very formal affair, though there was only enough diesel to run the generator lighting the room for two hours. A girls' choir sang a new patriotic hymn. A moment of silence for the dead. Greetings from dignitaries, including Izetbegović. I was the only foreigner present. At UNPROFOR's morning briefing I had handed out invitations to the entire foreign press corps. But there had been another massacre the day before. UNPROFOR was organizing a media trip to the site. The journalists went there.

Mine were the last greetings. How foolish I felt standing before these people who had endured so much. And what I said was even more ridiculous: "This is the most thoroughly documented war in history. We in the West have seen the siege of Sarajevo from its beginning, and still we have not intervened. Perhaps it is up to you, the poets and fiction writers, to give us the true meaning of this tragedy so that we may know in our bones what you have lived through."

The audience stared at me in silence.

Nedžad Ibrišimović returned to the lectern and joked: "Before the war in Yugoslavia there were more writers here than in China—that was the problem!"

He continued: "We used to have the Sarajevo Days of Poetry. Poets and writers came from all around the world. Except for Susan Sontag, no one has come or spoken out—why?" Nevertheless books were being written and published in Sarajevo: a testament, according to the declaration issued by the congress, "that the writer and his work are the lights in the darkness of civilization's changes."

Then the room went dark. Into the upstairs lobby we trooped, where windows provided some light. Within minutes the generator was running again, and when the congress resumed someone asked, "What is nationality to a poet? A dead Croatian poet, a dead Muslim, a dead Serb—they're just poets." With shells falling outside (the front was a hundred fifty meters away), a Montenegrin poet read a comic poem about his former colleagues, now in Pale, writing Serbian poems with

Serbian pencils on Serbian paper. Their subject? Serbian birds, which fly higher than other birds.

The diesel for the generator was running low, so the organizers skipped over some of the texts on a writer's relationship to his homeland and instructed Ferida to conclude by reading the declaration of the War Congress. Asserting that the writer exists to face evil, and determined to "restore power to the word, though the barbarians have destroyed libraries, cultural monuments, everything made out of words," the writers vowed to make the Bosnian language their only homeland.

At the reception, Ferida, Goran, and I ate our fill of *ćevapčići,* the spicy grilled hamburger selling for forty marks per kilogram on the black market, then left. Halfway across the wasteland next to the Holiday Inn, Ferida stopped. "Since the war began," she said, "we've learned to treasure every meal, every conversation, every poem we read or write, because we know it may very well be our last." She stretched her arms above her. "I just want to feel the sun on my body," she murmured. We were in the plain sight of snipers, and I just wanted to get to the shelter of the buildings a hundred meters away. But Goran and Ferida preferred the sun. I admired their devotion to pleasure. What Rebecca West observed in the 1930s remained true: "The air of luxury in Sarajevo has less to do with material goods than with the people. They greet delight here with unreluctant and sturdy appreciation, they are even prudent about it, they will not let a drop of pleasure run to waste." And as we walked toward the center of town, with shots ringing out all around us, Ferida and Goran did not even flinch.

"If you hear the shot," said Goran, "you're still alive."

His flat was on the top floor of a four-story building; the bullet holes in the roof, he said, were greetings from his brothers, commemorating articles he had published in *Oslobođenje*. His wife, Amela, and their two children were entertaining an elderly Austrian diplomat. The fire burning in the wood stove was fed with feathery bits of roots. Into the evening we drank Serbian brandy bought on the black market for a hundred marks and talked about how the siege had changed their lives and work.

"Before the war I didn't really like Goran's poetry," said Ferida. "It was too hermetic. But now it's so clear and direct. Now he only writes about what's important."

Amela lit makeshift candles—a little cooking oil laid on water in small glass cups, with string wicks worked into bits of cork—and served a dinner of thin vegetable soup. When the brandy was gone, Goran brought out a bottle of vodka saved for months. We ate and drank, joked and laughed, discussing the poetry of Rilke and St.-John Perse, Czesław Miłosz and Zbigniew Herbert.

The diplomat explained why Václav Havel had not attended the war congress. "Havel the President of the Czech Republic is not as free as Havel the writer.

That's a choice he made. You may not like it or understand it, but that's the way of the world. He is no longer free to do what he likes."

Ferida whispered to me, "I hate all politicians. They speak such bullshit. They're the ones who got us into this mess." She looked around the darkening room. "I don't want to talk about all the horrors I have lived through," she sighed. "I can tell that to journalists any time. With you I just want to talk about poetry and life. What else is there?"

Stupni Do was a Muslim village in central Bosnia, with a population of 250. None of my Bosnian or humanitarian friends had ever heard of the place until Croatian forces overran it one night during the film festival, dynamiting and torching every building. The masked soldiers shot scores of villagers, cut their throats, and burned them in their houses and gardens. Some villagers, fleeing into the woods, were gunned down; many bodies were never recovered. For three days the HVO prevented UN military monitors from inspecting the village, and not until the day of the War Congress was the press allowed to visit Stupni Do. Why did it not even occur to me to join the pool of journalists escorted there by UNPROFOR? I cannot say. But I am haunted by that failure of nerve and imagination.

17

Mostar

August–September 1994

An unscheduled stop in the middle of the night, in the mountains outside Zadar. When the bus driver, failing to fix the broken generator, advises us to get some sleep, the British journalist across the aisle from me deadpans, "Not a good place to stop." Parked on the side of the road, we are an easy target for any Serbs occupying the ridge above us. How I wish I did not have a window seat. And how I regret my late nights at Vilenica, now that I cannot sleep. Oh, the vanity of writers! This year at the literary festival the Slovenians honored a minor Sarajevo poet who is sitting out the war in Ljubljana. His citation was read in Slovenian—an English translation would follow—but the poet, who spoke neither Slovenian nor English, was so excited that as soon as he heard his name he marched to the front of the auditorium to receive his award. Someone clapped uneasily. Others joined in. So much for the translation. The poet was bowing to the crowd, which for the most part knew only that he had left Sarajevo before the war. What a change from the days when the Vilenica Prize went to writers like Czesław Miłosz and Milan Kundera! This is a time of dubious honors: Radovan Karadžić has just bestowed upon himself a new Bosnian literary award, and the Greek Orthodox Church has appointed him to the 900-year-old "Knights' Order of the First Rank of Saint Dionysius of Xanthe," calling him "one of the most prominent sons of our Lord Jesus Christ working for peace." He must work in mysterious ways, then, having once said that before long Sarajevans would count the living, not the dead. And his war is entering a critical stage. This week, another poet and more sincere peacemaker, Pope John Paul II, is planning to visit the Bosnian capital, where the siege is now in its twenty-ninth month. A cease-fire has been in effect there since February, when a mortar landed in the market, killing sixty-nine and wounding more than 200, a massacre that tested Karadžić's powers of invention. He insisted, of course, that the Muslims had shelled themselves, and then he added a poetic touch: the dead and dying Sarajevans rushed to the hospital were old corpses hauled out for television crews to film. No wonder he is the poet laureate of Republika Srpska! But that was the day the international community said, Enough poetry! A genuine threat of air strikes forced the Serbs to withdraw their heavy weapons twenty kilometers from Sarajevo, then American diplomats brokered an agreement between the Bosnians and Croats, ending their bitter war so that they might work together to roll back Serbian advances. The UNHCR Special Envoy described the Bosnian-Croat Federation as an imposed marriage for the Croats. "That's not Balkan," he told me in Zagreb. "For the Bosnians it was an arranged marriage, with mother or mother-in-law being in Washington: that's very Balkan, but not necessarily durable." Durable or not, the Federation means that more weapons are being smuggled into Bosnia, though the Croats take a cut out of every shipment. It also clears the way for the papal mission to the region, this week's media focus and the reason why I took the bus from Zagreb to Mostar, unwilling to wait with hundreds of journalists in Ancona for a place on a UNHCR flight into Sarajevo. Now a night at the Hotel Federico does not sound so bad: at least I would get some sleep. Nor is my journey devoid of religious elements: I am praying that no Serbs have us in their sights.

The destruction of the Old Bridge in Mostar on 9 November 1993, four years to the day after the Berlin Wall came down, completed a circle: the future met its herald. What hope had greeted the sight of Germans dismantling the most potent symbol of the division between East and West! Here was the promise of a new beginning: the end of history, as Francis Fukuyama said. But Misha Glenny was closer to the mark when he called it the rebirth of history. And nowhere was history's rebirth or revenge more evident than in Mostar, once the most ethnically mixed city in Bosnia and Herzegovina (34 percent Muslim, 33 percent Croat, 19 percent Serb, 12 percent Yugoslav). During World War Two, Glenny writes, Mostar's Muslims, Croats, and Serbs resisted "the temptation of mutual loathing which gripped the rest of western and eastern Herzegovina and the Neretva valley," a temptation they succumbed to during the Third Balkan War. In the first month of fighting the Serbs set out to destroy Herzegovina's capital (Glenny awarded the mad JNA commander Miodrag Perušić "first prize in the keenly contested 'Most Bloodthirsty General' stakes of the Yugoslav wars"), relentlessly bombarding its graceful Ottoman architecture, including the famous old footbridge arcing high above the Neretva River. The civilian population, Muslim and Croatian, was cruelly starved. That was not the end of it. In April 1993, the Bosnian Croats turned against their erstwhile allies, training their guns on the Muslims across the river. Before long the Old Bridge lay in pieces on the bottom of the river. Built of white cobalt in 1566, it had survived thirty earthquakes, the fall of the Ottoman Empire, the Balkan Wars, two World Wars, and the initial Serbian onslaught on the city. Now this symbol of connection is gone. "Finally—who did it?" the Croatian writer Slavenka Drakulić asks. "The Muslims are accusing the Croats, the Croats are accusing the Muslims. But does it even matter?

> For four centuries people needed that bridge and admired its beauty. The question is not who shelled and demolished it. The question is not even why someone did it—destruction is part of human nature. The question is: What kind of people do not need that bridge. The only answer I can come up with is this: people who do not believe in the future—theirs or their children's—do not need such a bridge.

Mine is a journey, then, to a place that has given up on the future. Was it a coincidence that the destruction of the Old Bridge also fell on the fifty-fifth anniversary of *Kristallnacht*?

It is still dark when the bus driver guns the engine and starts down the mountain road, without lights. By daybreak we are barreling along the Dalmatian coast, the sea glittering in the distance. Then the long climb to Herzegovina, and a new border. The guards interrogate an old woman smuggling shirts; three large duffel bags covered with mesh, which may be filled with rifles and grenades, are not examined. The mood lightens considerably once we enter Herceg-Bosna, as Bosnian Croats call the parastate they have established in defiance of the Federation. "They're going home," the British journalist says of our fellow passengers, some of whom are returning refugees. And some are pilgrims who get out at Medjugorje, where a vision of the Virgin Mary was supposedly vouchsafed to six Catholic children in 1981. Medjugorje (literally, between the mountains) remains a popular religious shrine, even in wartime. Who can judge the validity of the daily apparitions—a shimmering light on the hillside, messages of peace heard only by the visionaries—that draw the faithful from around the world? The Pope refuses to acknowledge *Gospa*, the Croatian name for the Madonna of Medjugorje, in part because it has acquired such political significance. Croatian nationalists seized on the vision and turned it to their advantage, despite the efforts of Mostar's bishop to have it condemned as a fraud. *Gospa* became not only a major source of income for an economically depressed area but also a rallying point for the Croatian people, not unlike the Slovenian dedication to Mount Triglav or the Serbian veneration of Lazar's relics. It was no coincidence, the British journalist remarks when the bus pulls out of Medjugorje, that Croatia declared independence on the tenth anniversary of the first apparitions, 25 June 1991. Soon we are driving down the winding road toward Mostar. "It looks so peaceful," the British journalist says of the medieval city nestled in the Neretva River valley. "You would never guess they've been slaughtering one another for three years now."

A Croatian friend has arranged for me to stay at the Franciscan monastery. I have no illusions about the Franciscans' record here. They came in 1260 to wipe out the Bogomil heresy, but after the Turkish conquest in 1463 they settled into reasonably harmonious relations with their Muslim and Orthodox neighbors. All that changed with the assassination of Franz Ferdinand. The friars who encouraged Catholic mobs to lynch Serbs set an example for the Ustaše, and when Ante Pavelić swept into power he found strong support for his "cleansing" measures among the Franciscans. As a matter of fact, the center of Ustaša operations in Bosnia and Herzegovina was the monastery at Široki Brijeg: one graduate won a competition for cutting the throats of 1,360 Serbs. His prize? A gold watch, a silver service, a roast suckling pig, and wine. The Franciscans backed the Ustaše in burning down or dynamiting a third of the Orthodox churches, often with the faithful inside. Catholic priests were condemned to death for telling their flocks "Thou shalt not kill." And historian Richard West quotes one particularly loathsome friar saying to his parishioners: "Brother Croats, go and slaughter all Serbs, and first of all slaughter my sister, who is married to a Serb, and then kill all the Serbs in a row. When you have finished, come to me and I'll hear your confession and give you forgiveness for your sins." Thus tens of thousands of Serbs were murdered in these mountains, the Neretva River brimmed with Serbian corpses, and less than two miles from Medjugorje is a ravine into which six wagon loads of Serbian women and children were thrown; the names of three Franciscan Ustaše from Medjugorje, killed alongside the German SS, are engraved on a plaque at Široki Brijeg commemorating those who died fighting the Partisans. Why stay with the Franciscans, then? Because Mostar's bishop was the only senior Croatian churchman to protest the Ustaša methods; because the current bishop has consistently counseled peace; and because I am low on money.

When the bus lets us out in the western half of the city (a banner hanging above the street reads, "*Dobro Došli u Hrvatski Mostar*"—"Welcome to Croatian Mostar"), I walk to the monastery, located on the front line. The church has been destroyed; long thin logs propped against the back wall mark the front. Three policemen in shirt sleeves are drinking beer at a picnic table. They do not flinch at the blast of a grenade: unexploded ordnance set off in a dumpster next to the church. One policeman leads me down a set of stairs, protected by logs blackened in a fire, into the basement of the monastery, where two monks in work clothes are finishing lunch. It turns out that my friend did not succeed in contacting the monks she knows: they are on their way to see the Pope in Zagreb. To meet him in Sarajevo would mean far less travel, and while such a journey is fraught with danger, its symbolic value would be immense. But the Franciscans prefer to go to Croatia. Anti-Muslim feelings are running high on this side of the city, according to the monks who stayed behind to guard the monastery. Little remains intact. Most of the cells are damaged, the red-tile roof is covered with holes, the library is gone; walls are propped up with sandbags, and in the courtyard, where forty grenades landed, a charred cross with a plastic date nailed across it—1992—is surrounded by palms and roses. Mass is now said in a room the size of a walk-in closet. The monks shake their heads. There is no place for me to stay. From an upstairs window they point to the Bosnian lilies painted on the sides of all the gutted buildings; and when I ask them about the Old Bridge the monks click their tongues. "Turks!" they mutter with contempt. Before I leave, one monk gives me a postcard displaying photographs of the church taken before and after the fighting. "Crazy," he says.

The EU has opened offices in the Hotel Ero, establishing a provisional government under the leadership of Hans Koschnick, the former mayor of Cologne. A young Croatian woman working for the EU rents me an abandoned flat for thirty-five dollars a night. The owners are not coming back, the local mafia having forced them out. I can stay as long as I like. Three Italians, two young men and an older woman separated from a Catholic group driving to Sarajevo, sweep into the hotel, needing directions to the capital. They *have* to see the Pope, they tell the Croatian woman; their idealism is touching, their naïveté frightening. Once installed in my flat, I take a walk around West Mostar, inspecting the damage—the houses, apartments, and municipal buildings destroyed by artillery fire or dynamite; the shelled stadium and shrapnel-pocked walls; the streets gouged by grenades and missiles. The traffic lights do not work, and cars careen around intersections, honking their horns at white UN and EU vehicles. Serbian forces are behind one mountain range, Croats behind the other. "The Serbs can shoot us any time," says the bartender who serves me a beer late in the afternoon. *Zdravlje!*

In the early evening I stroll to the bread factory overlooking the river, where I meet two Croatian special policemen. "This is a Disney war," says one. "You can rent a tank over there. I don't think it's a good idea to give the Muslims arms, but we're giving them everything—arms, food, clothes." Then he proceeds to show off his own equipment, most of which—his uniform, his knife, his radio—comes from America. Only his pistol was manufactured in Croatia, as it says, in English, on the handle: **MADE IN CROATIA**. His radio picks up twenty different bands, so we listen to Serbs and Muslims taunting one another. For the moment peace is holding, though within the week Croatian forces will fire on the Hotel Ero and the Serbs will resume shelling the city. The other policeman—only months before they were fighting in the Croatian Army—point to the former JNA barracks on the other side of the river, less than a hundred meters away, where Muslim forces are now housed. "You see that building? Those houses? They're filled with gold. Many people got rich off this war. We will never get along," he confides. "We're too different, maybe because of our religion." "What do you think of the Pope going to Sarajevo?" I say. They shake their heads. "I have a bad feeling," says the first policeman. "We listen to the Serbs and Muslims on the radio, and both sides hate the Pope. There are lots of people who would like to go down in history as the one who killed the Pope: they're the only criminals anyone remembers, like Oswald or O. J. Simpson. And these guys have Strelas [surface-to-air missiles]!" On either bank of the river teenage boys in bathing suits are fishing and throwing rocks into the water, cheerfully calling one another names—the Croats are Ustaše, the Muslims, Titoists. "The Muslims didn't arm themselves," says the first policeman. "They thought the West would save them. You see what they did to us? Nothing compared to what we did to them. We flattened their buildings, they just threw crap at us. But the Serbs did the real damage," he says with awe. "Who blew up the Old Bridge?" I ask. "The Muslims care more for bridges than for people," the first policeman sneers. "It was an old bridge, like an old person. It was time for it to die," says the other. "Who blew it up? We did, they did, but we finished it off," he says proudly.

There is no gold in East Mostar, as I discover in the morning when I cross a temporary bridge. What a change from West Mostar, where the cafés and bars are crowded, the shops and kiosks overflowing with goods; there is even a carwash operating. In East Mostar rows of lockers stuffed with bricks and debris line the streets; men and women wait by spigots or water trucks to fill their plastic containers; 16,000 buildings and flats are damaged or destroyed. Nothing was spared. A poster bearing a plea to drive slowly partly covers a sniper warning sign, but the graffiti tell a different story: on the strafed walls of many buildings are the names of children killed in the war—Saša, Ana, Hara. Here is the true horror of the war: twenty-seven of the twenty-nine historical monuments destroyed, every bridge down, every church, mosque, and cultural monument gutted. Even so the talk is of reconstruction. The remains of the Old Bridge in the river will be hauled up, and rock from the original quarry will be excavated to replace what was shattered. The joke is that the Old Bridge will be rebuilt exactly as it was, only older.

Green Islamic flags hang from balconies and above the streets; in a new graveyard, where there are scores of fresh graves and holes dug for the next wave of casualties, the wooden markers feature the same insignia—a star within a crescent moon; the damaged mosques are full of old men and boys. The European fear that Islamic fundamentalism will take hold in Bosnia, which hitherto seemed farfetched, is becoming, day by day, a reality in East Mostar—another blow to moderate and secular Muslims around the world, already hard pressed to explain to Muslim extremists and their followers the West's inaction in Bosnia. It will be an odd twist of fate if a tradition of religious tolerance gives way to fanaticism. A brief history of Islam in these lands is in order. Unlike other Balkan nations, before the Turkish conquest Bosnia had no single strong church but rather three weak confessions, all of them short on priests. The ground was fertile, then, for the growth of a new faith, which is why some historians describe Islam's advances here as a process of acceptance instead of conversion. The well-organized Ottoman conquerors offered Muslim converts lower taxes and better career opportunities. Yet the Orthodox Church also grew through the conversion of Bosnian Church members (who converted in larger numbers to Islam and sometimes to Catholicism) and Serbian migration into lands abandoned by Catholics fleeing to Croatia. Indeed the Ottomans preferred the Orthodox to the Catholics, partly because the Patriarch in Constantinople was easily controlled, partly because of the threat that the papacy might launch another crusade. (Hence the fear that the Franciscans would form a fifth column, notes historian John V. A. Fine.) But the Muslims were the chief beneficiaries of Turkish rule. Theirs was the military class, their descendants made up the landed elite, and their Christian serfs toiled for four centuries with little say over their affairs (though they were free to practice their religion) before they rebelled in 1850. The uprisings against landlords and local officials spread until 1878, when the Great Powers intervened, and Bosnia was assigned to Austria at the Congress of Berlin. Interestingly, many Muslims held onto their lands and positions, even after the Ottomans left (this was not true elsewhere in the collapsing empire), because they were first and foremost Bosnians. Still, thousands fled to Constantinople, and now the descendants of the Christian serfs occupying the mountains around Mostar would like to see all the Muslims gone. The peasant uprisings are not over.

In the Koski, Mehmed Paša's riverside mosque built in 1648 and all but destroyed last year, I meet with Jadran Jelin, the man in charge of reconstructing Mostar. A professor of civil engineering at Mostar University, he spent ten months in a Croatian concentration camp, where the guards were his own students. But he prefers to remember another Croatian student of his who helped him move his belongings out of his flat in West Mostar before the HVO blew it up. "There are still good people here in Mostar," Jelin insists. "They'll rebuild this city." I ask him about the Federation. The stories I heard in West Mostar, which now regards itself as the capital of Herceg-Bosna, and the divisions I feel between the two sides of the city, the mutual suspicion and grief built up during the war, lead me to doubt the enduring viability of this agreement. Two old women in East Mostar who invited me in for a glass of juice expressed what seems to be a common Muslim sentiment: they could probably live with the Serbs, they said, but not the Croats—not after what the HVO did to their side of the city. "You have been listening to evil people," says Jelin. "We are good people. The peace will hold."

Through the streets of East Mostar I wander, past hundreds of gutted buildings, and cross the makeshift bridge of wooden planks and cables that replaces the Old Bridge. The Pope has decided not to go to Sarajevo after all. Karadžić said he could not guarantee his safety. The Orthodox Church in Belgrade has already barred John Paul II's visit; the religious figure that future historians will associate with the downfall of Communism has met his match in these nationalists. A sad day for the world, I am thinking when I chance upon a burned-out flat in which a Muslim artist is painting, his subject always the Old Bridge. Not once does the young man look up from his canvas as I walk around the charred room, studying his primitive versions of the bridge built, legend has it, of mortar mixed with egg whites. There are no prices on his work. He has no interest in talking to me. Outside, in the sunlight, it is quite warm, a perfect Indian-summer day, yet I am shivering. "The bridge," writes Slavenka Drakulić, "in all its beauty and grace, was built to outlive us; it was an attempt to grasp eternity. Because it was the product of both individual creativity and collective experience," she adds, "it transcended our individual destiny. A dead woman is one of us—but the bridge is all of us, forever." The artist in his dark flat is determined to preserve some vision of that eternity.

Three Bosnian policemen lounge by a pile of rubble. The front is only fifty meters away, they say, directing me toward a row of deserted buildings that will provide cover, and here I find the British journalist I met on the bus. He is in Mostar on behalf of a German organization planning to build a community center here. What he must decide is where to house it. This flattened Muslim section of West Mostar, a kind of no man's land, may be the best place. "This is the real crime," he says with contempt, pointing at the rubble. The irony of the Holocaust Museum opening in Washington, D.C., its message—"Never again"—proclaimed in every possible way, even as genocide was being carried out again, this time in Bosnia, is not lost on him. Here is indisputable evidence of our failure to act against aggression. Yet this was where the international community could have made a difference, he says, poking his head into a storefront filled with debris. "Your president should have told the Croats, 'Look, you will not be part of Europe for the next twenty years if you don't stop shooting right now.' That wouldn't work with the Serbs, because they're not interested in Europe. But the Croats had something to lose: they want to belong to the West. Clinton let them get away with this. Should be enough to cost him the next election, don't you think?"

Where the Marshal Tito Bridge once stretched across the river, British peacekeepers are in the last stages of assembling a temporary bridge. Barbed wire is coiled around the construction site; cranes pivot on both sides of the Neretva. In three days these troops have built a bridge big enough to handle trucks filled with food and other humanitarian supplies. The British journalist and I watch them place a long green girder across the river. "This is just practice for another Falklands," he mutters. "They have thousands of these bridges lying around." But then he smiles. "What if we were to build our community center in three days?" he says. "It could be done, couldn't it?"

Midnight. I lie awake in the dark, in the well-appointed flat of a family that must have left in a hurry (clothes in the closets, toys in the children's bedrooms, photograph albums on the bookshelves). I am remembering my walk across the wobbly cableway strung between the pillars of the Old Bridge—how I kept losing my balance when the Bosnian boys dove from it into the river twenty meters below. The planks of wood shift underfoot, and I reach for the cables, embarrassed by my clumsiness and afraid of falling. I stand there as old men and women cross without a flinch. It occurs to me that this is how connections are reestablished between warring sides: with a makeshift bridge, which will be replaced by a permanent structure. Eventually the Europeans will close their offices in the Hotel Ero, and leave. The people of East and West Mostar will have to find ways to live together. I recall a conversation with a military monitor in Dubrovnik. In fifty years, he predicted, there will be another Yugoslavia. "They'll have to live together, won't they?" he said. I told him about the schoolteacher I knew from Vukovar, who was quarreling with her aunts because they had traded Christmas cards with their former neighbors, Serbian women living in refugee centers in Belgrade. The schoolteacher could not understand why the older women would reach out to their enemies. "Those Christmas cards are the best way to the future," said the monitor. "Lasting peace will come only with small gestures." Bridges are built with such gestures. One day this summer, for example, thousands of spectators turned out for the resumption of the diving competition from the Old Bridge, an annual event from 1566 until 1992. This rite of passage for seven-year-old boys, some of whom are now leaping, one after another, toward the emerald water beneath me, causing the cableway to swing in the wind, is called "the test of courage." No Croats took part this year.

18

Sarajevo III

March 1995

I had an excuse for missing my opening lecture at Sarajevo University. One morning I boarded a military jet in Zagreb, an Ilyushin 76 carrying peacekeepers, humanitarians, diplomats, and journalists to the besieged capital. After an hour's flight, we were on our final approach—the landing gear was down—when small arms fire sent us climbing steeply away from Butmir Airport.

"We are returning to Zagreb," the Russian pilot said over the intercom. "There is shooting."

"That must have been close," cried the UN radio journalist sitting next to me. Large white containers of frozen meat filled the cabin, leaving us with no leg room; the jet engines were so loud the journalist had to shout in my ear. "These Russians will usually fly through anything."

The navigator emerged from the cockpit, grinning nervously. One of his Orthodox "brothers" had shot at us, violating international law, a commonplace in this war. The UN's response was swift and predictable: the airlift was suspended, tightening the siege of the city. The cease-fire that former President Jimmy Carter had negotiated in December was falling apart.

"I shouldn't say this since I work for the UN," the journalist muttered when we landed in Zagreb, "but the Serbs are calling all the shots. Maybe we should start firing back."

This man was not alone in thinking wider war inevitable. "Our aim remains the unification of all Serbian lands," Ratko Mladić had said last fall. "Borders are drawn with blood." Nor had he changed his mind over the winter. Two days ago, his gunners locked a surface-to-air missile on a humanitarian flight, causing the Turkish president to cancel a visit to the capital. Cease-fire violations were increasing. And Bosnia's third anniversary of voting for independence was balanced last month by Sarajevo's marking its 1,000th day of siege, the longest in modern history. The Bosnian spring offensive, an open diplomatic secret, would begin well before the cease-fire ended on 1 May.

Tension was rising in Zagreb, too. Franjo Tuđman vowed to send the peacekeepers packing when their Security Council mandate ran out this month: their buffer between the warring sides merely preserved the gains won by the Serbs on the battlefield. But if the UN was forced out of Croatia it would also close down its Bosnian operations. NATO, the UN, foreign embassies, relief agencies, all were drawing up evacuation plans. The warring factions would probably try to seize hostages and equipment. Casualties would be high. The Bosnians would have to fend for themselves.

We shuffled out of the plane and stood around on the runway, waiting for our luggage. The navigator climbed up onto the wing to look for bullet holes. The sky was overcast; a cold front had swept in since our departure; snow was imminent.

"How long will the airlift be off?" I asked the journalist before we parted.

"As long as the Serbs want it to be," he grinned.

When I checked back into the hotel, I learned three things: 1) the Bosnian Serbs had rejected the latest peace proposal put forth by the Contact Group (diplomats from the United States, Britain, France, Germany, and Russia who had replaced the EU-UN negotiating team); 2) an unsubstantiated report of United States–led covert actions on behalf of the Bosnian Army had opened another rift between the UN and NATO; and 3) the hotel would not credit my frequent flyer account for my stay, because as a humanitarian I was only paying half-price. I went up to my room and unpacked.

I was traveling under odd circumstances, the Open Society Institute having hired me to teach English literature. Waiting for the airlift to resume, I read the poetry of Wyatt and Wordsworth; for diversion I monitored what passes in our time for "the still, sad music of humanity"—newscasts of Marines evacuating UN peacekeepers from Somalia (a dry run for the Balkans?); of the O. J. Simpson trial; of the arrest of Nicholas Leeson, the British stockbroker who had single-handedly brought down Barings Bank. Another broker described Leeson's trading tactics in familiar terms: "It started to go bad, and he must have kept hoping it would turn around, and then it was too late."

The same could be said of the international community's response to the war in Bosnia.

I made it into Sarajevo on Bairam, the holiest day of Ramadan. Rain fell on this day of thanksgiving, and sporadic bursts of gunfire accompanied the feast marking the end of the month-long fast, the beginning of the Muslim New Year. Since my last visit to the capital, the government had taken on a distinctly Islamic hue. The new minister of culture was speaking out against mixed marriages and Western influences; one of the two Bosnian television stations had begun to show Arabic films. (The second station continued to screen pirated American movies on video, some with the FBI warning not to rebroadcast them blazing across the airways.) At sunset I heard a sound new to the part of town I was staying in: a muezzin chanting over a loudspeaker, calling the faithful to prayer, among them thousands of refugees from the eastern enclaves of Goražde, Srebrenica, and Žepa.

There were other changes, prominently the restoration of electricity pumped through the tunnel dug under the airport. Through my window I was surprised to see lights in the houses and buildings. Dozens of restaurants and cafés had opened; more cars were on the streets than UN vehicles. Sarajevans, too. During the cease-fire more people ventured outside, though snipers were working again. This day two pedestrians were killed near the university building in which I would teach, and tram service was suspended when Serbian gunners fired on a tram, wounding several civilians.

My host at the IRC house, Eric Shutler, lived alone, my friends in the relief community having moved on to Chechnya and Rwanda. The ex-Navy pilot and former banker was restoring Sarajevo's schools, and over dinner at a nearby restaurant he complained of corrupt local contractors. Nor were they the only ones to take advantage of the anarchy. "The best and brightest are gone, except those who realize war's a good way to get rich. They're the new robber barons, and the government's with them." Meanwhile, trucking materials into the city was very difficult, since the Bosnian Serbs had closed the "blue routes" used by humanitarian convoys, despite the UN's mandate to keep them open.

But food was getting in. Cars, buses, and trucks loaded with meat, produce, and other goods (including arms) made the hazardous journey over Mount Igman, often traveling through the dark without headlights so as not to attract the attention of Serbian gunners. On a narrow, unpaved road these caravans wound along steep cliffs through the mountains west of the city before dropping down into the suburb of Hrasnica; from there it was a short dangerous run to the tunnel in Butmir, through which all manner of supplies came. You could even buy a satellite dish—one was propped in a window of the IRC house. And the construction of a second tunnel big enough for trucks was underway.

We feasted on shish kebab, roasted potatoes, and pickles: a welcome change from humanitarian fare. The restaurant had new windows, potted plants, a neon sign. An amplified four-piece band played Bosnian folk songs; at the next table a soldier sang to three young women. The ten o'clock curfew was relaxed because of Bairam. By the time we left the restaurant was crowded.

From Eric I learned that the program I was initiating—the woman who hired me said I would be their guinea pig—had just been canceled: the Open Society Institute (OSI), funded by international financier George Soros, was retrenching in Bosnia. Mine was one of dozens of projects, including Eric's, to fall under the ax. The prospect of teaching poetry in a city under siege was becoming ever more bizarre, not least because most of my Bosnian friends had left or suffered nervous breakdowns. I doubted that I would see them again.

"The city is full of despair," Eric said. "When this so-called peace came, the old people started getting sick and dying and killing themselves. They held on through the fighting, but now that they see what's happened to Sarajevo they've given up hope." The cease-fire was the calm before the storm: no one wanted to fight in winter with the roads impassable, so all sides rearmed, trained, and deployed troops for the spring. "Maybe the outside world should have let them fight it out among themselves right from the start," he sighed, "instead of trying to feed everyone. At least it might be over by now."

Zvonko Radeljković walked with a slight stoop, a legacy of a life spent hunched over books. The gravel-voiced professor of American literature had published essays, in English, on Hemingway; his translations included editions of Emerson and Whitman, as well as episodes of "China Beach" for state TV. He liked to quote T. S. Eliot; Hemingway was "Papa" to him. A prodigious reader, in the first year of the siege he went through a hundred new titles the British Embassy had donated to the university. No doubt he was better read in contemporary British letters than some critics in London.

Of eighteen professors working in the English department before the war only Zvonko taught full-time now; the rest had fled the city or else avoided the Faculty of Philosophy, which stood on the front line. Though most of his students were gone, some having escaped, some fighting in the army or working for relief agencies or the UN, he conducted tutorials with students living near him when school was closed; when it reopened he taught the entire curriculum, Beowulf to Virginia Woolf. How could I turn down his invitation to deliver a series of lectures? Our friendship dated from the film festival, when we had spent an afternoon discussing, of all things, country and western music. He was humming when I met him before my first class. It was bitterly cold that morning, and on the walk to school he said, "I must warn you that you are heading for a very dangerous place. Three people have been shot here in the last two days. But it's nice to be in the sun, isn't it?"

I took no consolation from the knowledge that Zvonko and his students made this journey daily down Sniper's Alley. Walking with my shoulder hunched, as if to protect my neck, I wished that Zvonko would pick up his leisurely pace. But he said he was not interested in running into a bullet. We passed behind a gutted high-rise, and my head began to spin: an APC was parked between us and the Faculty of Philosophy. This was the setting for dozens of sniping incidents, many recorded by cameramen, one of whom was waiting ten steps away, sheltered by the building, his camera rolling.

"You know why he's there," said Zvonko, signaling me to wait behind the APC.

"I can guess," I said.

A French peacekeeper trained his binoculars on the hill above. A small crowd had gathered behind the APC. The cameraman, shielding his eyes from the sun, stared at us.

"He just wants a little blood for the evening news," Zvonko said. Then we headed toward the cameraman. I held my breath until we were safely past him.

By the entrance to the faculty, a squat four-story structure wedged between the high-rise and the National Museum, was a large pile of glass; hundred of windows had been shattered. Unlike the high-rise, looted of every piece of wood including door jambs and window frames, the faculty had not been plundered, thanks to the caretaker, a clever man. He discovered that the plastic foil used to cover broken windows burned for hours; figuring that the most dangerous time for looting was between twilight and curfew, he would stuff the sheeting into a metal bowl in the lobby and light it when he went home at six. A ghostly blue flame reflected off the ceiling. Meanwhile, Zvonko let it be known that peacekeepers were using space in the faculty, and convinced a lieutenant colonel to drive his APC around the building to lend credence to the rumor. Shells had destroyed the roof and rainwater ran downstairs, but the library was intact and classes met, without electricity or heat; three hundred students (out of 3,000 before the war) were studying for their degrees.

"This emptiness is spooky," Zvonko said.

The window in his fourth-floor office looked out on the Jewish Cemetery. Government forces were on one side of the graveyard fifty meters up the hill, Bosnian Serbs on the other. Beyond the building in front of us was Serbian-occupied Grbavica. From the National Museum next door Bosnian snipers shot at Serbian civilians just as Serbs across the way fired into Sniper's Alley. There was a bullet hole in Zvonko's window, and when he beckoned me to look at the combatants on the hillside I hugged the wall, glancing at the graves out of the corner of my eye, then edged back to my seat.

"I don't know how you do this every day," I said.

"What do you mean?" He smiled. "It's normal."

Thirty-five women and one young man were waiting in the classroom. The plastic sheeting over the windows kept out neither the cold nor the sounds of passing vehicles and gunfire; the snipers were hard at work. Zvonko needed lectures on the English Renaissance, Romantic poetry, and the Civil War writings of Walt Whitman and Emily Dickinson. The important thing, he said when I raised an eyebrow, was for his students to hear me speak English. And he introduced me to his class without mentioning the fact that my program had been canceled. Zvonko, the eternal optimist, was one of very few who thought the spring offensive would not occur; peace might even be in the offing now that the Bosnians were nearing military parity with the Serbs. Not that he had illusions about the war. When I began my lecture on Wyatt I remembered his warning to anyone looking for heroes in Bosnia: "You guys always want to find the white. There is no white in Bosnia. All sides are guilty now."

I recited "They Flee from Me" with all the passion I could muster. I had to shout to make myself heard.

"Any writer or journalist who wants to retain his integrity finds himself thwarted by the general drift of society rather than by active persecution," George Orwell wrote during the Spanish Civil War. Into my notebook I had copied these lines from a newspaper article on films about the war in Bosnia, along with the reporter's assertion that "The drift in European capitals and in Washington has been twofold: politically it has been toward the cynical containment of the conflict; culturally it has been toward the triumph of narcissism." Had I returned to Sarajevo to combat my own cynicism? What a thought. But one night in Zagreb, waiting for another plane, I had lost my nerve and called a relief worker to ask her what purpose I would serve by teaching poetry in Sarajevo.

"Get real," said my friend. "What else *can* you do?"

In Zagreb I also called Jadranka Pintarić. The art critic believed the Bosnian-Croat Federation was doomed, and she said the peacekeepers had to leave, the Russians especially: they were corrupt, they were siding with the Serbs, and so on.

But the UN has kept humanitarian supplies flowing in Croatia as well as in Bosnia, I said.

"What you will do in Sarajevo," she said, "is no different than giving chamomile tea to a dying man. What he needs is medicine. Force and diplomacy. You will leave, and he will die anyway."

My next question to her was the one I least I wanted to pose. Before my first trip to Sarajevo she had told me I would meet with misfortune, a prediction I had not taken seriously. Accordingly, before each of my subsequent journeys, I listened more closely. What do you foresee this time?

There was a long silence.

"Jadranka?" I said.

"I have an unsure feeling," she said. "Maybe it will not be a disaster."

Zvonko Radeljković was attending an American studies conference in Seville when fighting broke out in Sarajevo. He called his wife, Ivanka, who told him not to worry. Indeed many Sarajevans mistook the initial Serbian onslaught for gunfire from the celebration of Bairam. But the barricades erected around the city had nothing to do with the beginning of the Muslim New Year. No one knew what was going on, Ivanka said the next time Zvonko called, but she feared the worst.

6 April 1992. In Seville, Zvonko was talking with a Serbian historian when the news reached them that the EC had recognized Bosnia's independence. The historian casually remarked, "There are not many examples in history of statehood being achieved without bloodshed."

"How right he was," Zvonko would later say. "How naïve I was."

Zvonko's return plane ticket included a layover in Paris, and while waiting for a flight to Belgrade it occurred to him that he could go anywhere in the world. He had money and an American Express card, but with his family in Sarajevo he traveled on to the Serbian capital, knowing he would not be able to fly home from there—the JNA had already captured Butmir Airport. That night in Belgrade the wife of his historian friend fed him dinner and offered to put him up for as long as he liked, but in the morning he decided to take a bus to Sarajevo—the last one, as it turned out.

In an essay he recalled the dreamlike nature of his journey home: the sunlight; the cab driver who said the war made no sense; how he followed his habit of buying Italian sausage and Emmentaler cheese before boarding the bus; the way he hid 2,000 marks in his flight bag; the paramilitaries with AK-47s who forced the bus to the side of the road and took away two young men. "Dirty swine," the Serbs called them; traitors fleeing their mandatory JNA service to join the Muslim Green Berets.

A Serb from Hrasnica introduced himself to Zvonko and said, "Did you see that? Poor boys! They could have been my sons."

The bus continued on only to be stopped again at the bridge over the Drina. The passengers were ordered to get out, and the bus drove off to deploy troops massing for an attack on Zvornik, the Bosnian city across the river. JNA soldiers with white ribbons tied around their arms swarmed in the road, carrying mortars and mortar shells. Tanks rolled toward the bridge.

Zvonko thought: this is not possible. But indeed Arkan's militiamen were about to ransack Zvornik, where they would kill hundreds of Muslim civilians and dump their corpses into the Drina.

Yet when the bus returned and the journey resumed, Belgrade radio was airing warnings about Green Berets attacking defenseless Serbs, though every checkpoint between Zvornik and Kladanj was manned by the JNA. The road was empty after Kladanj, and when the bus arrived in Sarajevo the streets were deserted. Zvonko walked home to his wife, two sons, and mother.

That first summer the windows in his flat near the center of town were blown out by a shell striking next door; in the coming years his building was hit on five separate occasions; within fifty meters of his front door were the remains of three public buildings: the Supreme Court, the Red Cross, and the Oriental Institute with its priceless library.

The next summer, offered the chance to host a weekly show on a new private radio station, he decided to play his favorite music and talk about the land and literature that had shaped his thinking. Country music became his way to introduce ideas from the place that in his view personified freedom: life, liberty, and the pursuit of happiness. He called his show *Sarajevo Country Club.*

"I used to say the function of the English department was to change our political thinking," he told me during the film festival, "to make a democratic future possible. My great joy was preaching American values through American writers. I cannot give formal lectures in our building, not while it's on the first firing line. But on radio I can talk about freedom, democracy, and individualism.

"Country music has no national determination," said Zvonko, whose own lineage was a mixture of Czech, Croatian, and Slovenian. "It's down-to-earth music. The music closest to life, whatever life is. It reveals the hollowness of intellectual justifications for nationalism."

His audience, including some besieging Serbs, numbered 15,000—when there was electricity; when the power lines were cut, 5,000 heard him on battery-operated radios. Some days he walked to the station under shelling and sniper fire. Once a grenade landed less than fifty meters from him.

That day I asked him to assess the prospects for his country.

"I feel the destiny of the world is being debated in Bosnia," he said. "After all the bad things there must come something at least a little better, something we can finally be proud of. As I say at the end of each show, Don't ever give up hope. That's what the people shelling us want you to do."

Certainly Zvonko had reason to despair: his oldest friend, Nikola Koljević, a Shakespeare scholar in the English department, was vice president of Republika Srpska. But for twenty years Zvonko and Nikola had met daily for a glass of *šljivovica*; twice a week they played table tennis in the Literature Institute loft. Their families spent their vacations together, and celebrated Christmas at each other's houses, now singing Croatian carols, now Serbian. Even after Nikola rose to power in the Serbian Democratic Society and became a member of the collective Bosnian presidency, where he made increasingly hostile speeches toward Muslims and Croats, Zvonko tried to reason with him. But Nikola suffered from what Zvonko called "Serbianism"—the fear that Serbs were "imperiled," though in most Bosnian institutions they held more key positions than their numbers warranted.

"You cannot understand," was Nikola's customary response to Zvonko's arguments. "You are not a Serb."

In the months leading up to the war they quarreled, then broke off their friendship.

"Once I met him, and he said, 'Let's have coffee together,'" Zvonko told me. "'Fine,' I said. Everyone but the government knew the Serbs had stationed rocket launchers all around Sarajevo. So I said to him, 'Did you put those guns in the hills to kill my sons?'

"'It's a long story,' Nikola said. 'We'll talk about it after the war.'

"'No, we won't,' I said.

"'Why can't you trust me?' he said.

"'How can I trust you,' I said, 'when I saw you on TV saying half the population of Dubrovnik is Serbian? What did you do—implant Serbian genes into them when they were asleep?'"

Zvonko sighed. "I told him he betrayed three things we had in common—literature, music, and the middle class. Propaganda is a negation of literature, *gusle* as a form of propaganda negates the jazz we used to play together, and by opting for Greater Serbia he destroyed the middle class in Serbia and Bosnia. I always thought the survival and flowering of the middle class was the only hope for the Balkans. Otherwise we'll just have a series of peasant wars."

He shook his head. "Why did he do it? When he was a boy he was taught that true Serbian values live only in peasants, the same peasants who want to exterminate any trace of Turkish rule in the Balkans. He used to speak out against capital punishment and state violence. Look at what he's accomplished: more than 10,000 people killed in Sarajevo, including Serbs, in a war he helped start."

Sarajevans had another explanation for Koljević's change of heart: the death of his teenaged son in a skiing accident plunged the scholar into a severe depression, for which he sought psychiatric treatment from Radovan Karadžić. The cure was Serbian nationalism. Zvonko shrugged. "One of my neighbors said to me, 'Koljević was your friend, please explain to me his motivation.' I exploded. I really cannot. I don't understand him. I don't feel I know him at all."

Money, money, money. The Bosnian government had "lost" a $250 million donation from Turkey. Some believed the money went to arms purchases. Others thought politicians had tucked it away in Swiss numbered accounts. Meantime, UNHCR had closed its operation in Split because the Croatian government had started charging exorbitant landing fees at the airport. And it was said the new tunnel being dug under the airport would hurt the French peacekeepers rumored to be stealing supplies from UNHCR flights, in the same way that goods delivered over the blue routes cut into their profiteering.

Even a story that caught the world's imagination was not immune from this contagion: it was said the lovers killed on a bridge while trying to escape from Sarajevo were carrying cash earned on the black market, perhaps as much as $100,000. Their corpses lay on the bridge for a full week, riddled with incendiary bullets, because the Bosnians and Serbs were determined to keep each other from getting the money. Had Goran Simić known this when he wrote "Love Story"?

> The story about Boško and Amira, who tried to cross the bridge out of Sarajevo believing the future existed on the other side where the bloody past was already underway, was a central media event in the spring. They were already dead in the middle of the bridge and the one who pulled the trigger wore a uniform and was never declared a murderer. All the world's newspapers wrote about them. The Italian newspapers published stories about the Bosnian Romeo and Juliet. French journalists wrote about a romantic love which transcended political borders. The Americans saw in them the symbol of two nations on the divided bridge. The British explained the absurdity of war on their bodies. The Russians were silent. The photographs of the dead lovers floated into peaceful springs.
>
> Only my friend Pršic, a Bosnian, the soldier who guards the bridge, had to watch each day as worms, flies, and ravens finished off the swollen bodies of Boško and Amira. I can hear him swearing as he places a gas mask over his face when the spring wind from the other side of the bridge brings the stench of their decaying bodies. No newspapers wrote about that.

Goran and his family were luckier. They were in Italy, about to emigrate to Canada. Ferida Duraković was at an artist's colony in Oregon, insisting she would return in time for the spring offensive. Meanwhile, Sarajevans were selling their belongings on the street. Crones picked through piles of garbage. And the old man sprawled on the sidewalk behind the Presidency building? He had not been shot. He was drunk and slurping water from a puddle.

The British Council had a full house for its production of *Alice in Wonderland* by a troupe of Bosnian children. Musicians—a guitarist, a female singer, several girls with shakers—began to play. The first act was a pantomime: a boy tried in vain to get five girls to pay attention to him. Then a dozen girls, the jurors, danced solemnly around the stage, holding red candles. Next came the trial over the stolen tarts, only this man was condemned to death, without judge or jury: "Sentence first—verdict afterwards!" a child cried. Puppets followed: a monster engulfing a turtle and a hare. It was, in Alice's words, "Curiouser and curiouser!"

The lights came up. On the stage was an empty chair. Only the musicians knew the show was over. When they began to twitter, the audience clapped.

Britain's first ambassador to Bosnia took the stage. He was deeply moved by the children's performance. As a matter of fact, he was crying. It seemed he had gone around the bend. His tour of duty was over; the embassy in Zagreb had sent a delegation to escort him home.

"I shall be leaving you with love in my heart," he said.

"Not a moment too soon," one of his colleagues whispered to me.

Three British truck drivers were spending the night at the IRC house after being trapped for a week in Serbian territory. They said this would be their last delivery for some time. And they were glad to be leaving in the morning even though they would have to drive over Mount Igman. Last week, the Serbian Navy had gone on maneuvers in the Adriatic Sea. This week the U.S. Navy's Sixth Fleet moved into position. Butmir Airport was closed. Total war was coming.

"The Serbs know the Bosnian Army's growing, and they won't let that happen," said one driver. "They'll cut off the head of that monster before it gets too big."

Radio Zid (Wall) was housed in a ground-floor apartment at the base of Koševo Hill. Across the plaza was a UNHCR bread distribution center; further on, the remnants of a tram that used to ferry people up the hill. Sandbags were piled inside the station. Posters hung from the walls: *Disunited Nations of Bosnia and Hercegovina*; *UN = United Nothing*; an ad for Coca-Cola reading *Enjoy Sarajevo*; *Welcome to the Winter Olympics*, with coils of barbed wire replacing the Olympic rings.

The station's country music library belonged to Zvonko. Journalists and humanitarians brought him tapes and appeared on his show. When I joined him on the air we discussed the debate in American universities over multiculturalism, in between songs by Bonnie Raitt, Willie Nelson, and Emmylou Harris. He thought the debate resembled the arguments that had torn his country apart.

Do you find it paradoxical to hear this music in Sarajevo? he asked me.

I said I had come to expect the unexpected in Bosnia.

"Very well," he grinned, signaling the engineer to start Elvis Presley's "Heartbreak Hotel."

"Then what are your feelings about Elvis?" said Zvonko.

"In Sarajevo," I replied, "it's obvious Elvis lives."

Zvonko laughed. "You believe in hope," he said, and closed his show with his refrain, "Don't ever give up hope. That's what they want you to do."

I looked at my watch: eleven o'clock. "What about the curfew?" I said to Zvonko and Ivanka

Their kitchen table was littered with plates, empty wine and brandy bottles, overflowing ashtrays. Smoke filled the air. We were quite drunk.

"You should experience everything in Sarajevo," said Zvonko. "Let's see what happens."

Out into the drizzle I stumbled, vaguely remembering that Serbian snipers were armed with night vision scopes, supplied, it was said, by Russia. I recalled a joke about the curfew. Two soldiers at a checkpoint see a man in the street at 9:45. One soldier shoots him, and the other asks why.

"I know where he lives," says the first soldier. "He'd never make it home by ten."

I walked up Đure Đakovica, past the new American Embassy (in its courtyard I had once spent forty minutes behind a tree, under sniper fire), and when I turned the corner by the London Café two Bosnian soldiers asked to see my papers. I fumbled for my passport and blue card. I told them I had only a few hundred meters to go. Suddenly they were laughing.

"Thank you very much," they said in a mock English accent.

Up the stairway by the destroyed tram I hurried. Firefights began in the mountains; tracers lit up the sky. Eric was watching TV when I walked in. While state TV showed troops cleaning their weapons interspersed with explications of passages from the Koran, a satellite dish could rein in Italian movies, MTV, or coverage of the O. J. Simpson trial. Eric tuned to the NBC Superchannel to catch the beginning of "Real Personal," a talk show about sex. A black man and a white woman had just returned from a week in the Caribbean, where a sex therapist had taught them how to masturbate. To save their marriage, the therapist said with a straight face. Hers was a severe regimen: her patients were not to make love until their last day at the resort. They had not disobeyed her.

"You see," the wife explained, "this is our third marriage."

"And I must confess," the husband said solemnly. "I came to masturbation rather late in life."

"Did I just hear that?" I asked Eric.

"I thought that was the one skill we were born with," he said.

Outside, gunfire filled the air.

Zvonko's study was rented out when he was a boy, first to a pair of Serbian prostitutes, then to a secret policeman; the government decree that a family's "fourth room" go to the refugees streaming into the city after the war was not rescinded until Zvonko was a teenager. Now used for entertaining and as the master bedroom, it remained Zvonko's studio. Photographs of Thoreau and Hemingway hung above his desk, along with an Eliot-inspired painting titled "The Unreal City," completed by an artist friend before the war; on his bookshelves, among recent *Best American* anthologies of poetry, fiction, and essays, were plane ticket stubs from trips to the United States, a sign reading "Don't just sit around doing nothing: Get Drunk!," and a gift certificate for a winery near Dubrovnik that was destroyed in the war. The window by his desk was covered with plastic sheeting. Zvonko was reading on the sofa the day the flat next door was blown up. Shattered glass fell around him. The Serbs were aiming for an army printing press rumored to be in the next building.

His mother's side of the family had lived in Sarajevo since the nineteenth century, when his great-grandfather, an Austro-Hungarian Marxist, came from

Bohemia to work on the railroad. His grandfather was an architect whose designs included the façade of the National Museum. His mother, a teacher of German, married a man from Travnik, where Radeljkovićs had been merchants for 250 years. But that was all that Zvonko knew about them: his father, a Partisan leader, was killed in mysterious circumstances in the last days of World War Two, either by the Nazis or by other Partisans.

Hence his dim view of politicians and intellectuals claiming to have special knowledge about right and wrong, the certainty once promulgated by Communists and now by nationalists. Raised by his mother and her parents, Zvonko could not remember a single argument from his childhood about ethnicity or faith. His friends belonged to all Balkan nations—Serb, Croat, Muslim, Jew—and many were of mixed origin; in those days quarrels were ideological, revolving around support for the Soviet Union or the United States. Zvonko's sympathies lay across the Atlantic.

On Sunday afternoons he would listen to "Country Countdown" on Voice of America; as a teenager he had his own combo, which played British and American music at dance halls; late at night he and his friends tuned in Radio Luxembourg to hear new songs by Conway Twitty, Marty Robbins, and the Everly Brothers. His undergraduate degree at Sarajevo University was in English literature; Thoreau was the subject of his masters thesis at Indiana University and of his doctoral dissertation in Sarajevo. He treasured his visits to America, and it pained him that on a 1987 USIA tour he could not convince his colleagues to go to Graceland. But he did spend three days with Nikola Koljević, who was a visiting professor at Hope College in Michigan. Zvonko lectured there on Hemingway.

"We had crazy times," Zvonko remembered. "Parties lasting till four in the morning."

But now several of his former colleagues were members of the Bosnian Serb Parliament. He could say without a trace of irony that some of his best friends were Serbs, almost all of them, in fact. He recalled the night an official from UNHCR visited him. During a break from intensive negotiations that afternoon in Pale the official had found himself alone in a room with Koljević.

"Zvonko Radeljković sends his greetings," he said to the vice president.

Stunned, Koljević said, "Zvonko has a happy family," and walked out of the room.

"Why did you go there—to negotiate or make trouble?" Zvonko asked the official. "All you've done is remind him of what was."

But Koljević was right: Zvonko had a happy family. His mother was still alive. He and Ivanka were proud of their long marriage; his older son, Nikola, following in his grandfather's footsteps, was studying design in Zagreb (after a stint in the army); and Gigo, the youngest, was a guitarist in a popular blues band. When I asked what the band was called, Zvonko said, "*Grateful Deaf*."

Gigo immediately corrected his translation. "*Blessed are the Deaf*," he said with a smile.

I did not know why the audience was laughing. For Winter Carnival I had agreed to give a reading, but when I arrived at the Writers' Association one rainy afternoon I learned that my part in the program would be small. Zvonko and Tvrtko Kulenović would deliver disquisitions on my work, an actor would recite three poems in translation, then I would read the same poems in English. I sat at a table between Zvonko and Kulenović; the actor was off to the side, where the TV camera could get the best shot of him. The wall behind us faced Mount Trebović. I wished it was not made of glass.

In his writings Kulenović drew parallels between this war and the Trojan War, another example of genocide and urbicide: "In genocide you kill those you do not like. In urbicide you do not discriminate: you destroy everything." And while he spoke about me—it was unnerving not to know what he said—I had my notebook open to a page on which I had copied down figures, from the Bosnian commission on Serbian war crimes, on detention camps, mass murders, and razed villages. The evidence—testimonies, photographs, videos, captured documents—could be used only to open investigations at The Hague, but it was more than enough for me to wonder, what good is poetry in the face of these atrocities? That was when Zvonko concluded his remarks with the joke I did not get. The audience was still laughing when the actor began his recitation, the cameraman filming him from several angles. The joke, Zvonko whispered, concerned a Russian poet named Eduard Limonov who went to Pale to lend support to the Serbs and was filmed firing several rounds from a sniper's rifle down into Sarajevo. The punch line was that for Bosnians there were only two kinds of poets, those who were shooting at them and those who were trying to help. With that I started to read.

"I did not shoot the wretched in the dungeons," was the line of verse Osip Mandelstam never tired of reciting to himself on winter nights in Moscow. Composed by his friend Sergei Esenin, who in the West is remembered more for his brief marriage to Isadora Duncan than for his poetry, this line was for Mandelstam "the symbol of faith, the true canon of the genuine writer, the mortal enemy of Literature." Clearly, Esenin's verse did not make the same impression on Eduard Limonov.

"In the years of Buchenwald and Auschwitz," Odysseas Elytis writes in "Chronicle of a Decade," much of it set in Nazi-occupied Athens, "Matisse painted the juiciest, rawest, most enchanting flowers and fruits ever made, as if the miracle of life itself discovered it could compress itself inside them forever. Today, they speak more eloquently than any macabre necrology." The Greek poet goes on to suggest that "An entire contemporary literature made the mistake of competing with events and succumbing to horror instead of balancing it, as it should have done."

This is what I said, more or less, in my talk at the European Club. In an office overlooking Marshal Tito Boulevard thirty students had gathered to discuss the writer's role in war. They sat in overstuffed chairs, some watching MTV, others reading magazines. I quoted from Czesław Miłosz's "The World," a suite of poems celebrating the natural order, composed during the Nazi occupation of Warsaw. But what language should we write in? one student asked. Many of us have given up on Bosnian. I offered the example of the Romanian poet Paul Celan: after Nazis murdered his parents, he chose to write in German, which he said "had to pass through the thousand darknesses of deathbringing speech." He wanted to redeem the language used to justify the most heinous crimes against humanity. Yes, said another student, staring at the television, and he was a suicide.

Then I made the mistake of quoting Zvonko: "These are the best years of your life, wherever you are, even if it's in Sarajevo. You have to make the most of your fate."

"We're living in a prison camp!" cried Amra Pandžo, the organizer of the event. "Don't talk to us about choice."

After my talk, a young man from Dobrinja came up and introduced himself as the custodian of one of Sarajevo's biggest libraries: the American Center's, which he had saved early in the siege. He asked for a copy of my book, but it was not my poetry that interested him. When I turned to talk with Amra Pandžo, I saw him reach into a cardboard box and withdraw a dozen condoms. These he stuffed between the pages of my book, then stole out of the room. I mentioned this to my hostess.

"What do you expect?" she said impatiently. "What else is there to do in Dobrinja?"

And Auden said, "poetry makes nothing happen."

Early Sunday morning I woke from a dream in which someone was knocking at my door to hear a 120-millimeter shell crash in a nearby park: the Serbs were aiming for a mortar near the Residency. I went downstairs to sleep on the floor. The gunners were determined to keep the city awake. Large shells landed every hour until dawn, killing one man and wounding several others: retribution for the sniping deaths of two Serbian girls in Grbavica the day before.

My lectures complete, after breakfast I packed and went to the airport to see if I could get on a flight to Zagreb. I did not feel hopeful. The airlift had been suspended during much of my stay, the clerks in charge of flight manifests took pleasure in denying journalists and humanitarians places on the planes, the line of peacekeepers ahead of me was long. When it was my turn to hear a clerk say he had no record of my requests to get on a flight, a peacekeeper ran into the room with news of Serbs firing on the plane landing with UN Special Representative Yasushi Akashi and two military officers on board; a bullet had ripped through one officer's garment bag, leaving a hole in every shirt. The airport went on red alert. The airlift was suspended. I hitched a ride back into town.

"We have such an advantage, technologically speaking, over the Serbs," the UN Civil Affairs officer was telling me over lunch at a pizzeria near Koševo Hospital.

"We can hear and see everything they say or do. You wouldn't believe what the photographs show. And we have any number of military experts paid to guess what the Serbs will do. The problem is, we don't do anything with this information."

A soldier walked up the street carrying a bag of groceries.

"But I can tell you one thing," said the UN officer. "The spring offensive won't start today."

"How do you know that?" I said.

He smiled at the soldier. "The Bosnians can't do anything without their logistics man."

The air-raid signal blared. All afternoon I had worked at a furious rate, digging up first the vegetable garden then a flower bed at the IRC house. Eric was pruning apple trees. The guards trimmed the hedge and burned debris. My frenzy had nothing to do with Martin Luther's vow to plant a tree on the last day of the world. Spring was in the air—this was the first sunny day in over a week—and I had nervous energy to work off. The warmer weather brought others out. The hill below the mangled radio and television tower, riddled with new underground munitions factories, was shrouded with smoke from burning trash. There was the sound of hammering across the valley. Then shelling in Stup. Then shooting in the Jewish Cemetery. Eric chose this moment to tell me the average firefight lasts fifteen seconds. I asked him when he thought the blue routes would be reopened.

"In about eight months," he laughed, "when the spring offensive and then the war are over."

Bowing to international pressure, Tuđman had decided to let UNPROFOR stay in Croatia, in reduced numbers, but peace was not at hand, in Bosnia or Croatia. Fifteen Bosnians had been shot this week in Sarajevo alone (including one this day at the Faculty of Philosophy); and in the last several days the Serbs had made a number of helicopter flights over Mount Igman—in the no-fly zone—to supply troops; stolen two UN vehicles and 120 flak jackets; blocked aid convoys to the eastern enclaves; and arrested five humanitarians who made a wrong turn near Sarajevo. The aid workers were accused of providing logistical support to the Bosnian Army. When they would be released was anybody's guess.

Muha, the technician who used to live in the basement, came by with a letter for me to mail to his wife, Mirna. A year ago, he had gone to Germany to join her and immediately returned, a shattered man. Mirna was living with someone else. Muha lost his mind. His in-laws were taking care of him now. His father-in-law appeared in the driveway. He shook his head when he saw me.

"Always trouble when you come," he joked. Then he took Muha's arm and led him home.

A dance performance at the National Theater, an exquisite Austro-Hungarian building with 5,000 lights, red velvet curtains, mannequins in baroque dress—and no heat. A Renaissance group from Brussels performed on period instruments, which they could not tune in the cold. Then a woman performed a long dance around a tub. She would climb up onto its rim, sway back and forth, as though trying to decide whether to jump in, then run away. At last she made the plunge, splashing water on the stage. She stayed underwater for a full minute. The audience howled.

"She's had one bath more than anyone here has in the last week," said a humanitarian.

The dancer took her bow, flinging water from the tub toward the crowd.

The old man was selling books on Marshal Tito Boulevard: Richard Howard's *Alone with America*, Randall Jarrell's *Poetry and the Age*, John Ashbery's *A Wave*. A curious assortment for a bookseller who spoke no English. Zvonko opened a book and read the inscription. These were the remains of a well-known Bosnian poet's library. This poet had made a bold declaration early in the war: if George Bush did not intervene to lift the siege of Sarajevo, he would kill himself. Then he had fled to America. His neighbors had plundered his apartment with uncommon zeal.

"A dirty business," Zvonko muttered. "Do you need a copy of *A Wave*? Look: only three marks. What a deal."

We walked to the covered market. With the airlift suspended, Zvonko was showing me some of his favorite places. The weather had turned cold and overcast again. The cease-fire was over. Battles raged in the Jewish Cemetery; a sniper had just killed a woman near the Holiday Inn.

"They've decided to close the university this week," Zvonko was saying, "because of all the sniping *last* week. Now we'll have a week of peace. It always happens like that."

In the market he wanted to see if prices were rising now that the blue routes were closed for good. But only brandy and gasoline were more expensive. Here were potatoes, onions, garlics, leeks, apples and oranges, cigarettes and beer, various cuts of meat, a wooden pallet piled high with bags of stolen flour, each bearing a picture of an American flag and a warning—in English—not to resell it.

"Quite reasonable," Zvonko said of the prices. In the first year of war he had lost more than fifty pounds, and while he had gained some back with the help of the airlift and the blue routes he did not think he would soon grow fat. "Unfortunately, we have no income."

With the sound of gunfire in the distance, we headed for the old part of town, passing the central bank which had lately received a fresh coat of paint, courtesy of Middle Eastern money. The bright green façade stood out among the ruins (an estimated 600,000 shells had struck the city, damaging or destroying 60 percent of the buildings), like the refurbished Iranian Embassy along the Miljačka River. Rebuilding was underway in Sarajevo. Engineers were planning to begin restoring the airport in April, during the spring offensive.

"April *is* the cruelest month," said Zvonko.

On to the river, where we paused by the spot where Franz Ferdinand and his wife Sophie were assassinated on 28 June 1914. Gavrilo Princip and his comrades from *Mlada Bosna* (Young Bosnia, a group with ties to the Black Hand) chose Saint Vitus Day to commit this crime. It was a sacred date in Serbian mythology: on this day in 1389 Sultan Murad's troops had defeated the Serbian army on Blackbird's Field in Kosovo, and on the 525th anniversary of that battle Princip fired the most decisive shots of the twentieth century. Seven years later, Yugoslav authorities proclaimed their constitution on Saint Vitus Day, by which time the busts and plaque erected in Sarajevo to commemorate the royal couple had already been torn down. In 1930, on the façade of the building before us, these words were engraved: "On this historic spot Gavrilo Princip, on Saint Vitus Day 1914, heralded the advent of liberty." (Freedom from a different yoke was heralded by Yugoslavia's expulsion from the Cominform on 28 June 1948.) The footsteps pressed into the sidewalk to mark the place where the assassin fired his revolver became a popular tourist site (a monument to Serbian infamy, said Winston Churchill), which did not survive the siege.

"A Serb terrorist," was how the policeman standing guard described Princip.

Here was where the twentieth century began and ended. Historian John Lukacs suggests that ours was a short century, bracketed by Ferdinand's assassination and the Berlin Wall's dismantling. But his calculations were slightly off. A little remarked upon event from this war provided better closure. French President François Mitterrand had his eye on history when he made a secret flight into Sarajevo on 28 June 1992, a daring gambit that opened a humanitarian corridor and scuttled serious talk in the international community of using military means to lift the siege. That was the day the West prolonged indefinitely the suffering in Bosnia, placing its seal upon history's bloodiest century.

"Mitterrand knew he had cancer," said Zvonko. "He could be another Franz Ferdinand. Why else did he fly in on that date?"

Across the Mayor's Bridge we went to Saint Antony's Church, where all but three of the thirty Franciscan monks in residence before the war had gone to Croatia. "They have a lot of experience with moving like this," Zvonko said, deliberately understating their checkered history in Bosnia. And when we walked back across the river some minutes later he pointed at an old woman in a Muslim cemetery. She was laying flowers by a gravestone, making the sign of the cross.

"You see how Islam has been corrupted?" he said. "In a true Muslim graveyard there would only be white stones. The devout mix customs from every religion. Whatever works, I guess."

There was a larger point to make. In this place of shifting beliefs and alliances, many Bosnians traded one folk religion for another because, as Noel Malcolm writes, "where the basic attitude to religion is practical-magical, even the most important ceremonial elements of one religion can be borrowed by another—or rather, *especially* the most important, since these are thought to be the most powerful." The woman before us was calling on every divinity, every power, she knew.

Sarajevo offered no shortage of spiritual choices, even after so much fighting. Within five hundred meters of this cemetery you could find a mosque, an Orthodox church, a Catholic church, and a synagogue. For half a millennium four confessions had lived side by side in relative harmony, and the siege had not destroyed that tradition. Next month, Jews, Croats, Muslims, and Serbs would gather for a special Passover service, at which the Sarajevo Haggadah, a fourteenth-century illuminated prayer book telling the Exodus story of the Jews rescued from slavery in Egypt, would be opened for only the third time in the last hundred years.

Our next stop was the coppersmiths' street, a cobblestone block of shops which had thus far survived the war untouched. The windows were filled with pipes, coffee services, and jewelry.

"Ivanka and I used to walk here every night after dinner," Zvonko said. "That's what I miss."

And as we climbed a hill toward the Muslim barracks built in the Austro-Hungarian era he gazed wistfully at the mountains beyond. On one slope was a new cemetery, on the other, Serbian positions. Zvonko and his wife used to hike up there to pick mushrooms, return by way of the barracks, and stop at a café overlooking the city. All that remained of the café was its foundation; the wood had been sawed up and burned; likewise the trees that once bordered the road winding into town, which was no longer safe to walk down.

A Habsburg commander, Prince Eugene of Savoy, was the first to raze Sarajevo, during the last Christian crusade against the Ottomans. He reached the city on 22

October 1697 and, finding the Turks unprepared for battle, demanded their surrender. In his diary he recorded:

> On 23 October I placed the troops in a broad front on a height directly overlooking the city. From there I sent detachments to plunder it. The Turks had already taken the best things to safety, but still a great quantity of all sorts of goods remained behind. Towards evening the city began to burn. The city is very large and quite open; it has 120 fine mosques. On the 24th I remained at Sarajevo. We let the city and the whole surrounding area go up in flames. Our raiding party, which pursued the enemy, brought back booty and many women and children, after killing many Turks. The Christians come to us in crowds and ask for permission to come into our camp with their belongings, since they want to leave the country and follow us.

Some things never change.

The air-raid siren blared.

Zvonko laughed. "Probably the all-clear signal from the shelling yesterday."

Through one of the two ancient gates (both shelled by the Serbs) we wandered on to Markela Market, the site of the massacre in February 1994 that led the UN, backed by NATO, to create a heavy-weapons exclusion zone around the city. Zvonko was in no mood to talk about what he called "that savagery." He was looking for flowers for Ivanka—it was their anniversary—but only wilted carnations and plastic flowers were available. The plastic made him grin.

"An old literary theme," he said, "the difference between appearance and reality, which means something altogether new in Sarajevo. I'm afraid I'll have to buy her green onions."

Further on, though, was a newly reopened flower shop, where Zvonko found a single yellow lily for the exorbitant price of ten marks. Supplied with brandy, onions, and a lily, we walked on to the Presidency, which was surrounded by UN armored vehicles and soldiers with walkie-talkies, one of whom directed us across the street, into a crowd of men and women coming home from work.

"Looks like the Presidency's under siege," said Zvonko, hurrying toward his flat.

Inside, Yasushi Akashi was making no progress in his talks with the Bosnians: their offensive would begin within the week. The Serbs would counter by intensifying their shelling of the city, striking the Faculty of Philosophy, for instance, with wire-guided rockets, forcing authorities to move classes downtown, to the basement of the Law School. By June the capital would be on permanent alert, large groups of people forbidden to gather, classes canceled for the summer.

But today we were celebrating the Radeljkovićs' thirty-first anniversary. We sat at the kitchen table with Gigo and Vlado Sučić, one of two remaining French professors, and drank coffee until it was time to drink brandy. Wine and brandy bottles filled with water lined the cupboards and window sills; a makeshift gas burner attached to a length of tube resembling an intravenous drip dangled above the stove. Zvonko told stories into the night: of being seated next to Jacqueline Onassis at a dinner in the Kennedy Library, at the opening of the Hemingway room; of how the war had weaned him of his love of table tennis and chess; of Vlado's efforts to open the faculty each morning. But now Vlado and his colleague were being considered for positions at the Sorbonne. Vlado rose to his feet.

"If you go to Paris," Zvonko said when his friend started for the door.

Vlado nodded. "It will be the end of our department," he said, and walked into the dark.

Zvonko closed the door. "Everybody has the right to be afraid and to leave. But I will stay as long as I can," he said defiantly.

She wanted me to see her frescoes. On the walls of her apartment the old woman had painted pictures of what she missed during the siege—the sun; fruits and vegetables, tea and sugar, a box of cereal; a jungle scene from a Kipling novel burned for the sake of a meal; her daughter and granddaughter, who were living in a refugee camp in Holland; her husband, killed at the start of the war. Here was a vision of yearning in the style of Henri Rousseau: a petition to the gods of desire, counterpointed by the rattle of gunfire. Overlooking everything was a large eye: the Buddha's.

"I became a Buddhist," she said, "because there are no Buddhists among our nationalists."

The whole city was a graveyard. Parks, soccer fields, tennis courts: there were new white markers everywhere. With the airlift suspended indefinitely, one afternoon I accompanied Zvonko to the old Catholic cemetery, where his father and grandfather shared a plot next to a German monument cut in two by a grenade. Dogs prowled in the overgrown grass. Magpies scavenged for food.

"My ambition," Zvonko said, "was to create the language of freedom through my teaching and *Country Club.* From that may come the institutions. We haven't had democracy for fifty years, we don't even have the vocabulary for a free market. First you have to name the animals. That's why Emerson felt like a new Adam naming the New World, naming America. We have to do that here."

Nikola Koljević was still on his mind. "Do you know, he used to call Milica, his wife, *my Milicia*, which in our language means *my policewoman*. So you see: he took his bad marriage out on Sarajevo. I'm serious. He told a journalist he had to do something to interrupt his middle-class life. That's one reason why this war started. Unpleasant, yes, but then reality tends to be unpleasant."

"And yet you don't leave," I said.

"I would be debased in my own eyes to leave like a dog under hail," he said. "Even if one decided that it was absolutely necessary to leave, one would want to do it in a dignified way."

He was reminded of "A Letter from 1920," Ivo Andrić's story about a doctor from Sarajevo who moves to Paris after the war. "Bosnia is a country of fear and hatred," the doctor writes to a childhood friend. "Maybe in Bosnia men should be warned at every step, in every thought, and in every, even the most elevated, feeling to beware of hatred, innate, unconscious, endemic hatred. Because this backward and poor country, in which four different faiths live packed together, needs four times as much love, mutual understanding, and tolerance as other lands." Believing Bosnians cannot do that, he refuses to return to his homeland. Yet in Paris he becomes known in the Yugoslav colony as "our doctor," treating workers and students for free. Inexplicably, when the Spanish Civil War breaks out, he joins the Republican army as a volunteer and sets up a hospital, where he is killed in an air attack. "And so ended the life of a man who had fled from hatred" is the last line of the story.

"The doctor was right," Zvonko said. "Bosnia *is* a country of fear and hatred. But then so is every other country. He proves that by going to Spain, where fear and hatred are what kill him.

"Think of your own Southern gothic novels—take Faulkner or Styron—and you'll see fear and hatred are characteristics even of 'this most benevolent republic,' as Hawthorne described it. These are general human conditions, which aren't overcome by running away. The doctor in Andric's story says, 'Only in flight is there salvation,' but that's philosophically wrong—and cowardly. You have to face these things. You have to sail through them any way you can. If you run away, you don't help yourself or the land. Both lose, you and whatever the country is."

"It's man against the elements"—that was how Andy Loescher, a British relief worker, described the drive over Mount Igman. "The most exciting work I've ever done," he said on my first day in Sarajevo. On my last day I learned that he would drive me over the route that claimed, on average, one life daily. Three OSI administrators, unable to fly to Sarajevo, had gone to Split; with luck we would meet them in Konjic, and Andy would drive them back to the capital.

He was a short, barrel-chested man with a gold front tooth. A roofer by profession, he gave as his home address a pub in Sussex, which he did not expect to visit soon. Thirty times he had made the trip over Igman, but with the coming offensive he expected to find himself trapped in Sarajevo.

This day was overcast and blustery. In the leafless birch outside the office where I met him, next to a mosque whose minaret lay on the ground like a felled tree, a crow was cawing. I took it as an omen and followed Andy and Menja, a young Bosnian, into the APC. We rolled through the deserted streets and came to the checkpoint in Dobrinja, where two Italian journalists hitched a ride. The airport was on red alert. At the end of the runway was an Ilyushin 76 with its nose buried in the macadam; on the far side was a checkpoint manned by two peacekeepers who made us sign waivers of responsibility, despite the UN's mandate to protect Mount Igman. Around a corner we stopped in the village of Butmir. Every house was leveled, and at the checkpoint a Bosnian soldier emerged from one dynamited place to warn Menja of heavy shelling and sniping ahead.

"You're a dead man without me," Menja grinned, punching Andy. "Everything's on red alert."

A hundred meters on were the new and old tunnels, and a line of people of all ages loaded down with goods. Then we were bouncing along a rutted road into the ruined village of Hrasnica. In the schoolyard two veiled women sat by a fire; a small boy warmed his bare feet over the open flame. An old man was leading a pair of horses up the road, his wooden cart filled with hay. Soldiers stood around in twos and threes. At the next checkpoint, at the base of Mount Igman, we were told it was too dangerous to get out of the APC anywhere on the road.

"You're a dead man without me," Menja repeated solemnly as we started up the mountain. Nearing an exposed corner protected by a concrete oval, he cried, "Fast here, Andy!"

We wheeled around the corner, tires spinning in the mud, and headed east along the side of the mountain, climbing slowly toward thick clouds. Two girls walking down from the pass huddled by a rock. At the next switchback we began to rise through fir trees. Branches and tree limbs were strung together, in crude imitations of snow fences, to protect against snipers and shells. Andy pointed out the wrong turns that would take us directly to the Serbian side. In the forest, snow covered the ground and fog enveloped us. The muddy road turned to ice. A light snow was falling. Woodcutters were lopping branches off felled trees, and from the woodcutters' cabins soldiers kept watch; in a hunter's hut was a dead man slumped in a wheelbarrow.

"Can you light me a cigarette?" Andy asked me.

An old man was leading a horse with a pair of wooden boxes on its back. It was as if we had entered a medieval world. I remembered hiking in the Pohorje Mountains: the wind in the trees by the woodcutter's house; the couples arguing over the origins of folk songs; the sound of a man sharpening his scythe. At the bottom of the steep cliff, four hundred meters below, were rusting buses, trucks, and cars. The one-lane road was a sheet of ice laid on packed snow. Andy was going less than ten kilometers an hour, keeping the front tires in the ruts so as not to spin out of control. He pumped the brakes constantly as the road sloped downward.

"I'm glad you're driving," I joked, and then the APC was skidding sideways.

Any skid seems to last forever. The APC picked up speed, the extra weight of its armor now a liability. Trees flashed by. No one said a word. The only sound was

Andy grunting as he tried to steer the APC into the bank. Roots and rocks, icicles and snow, patches of grey sky—the world was passing quickly. What a way to die, I was thinking, when the front bumper crunched into the bank. The APC, fishtailing, bouncing in and out of the ruts, finally slowed to a stop.

"We're on the edge!" cried one Italian, peering out the back window. "It's a *long* drop."

"Then move to the front," our driver advised.

He took a deep breath. He rocked the APC away from the cliff and straightened the wheels. He turned off the engine. We got out to inspect the damage: only the bumper was dented.

"Not too bad," I said hopefully.

"I know someone who can pound that out," said Menja. "You're a dead man without me."

Andy shook his head. "It's an expensive machine," he groaned.

We stared at the APC, as if it were a bad dog. The wind had picked up; snow was swirling around us. I wondered if we were in anyone's sights. Andy took out the tire chains, laid them on the snow, and cursed. The chains were hopelessly tangled. A half-hour later, Andy threw the chains behind the seat and off we went, even slower than before. I popped a Pink Floyd tape into the tape deck. Soon we were singing, at the top of our lungs, "The Dark Side of the Moon."

We passed more huts and cabins, barracks filled with soldiers, stacks of split wood. The road widened slightly, and once we pulled over to the side to let a UN convoy by. Staring into the eyes of one peacekeeper, I began to shiver. At the next checkpoint, a hut marking the halfway point, trucks, cars, and buses were lined up on either side; traffic was one way. Eventually the snow and ice turned to mud. A meadow came into view, filled with bombed-out buildings. Andy opened a Coke. Truckloads of Bosnian soldiers passed. One Italian, fresh from three days at the front, called the coming offensive "the biggest public secret in Bosnia."

The Soros people were waiting for us at the UNPROFOR compound in Konjic. I did not envy them their journey back. Nor was I pleased to learn, when the Italians and I climbed into a car, that our driver, a young man from New Jersey, had wrecked several vehicles in six months of service.

This son of Croatian immigrants sped along the Neretva River, passing cars and trucks without checking for oncoming traffic. He hated the Bosnians. "*Mujahedeen*," he sneered at the soldiers. Granite peaks rose steeply from the green water, and nearing Mostar we raced by dynamited houses. "Ethnic cleansing," our driver intoned monotonously. Forsythia and fruit trees were blooming on the outskirts of the divided city. We crossed the bridge I had watched British peacekeepers erect six months earlier, accelerated up the winding road toward the border, and entered Croatia at dusk.

This day the shelling of Sarajevo had resumed in earnest. The war was widening. We came to a ridge overlooking the Adriatic. Behind us, on Mount Igman, on the same stretch of road that had nearly cost us our lives, nine French peacekeepers in the UN convoy we had passed were killed when their APC pitched over the cliff and down the steep ravine. Pressure was mounting in Paris to pull the peacekeepers out of Bosnia. There was snow in the mountains along the coast. The *bura*, the spring wind feared throughout the Balkans, was gathering force. On we drove toward the sea.

19

Albania

30 June 1995

Beyond the cove, beyond the promontory, where the Adriatic meets the Ionic, the emerald water is warm and clear and I am floating on my back, gazing at the sky. The others watch from shore, or rest on the rocks at the base of the cliff, or swim farther out to sea, toward the island newly leased to the Americans. Bujar Hudhri coaxes his little boy through the ripples and small waves lapping against the shore, until the child, with a sudden shriek of terror, runs back to his mother. Very well, Bujar says, and begins to swim, flailing at the water in an awkward version of the crawl, his stiff arms revolving like paddle wheels. No one will be surprised if he sinks. I am so tired, he said one night in Elbasan. I am forty years old, and all I have ever heard is war against Serbs, against Greeks, against Bulgarians. People kill each other in Albania for a small piece of land—brothers, neighbors. Because there is so little land for so many people. The publisher, thin and dark and angular, has a deep, weary voice. If you are at war, he likes to say, you cannot worry about why there is no wine. That was how the government kept control—building bunkers, making rules. Only four people a year may learn English in Elbasan. Only one child from each family may finish high school. Do not talk to foreigners, not even to tell them where the hospital is. The one up the coast in Vlorë, for example—dirty, dilapidated, at the end of a dirt road lined with boulders. The patients stared forlornly from the open windows. Cracks in the ceiling and walls. Plaster and pools of dank water on the floor. Then an old peasant woman, leaning on her husband's shoulder, was standing in the entranceway, cupping her hand over the bloody hole that was once her ear. Whatever you do, said the novelist Mike Keeley, don't get hurt here. He let out a wicked laugh. I knew I would like him: I went to school on his translations of Cavafy and Seferis, Ritsos and Elytis.

We met earlier in the week, on the drive from the airport in Tiranë to the city of Elbasan, along a cracked and twisting mountain road only marginally better than the route the magical poet and painter Edward Lear, recuperating from an unspecified illness, had followed, on horseback, one hundred fifty years ago, journeying from Thessaloníki to Elbasan and Tiranë. Mike instantly charmed me with a story about Ritsos's imprisonment, during the last military dictatorship, on the islands of Yiaros and Leros. Ritsos was prepared for his incarceration. In his winter coat, the lining of which he had removed, he hid the poems he wrote each day on cigarette papers (the guards permitted him to receive unlimited supplies of cigarettes), and upon his return to the mainland he gave the coat to his publisher, with instructions to print what would be seven books. My daughter's dowry, he explained. But how to make a selection from this and other work completed under the colonels for an American edition? Mike went to Ritsos for advice. How can I select when they are all good? said the poet. Of course they are all good, said Mike, but perhaps some are better than others? Ritsos opened a book and read the first poem aloud. That's pretty good, he said. The next poem he read with more passion. Yes, that's good, he said. And so it went, for an hour and a half, until he had read the entire volume to his translator. There may be one poem, Ritsos allowed, to leave out. Mike wisely made his own selection. *The poem/ year after year/ searches for its reader*, Ritsos wrote during the dictatorship. Mike was such a reader. What luck to travel with him through Albania, the mysterious country that for the better part of a half century was sealed off from the outside world! And now our literary delegation to the land of Enver Hoxha, the most diabolical dictator of all, is coming to an end, along an isolated stretch of the coast, on the very day of the EU vote on Albania's application for associate membership. A strange turn of events. Hoxha's forty-year reign of terror, an exercise in national paranoia, was hermetic in the extreme. The United States and Britain, which in World War Two had helped Hoxha drive out Italian and German forces and consolidate his power, severed relations with him in 1946, when his antipathy to the West could no longer be ignored. The former Partisan made a practice of breaking with his Communist allies, too: with Yugoslavia, in 1948, when Tito (who had designs on Albania) defied Stalin; with the Soviet Union, in 1961, over Khrushchev's destalinization policies; and finally with China, in 1976, as a protest against improving Sino-American relations. Hoxha was the purest ideologue, a madman who condemned his people to isolation and economic ruin in the names of Marx and Lenin. Not until 1990, five years after his death, did Albanians come to their senses. Thousands fled the country, and some who stayed behind stormed Western embassies in Tiranë. They demolished the statues of Hoxha, and then, in tens of thousands, seized ships moored in the harbor at Durrës and forced them to sail to Italy: *la dolce vita*, televised images of which had begun to reach Albania in the 1980s (though you could go to prison for pointing your TV antenna toward Italy), had a stronger pull than memory and a half-century of screeds against the Fascist occupation. With the election of the cardiologist Dr. Sali Berisha to the presidency in 1992, Albania started to reach out to the world. The military base, for example, on the island toward which Mike is swimming (with considerably more grace, at nearly seventy, than our friend Bujar): it is said that what was once an outpost for *terra incognita* now houses American intelligence analysts monitoring developments in Bosnia. Albania, the poorest country in Europe (followed by Bosnia and Serbia), is enjoying a windfall, diplomatically speaking, since the international community has written Bosnia off—as a quagmire, a civil war, an ancient feud, et cetera, et cetera—and Serbia is still under UN sanctions. If all goes well today, Albania will become the first Muslim country to join the EU. And it *is* Muslim (65 percent, with 10 percent Catholic, 20 percent Orthodox), notwithstanding Hoxha's 1967 decree that Albania was the world's first atheistic state, accompanied by orders to destroy hundreds of mosques and churches. My generation had no luck, Bujar said this morning in a mosque rebuilt with funding from Saudi Arabia, in the city of Pequin. At the age we were supposed to learn to love God we were taught to hate him. The mufti, a wizened man with a grey mustache, had the thick hands of a bricklayer, his occupation under Hoxha. He was pleased to escort us to the top of the new minaret, as if it were a shrine he had built himself, not the place from which to call the people to prayer. It was an odd sensation—part forbidden pleasure, part devotional—to stand there in the sun, taking in the view: olive trees on the surrounding hillsides, horse- and tractor-drawn carts carrying cattle, Chinese trucks sputtering down the street, a Partisan monument beside the mosque, and next to that a concrete bunker. Another pillbox built by our great criminal to defend us against Americans, Bujar said sarcastically. But Americans do not even know where is Albania! The bunkers—700,000, by some estimates—are everywhere, along roads and ditches, in alleys and squares. Row upon row, hill upon hill. You can even find them, as we did, high in the mountains, at the base of a waterfall far from the nearest village or in the underbrush along an unnavigable river. Hoxha commanded his people to build bunkers instead of houses, a national project even Edward Lear—"one of the dumbs," as he described himself—could not have dreamed up, though he would have appreciated the uses that lovesick teenagers have found for the bunkers. Thank you, Enver Hoxha, for this bunker, the lovers pant. And there are privatization jokes about peasants dividing up the land—I have ten kilometers and three bunkers—which, I fear, Bujar does not tell very well. In the opposition between lightness and gravity, which Milan Kundera describes as "the most mysterious, most ambiguous of all," Bujar takes the side of gravity. Who can blame him? If our suffering had produced anything, he told us, maybe we could say it was worthwhile. But it was all absurd. It added up to nothing. We were eating a late dinner at a sidewalk café in Elbasan when he recalled the brief period, in 1972, when Albania looked to the West. Everything changed, he said. We had music, books, bars opening. But then our dictator explained the West was decadent. Art and literature—culture—would weaken Albanians. The avant-garde were sent to prison camps or killed. Dances were prohibited. Every Sunday we had "volunteer work" or learning to use a gun to be vigilant against Americans. Bujar went directly from grammar school to work in the large steel mill (Albania's second national liberation, said Hoxha; the new Coca-Cola plant outside Tiranë is the third, says Bujar) that dominates the river valley east of Elbasan. From the Balkans' tallest chimneys (Chinese-built) spewed enough smoke and toxic wastes to destroy what was once a fertile agricultural region. I was seven years in the steelworks, Bujar said, then three more in military service. So it seems I had no childhood. He pushed his plate away. Mike caught my eye. Bujar's culinary plan for our delegation—to feed us a succession of heavy meals—left us scheming for polite ways to eat sparingly in such hot weather, in the midst of such poverty. But our host was watching when Mike and I slid our plates to the center of the table. Eat, he commanded, and we obeyed him. For Bujar has an autodidact's authority (in the army he taught himself English, Russian, and Italian); his formal education began only after Hoxha's death. As I had published articles as a student, he said, in 1988 I went to work for a newspaper. I had nothing in common with the other journalists—they were Communists. When the mayor suggested I become the editor of an independent newspaper, I told my editor I will quit because I have other thoughts than you. He said, It hurts me that you think differently but I will let you go. For three nights I didn't sleep. It was dangerous to go against the Communists, but I was lucky. I could have ended up in prison or worse. Instead, his paper, *Free World*, distributed in Elbasan, Tiranë, and Durrës, thrived. Before the 1992 elections Bujar interviewed the leaders of every political party, a hitherto unheard-of practice in the Albanian media. And when the Communists were swept out of office they found in him an unlikely champion for their right to express their views. Our chief of police closed a shop that sold Communist newspapers, said Bujar. I asked him why. They're Communists, he said. Now we're Democrats. I wrote fourteen lines criticizing him for that, and he took my paper away! The printer in Elbasan was ordered not to print for me, so I went to Tiranë to publish another paper, explaining what I did, who I am, and what these others were trying to do to democracy. I said a new life has begun. I want to be a friend to my people, not to any party. Then another war was waged against me. In my own paper they printed my obituary: Bujar is dead. This was written by a poet I had published in those very pages! I went to the minister of justice and stood outside his office for a week, saying, I have a private newspaper. How can they take it away from me? Finally, the vice minister said he could not help me. I said, I came here to exercise my rights. Who can help me if not the vice minister of justice? He was silent. I am ready to die, I said, but I will fight for what is right. He said, Change the name of your paper. Two weeks later I started *Where Are We Going?*. This is how I answered the poet who wrote my obituary: Bujar is more alive than before. Let me tell you something about that poet. He's sixty years old, and he spent twenty years in a political prison. I knew he had a lover, so instead of his obituary I wrote a wedding announcement for him! I wished him marriage to a girl who would excuse his crazy acts. I went to war not with their weapons but with culture, with art, with words. With black humor, that is, Albanian style, which comes out at the oddest moments. On the drive to the coast, for instance, speeding by kiosks lined with slaughtered sheep hanging from hooks, Bujar was reminded of a law passed after Hoxha's death: a peasant could have two lambs only as long as they were of the same sex; litters went to the state. You see, he said, deadpan, the lambs posed a threat to the government. Soon after, a policeman pulled us over for speeding, and when our driver got out Bujar said, Do you know how Albanians buy such nice vans as this? Someone in France pays someone to steal their car, which is insured, and then they drive it through Kosovo to Albania. This $30,000 Renault, with Albanian title and registration, cost our driver $10,000. Sali or Lali, Bujar said to him when he returned less than a minute later. Lali, the driver said with a laugh. You can pay four dollars to Sali the president, Bujar explained, or two to Lali the policeman. Corruption is endemic—the government turns a blind eye to the black market, the growing drug trade, the pyramid schemes in which Albanians invest their meager life savings, the gasoline smuggling to Montenegro in violation of the UN sanctions on Yugoslavia, et cetera, et cetera. If there is no justice in Albania at least there is symmetry. This is the land of the vendetta: the police chief who closed down Bujar's paper was fired, along with the poet who wrote his obituary. By then Bujar had become a private publisher, opening the House of Onufre, which employs fifteen writers, editors, printers, and binders (who use forks instead of knives to score book covers) to publish everything from hospital documents to a translation of *The Tempest*. With a used printing press from Greece, ink from Turkey, paper from Italy and Germany, and American computers, Bujar is building an empire that may one day include a television station; soon even Berisha will pay him a visit. This week's lavish publication party for *The Walls of Loneliness*, a collection of poems by one of his editors, Flutura (which means butterfly) Açka, has only increased his stature. It was, in fact, more of a performance than a party. The setting was a military barracks—soldiers stood outside the door—and the script was Bujar's. Two university students, accompanied by synthesized music from *Love Story,* recited Açka's "A Song for Emigration," dedicated to the 400,000 Albanians who in the last four years have left the country (out of a population of three and a half million): *Voices/ They are going far away/ They will go.* Then there were interviews, for television, with the poet's first-grade teacher, with the president of the Writers' Union, and with Albania's most famous actress, a plump elderly woman with a large wart on her forehead. I was asked if I have a daughter, the actress told the crowd. No, I said, and that is bad luck, because a daughter is a woman's poem. So with our young women poets we now have poetry writing poetry! With that the dancing began. And singing. And testimonies from a dozen writers. I thought it would never end. The poet herself was the last to speak. Reciting a poem to her mother, Flutura burst into tears and wiped her eyes with a dramatic flourish, a gesture she had to repeat for the photographers and television cameramen who missed it the first time. The actress liked that, in a matronly way. She had just played Queen Elizabeth in *Richard the Third*, and her next role was to be the mother of an Albanian refugee in a Greek film. A part I don't have to prepare for, she said bitterly, then quoted Flutura: *I want to make the sky angry*. A common sentiment among her countrymen. During the revolution, Albanians went after everything associated with the state, looting and plundering and wrecking factories, shops, schools, even hospitals. It is no accident that Ismaïl Kadaré, Albania's greatest living writer (and the latest addition to Bujar's list), believes his finest novel is *The Palace of Dreams*, the work in which he invented a version of hell bearing an uncanny resemblance to life under Hoxha. Kadaré's Inferno is a vast empire, at its center a secret ministry charged with collecting, classifying, and interpreting all the citizens' dreams, the unconscious life of an entire nation. *For in the nocturnal realm of sleep are to be found both the light and the darkness of humanity, its honey and its poison, its greatness and its vulnerability. All that is murky and harmful, or that will become so in a few years or centuries, makes its first appearance in men's dreams.* Those who work in the ministry, then, are terrified of overlooking the dream foretelling the Sultan's doom, and to preserve his fragile empire they have assembled files containing *all the sleep in the world, an ocean of terror on the vast surface of which they tried to find some tiny signs or signals.* Their real-life counterparts, of course, the informers and censors and secret police who maintained Hoxha's dictatorship, still live here: one reason why Albanians have yet to shake off the nightmare of their modern history. Perhaps that is why they have so little to say about their Balkan neighbors. No one will discuss the war in Bosnia or the problems facing Albanians in Kosovo and western Macedonia, and when I bring up the longstanding dream of Albanians uniting to create a Greater Albania, as I have done with everyone from Bujar to the actress to the minister of culture and the mayor of Elbasan, I am met with blank stares. What I would give to hear someone say, Yes, yes, shaking their head (an Albanian mannerism), as if to disavow their own words, let me tell you something about our common future. But the habits of secrecy and circumspection die hard. I roll over onto my stomach, dive down until I come to cold water, and then float back up to the surface like a drowned man. The island is a tall dark blur in the distance. First there were Soviets, Bujar said of the island's residents, and now there are Americans. To be precise, first there were Illyrians and Thracians in Albania, and then there were Greeks, Romans, Byzantines, Slavs, Bulgarians, Normans, Venetians, Serbs, Turks, Serbs (again), Greeks (again), Italians, and Germans. Albania was last divided up after the Second Balkan War, when land was ceded to Greece, Montenegro, and Serbia. When World War One broke out, Austrian, Bulgarian, Greek, Italian, Montenegrin, and Serbian forces fought over this ill-starred country, and in World War Two Albania resembled a transit station, with one enemy army after another crisscrossing the country, sometimes at top speed. In Kadaré's *Chronicle in Stone*, a novel set in Gjirokastra, near the Greek border, one chapter opens with the city changing hands between the Italian and Greek armies six times in a single week. More symmetry: both Kadaré and Hoxha were born in Gjirokastra, a city carved from stone, about which the novelist writes, *The traveler seeing it for the first time was tempted to compare it to something, but soon found that impossible, for the city rejected all comparisons. In fact, it looked like nothing else. It could no more support comparison than it could bear the rain, hail, rainbows, or multicolored foreign flags that vanished from its roof-tops as quickly as they had come, ephemeral and unreal as the city was eternal and concrete.* The same holds for the whole country, as I learned when I made the mistake of comparing Kadaré and Kundera—two novelists from Communist countries living in exile in Paris. You cannot compare anything to Albania! Bujar corrected me. Thus: across a rope bridge, deep in the mountains, we came upon two young women, in bare feet and flowery dresses, who left off weeding a field to serve us cold water from a well and Mars Bars. Or the paper lanterns, fashioned out of pages from Hoxha's *Collected Writings,* that Bujar showed us in the living room of an old peasant woman; once upon a time, she had affixed bayonets to her grape vines to deter the American paratroopers the dictator said were always on the horizon. Or the Albanian-American impersonator of Elvis Presley we see each night on TV... I am remembering our excursion to Lake Ohrid, on the Macedonian border: how we stopped by the side of the road, just past an abandoned ironworks, to go for a swim. I took off my shoes and clothes and started down the steep rocky slope to the water. I had not gone far before I scared up a Balkan adder, a thick black rope that slithered into the underbrush. Not until I had waded free of the algae close to shore to dive into the clearest water imaginable did I calm down. Now, on the other side of the country, as I swim in Mike's direction, toward the island, it seems to me that the dream of Greater Albania is like that adder in the underbrush. One day it may strike, and those not killed outright will flee: the oldest Balkan story. The "blue grave"—that was what Serbian soldiers called this part of the sea when their comrades died off in the exodus of 1916. The occupying Axis forces had put to flight Serbs in the hundreds of thousands; the army, the government, King Peter,

Prince Regent Alexander, priests and teachers, women and children, all retreated through Kosovo, Montenegro, and Albania. Many froze to death in the snowy mountains, others died at the hands of Kosovar Albanians (the Serbian army made its last stand on the Field of Blackbirds), and still others succumbed to typhus and dysentery. Corpses lay among the carcasses of horses and oxen, food for the starving refugees. Of the soldiers who made it to Vlorë, where French forces were waiting to rescue them, thousands were sick enough to be quarantined on the tiny island of Vido; the bodies of those who died were thrown into this "blue grave." But more than a hundred thousand troops survived to fight again alongside the Allies in Thessaloníki; and it was on the nearby island of Corfu that in the summer of 1917 Serbian diplomats and members of the London-based Yugoslav Committee laid the foundation for the Kingdom of Serbs, Croats, and Slovenes, declaring they were "the same by blood, by language, both spoken and written, [and] by the feelings of their unity." In short, they created a fiction, which was not only at odds with the thinking of most Serbs, Croats, and Slovenes but also ignored the wishes of Albanians, Macedonians, Montenegrins, and Bosnian Muslims, who were not represented at the negotiations on Corfu. Indeed one Serbian diplomat said that as soon as his forces crossed the Drina River into Bosnia the Muslims would be given forty-eight hours to return to the Orthodox faith or be killed. "You can't be serious," said a Croatian diplomat. "Quite serious," said the Serb. The union of South Slavs was thus doomed from the beginning. Now the Serbs are fleeing again, the Croatian Army having recaptured all of western Slavonia in a blitzkrieg launched on 1 May. Three centuries of Serbian settlement there ended forty hours later, when 12,000 Serbs took to the road. Here was another dramatic turnabout in the Yugoslav wars of succession, presaged in some ways by the Serbian escape to this "blue grave," in which I find only pleasure. Mike, too, is exuberant. Incredible, he says before swimming back to shore. Just incredible. This *is* a momentous day. The excitement over the EU vote has pushed to the background any discussion of a potentially more significant vote on the Continent: the German Parliament's decision to send troops abroad for the first time since World War Two, to Bosnia, where the peacekeeping mission is falling apart. We will not learn the results of the EU vote until this evening, during our tour of the thirteenth-century Orthodox monastery of Ardeniça, which miraculously escaped major damage during the anticlerical campaign. After the army destroyed the library, the bishop imprisoned in the monastery begged Hoxha to save this national monument. It turns out that Skanderbeg, the fifteenth-century prince (and father of Albania) who waged a guerrilla war against the Ottomans, temporarily keeping them at bay, was married here. Honeymooning couples now stay in the monks' cells, which have been refurbished stylishly: sheepskins draped across the beds, wrought iron lamps, TV sets with remote controls. Cypresses on the hilltop. Streaks of blue paint showing through the high walls. Frescoes behind a barred window. Running his fingers over the rope grooves worked into a stone well, Mike will quote Seferis—*The grooves on the well tell us something our ancestors knew*—before we climb to the bell tower, on top of which are propped a wooden cross and a TV antenna pointing toward Italy, toward fields and olive trees, bunkers and hay piled in the form of bunkers, concrete tunnels (built in the 1970s to protect Albanians from a nuclear attack which, Hoxha said, would be launched jointly by the United States and the Soviet Union) in which farmers keep their livestock, and then the sea. As the sun sets over the Adriatic a roar will go up from the tourists dining in the monastery restaurant: the EU has voted in Albania's favor. Which means that our drive back to Elbasan, past burning fields, men selling gasoline in plastic jugs by the side of the road, and more slaughtered lambs hanging from hooks, will go by in a blur of *rakija*. We are in for a marvelous night. What we do not know, what the intelligence officers on the island may not know (if, in fact, they are even there), is that sometime in the next week the worst massacre of the Third Balkan War will occur. Greater Serbia, first glimpsed perhaps in the dream of a shepherd or a poet centuries ago, is within the grasp of the Bosnian Serb leadership, for Ratko Mladić, emboldened by the looming failure of the Bosnian Army offensive to break out of Sarajevo, is preparing to overrun the safe areas of Srebrenica, Žepa, and Goražde. A terrible crime against humanity is in the offing: thousands of Muslim men are about to be executed. Soon there will be a photograph, on the front pages of newspapers around the world, of a woman from Srebrenica, with a noose tied around her neck, hanging from a tree, and then reports of mass graves will trickle in—reports confirmed by aerial reconnaissance photographs. But as I swim farther out to sea I am only thinking about how clear the water is, how warm. I could stay here all day, if not for the voices calling me back to shore. You see, Bujar wants to take us to lunch.

20

Expedition

December 1995–January 1996

It was not easy to convince Jadranka Pintarić to drive me to Jasenovac. The village on the Sava's northern bank, 130 kilometers from Zagreb, was once home to the largest Ustaša concentration camp, and the art critic was not keen to recall her countrymen's butchery of Serbs, Jews, and Gypsies. But what a legacy. The Nazi puppets "scorned modern technology [read: poison gas] at Jasenovac and the other camps," Richard West writes. "They normally killed with knives, axes and clubs, or by hanging, burning in furnaces or burying alive." Among the executioners were six Franciscans, whose victims included Croatian and Slovenian priests. Archbishop Stepinac of Zagreb never publicly denounced the camp, one of many failings for which he was convicted on charges of collaborating with the enemy. Even so Croats were campaigning for his canonization. Not Jadranka. Her main concern on the drive through the flatlands of western Slavonia was the weather. The sky was overcast, six inches of snow lay on the ground, and her windshield wipers did not work. She pointed to a hawk perched in a bare tree, as if it were an warning sign.

"The Sava's flooding," she reminded me.

"We won't stay long," I said.

Jadranka gave me a dubious look. She was wary of any uses, literary or otherwise, to which this lethal symbol might be put. The disagreement between Serbian and Croatian historians over the number of victims at Jasenovac had caused enough grief. Under Tito it was said that 700,000 people died in the camp (though the best estimate was 80,000), but when Serbian nationalists argued that more than a million Serbs alone were executed there (the same as the number of Yugoslavs killed in the war), Franjo Tuđman countered with the figure of 28,000.

The fluctuating numbers—unlike their Nazi masters, the Ustaše destroyed their records—illustrated Stalin's cynical dictum that three people killed in a traffic accident constitute a tragedy whereas a million victims of a pogrom are statistics. And an argument over statistics deflects attention from the individual nature of tragedy. A debate was raging in American foreign policy circles, for example, over the number of Bosnians killed in the Third Balkan War: 200,000, the figure cited in most news accounts, or 25,000, as George Kenney, a former State Department official, asserted. Perhaps the truth would never be known.

The same held for the massacre at Srebrenica, the safe area overrun by the Serbs in July. How many Muslim men were shot? Eight thousand, said the Bosnian government. Five to seven thousand, said UNHCR. None, said the Serbs. Indeed Ratko Mladić himself was filmed guaranteeing the safety of thousands of Muslims who had sought refuge in the Dutch peacekeeping compound. Then scores of buses arrived, some to take women and children to Tuzla, others to deliver every man between the ages of seventeen and sixty to killing fields beyond the camera's range. A soldier confessed to taking part in executing 1,200 Bosnians, in groups of ten. Too slow, said the commander. Use the machine gun. "But it was not very professional: the bursts of fire wounded rather than killing off," said the soldier, who was sentenced by the War Crimes Tribunal to ten years in prison. "The wounded begged us to finish them off." Which they did, with bullets to the head. But there were enough survivors—wounded men who lay under the corpses until nightfall, when they escaped into the woods—willing to tell their stories and aerial reconnaissance photographs of mass graves for The Hague to indict Mladić and Radovan Karadžić on more charges of crimes against humanity.

The U.S. government picked a critical moment to release the pictures to the Security Council: during the Serbian exodus from Krajina in August, the most dramatic example of "ethnic cleansing" in the war. Operation Blitz, which cleansed the Serbs from western Slavonia in May, was the first step in Croatia's drive to recapture its occupied lands. Even as Bosnian Serbs murdered the men of Srebrenica the revitalized Croatian Army (trained by retired U.S. Army officers and armed with heavy weaponry acquired from abroad or manufactured at home) was preparing to mount Operation Storm. The offensive began on 4 August, with Croatian forces attacking Krajina at thirty separate points, including a well-executed pincers movement (an American military trademark) on Knin. It took them just eighty-four hours to destroy the Serbian Republic of Krajina and put its entire population to flight. Of the 600,000 Serbs who had lived in Croatia before the war only 150,000 remained, mostly in eastern Slavonia, which Croatia won back not on the battlefield but in the subsequent negotiations leading to the Dayton Peace Accords signed last week in Paris. The majority of Serbs in and around Vukovar would flee when Vukovar was returned to Croatia. The thousand-year-old dream of independence had come true. There was no place for Serbs in this new country.

I had at first shared Jadranka's excitement over the breaking of the military deadlock. But then the televised images of Serbs fleeing in cars and trucks, in horse-drawn carts and by foot, bringing to a close hundreds of years of continuous presence in Krajina, made me queasy. They had brought this on themselves, ethnically cleansing the area to start the war. And the perpetual cycles of revenge in the Balkans perhaps dictated that population exchanges precede a peace settlement. Milošević himself chose not to send reinforcements to the Croatian Serbs, abandoning them in a *quid pro quo* with his enemy counterparts: Bosnia's eastern enclaves went to the Serbs, Krajina and Slavonia to the Croats, territory in northwest Bosnia to the Bosnians. As Warren Christopher said of the Croatian offensive: "it always had the prospect of simplifying matters." But war's simplifications are usually tragic.

Witness the refugees living in cattle cars by the riverbank we drove along. And the empty cattle cars in Jasenovac, from which prisoners who did not survive the journey to the camp were dumped into the river, lined a meadow of raised mounds: mass graves. Above them towered a concrete monument, twenty meters high, in the shape of a six-petaled flower, designed by Bogdan Bogdanović, the acclaimed architect and former mayor of Belgrade. The adjacent museum was gutted, and since record-setting rains had caused the Sava to flood over the bank, turning the killing field into a frozen lake, the monument now stood as an island unto itself.

Jadranka parked by the Catholic Church, of which only the damaged walls remained, though a crane was in position to begin repairs; the façade of the nearby Orthodox Church was unmarked. We wandered along a muddy road, passing soldiers and one or two old men, and stopped at a curve in the river. Crows were cawing. A dog barked. The village on the Bosnian side of the Sava looked empty. Jadranka turned around and started to retrace her steps. What she remembered from her obligatory schoolgirl excursion to the museum was the film her class had to watch about the camp.

"I had never seen such footage," she said. "Bodies, bones, hair, like garbage. I was ten years old, and I got sick, but I was forced to stay until the end, even after vomiting!"

In the center of the village was a vacant grocery store. From the second floor came the sounds of revelry—drunken soldiers singing folk songs, ringing in the New Year two days ahead of schedule. They have nothing better to do, muttered Jadranka. She was resigned to seeing conditions worsen in her homeland. The war, refugees, corruption in the ruling party, mismanagement, all had left the economy in shambles. Worse, Tuđman had used his military success to consolidate his power in the recent elections, expanding the electorate to include Bosnian Croats and ethnic Croats the world over: the joke was that votes were still trickling from submarines (Croatia did not have a navy) and Antarctica. Young Croats were voting with their feet, moving in droves to Australia, Canada, and New Zealand. But Jadranka had abandoned hope of emigrating. What country would take a single woman reduced to editing Italian cookbooks? She cursed the Serbian unity graffiti on the crumbling walls of the houses, and her mood grew darker yet when a cold drizzle began as we circled the killing field. If Croats committed genocide in World War Two, she said, there would not have been enough Serbs left to start the war in 1991. And she had the numbers to prove her point.

I, too, had a figure in mind: that none of the major war criminals from the Third Balkan War—Karadžić, Mladić, et al—was in custody. Milošević and Tuđman, the war's architects, had not been indicted; murderous paramilitaries like Arkan and Šešelj seemed in no danger of being taken to The Hague. War crimes investigators had located 300 mass graves in Bosnia, yet few Balkan observers believed the victims' families would find justice, which was crucial to creating enduring peace.

The rewriting of history was already underway. Serbian high school textbooks blamed the Vatican for the war with Croatia; the Serbs fought back to prevent the Church and Ustaša descendants from orchestrating another genocide. Any mention of Jasenovac, meanwhile, had disappeared from Croatian textbooks. There was talk of turning the camp into a memorial for all war victims so that Croats killed in this war could lie alongside those murdered by the Ustaše. A journalist accompanying Tuđman on an official visit to Argentina the previous winter had even interviewed Dinko Šakić, the commander of Jasenovac, who not only said that no mass executions had occurred there but also that the Ustaša state "was the foundation on which today's Croatia is built"—a sentiment echoed by Tuđman when he called for the return of Ante Pavelić's remains from their burial place in Spain.

"Yes, this monument is a beautiful object," Jadranka was saying, "but it was built to make us ashamed of our past. And this flower is more than a flower: it was imposed on us. For a half century the Serbs concentrated only on three years of our history. We're fed up with it. You know, you can't feel guilty forever. Even Germans are tired of talking about the Nazis."

But I was thinking of a scene in Curzio Malaparte's surrealistic memoir, *Kaputt*, depicting Pavelić at his most demonic. What the Italian writer called his "horribly gay and gruesome book" includes his interview with the Ustaša leader or *Poglavnik* in Zagreb:

> While he spoke, I gazed at a wicker basket on the Poglavnik's desk. The lid was raised and the basket seemed to be filled with mussels, or shelled oysters—as they are occasionally displayed in the windows of Fortnum and Mason in Piccadilly in London. Casertano, an Italian diplomat, looked at me and winked, "Would you like a nice oyster stew?"
>
> "Are they Dalmatian oysters?" I asked the Poglavnik.

> Ante Pavelić removed the lid from the basket and revealed the mussels, that slimy jelly-like mass, and he said smiling, with that tired good-natured smile of his, "It is a present from my loyal *ustaše*. Forty pounds of human eyes."

This image was burned into Serbia's collective memory, whether it had actually happened or not, and it would only burn brighter so long as Croatia played down its Ustaša past.

"Have you seen enough?" said Jadranka.

Dusk was falling when we set out for Zagreb. The roads had iced over, which did not deter my friend from attempting to pass a long line of cars, each with a slip of white lace hanging from the trunk or door handle: a peasant wedding procession. On my first trip to Slavonia I had seen a peasant funeral, and I was thinking my Croatian journeys were drawing to a neat close when a car sped up behind us, flashing its high beams. But Jadranka could not get past the truck carrying the bride and groom. The high beams flashed again, a horn honked, and when we finally slipped into the right lane Jadranka gave the finger to the impatient driver, who sped ahead of us in his new Mercedes. Jadranka was shaking her fist at him, crying, "Lunatic," when he pulled into our lane, slammed on the brakes, and sent us into a skid. Across the highway we fishtailed in front of the wedding party. Several harrowing seconds passed before Jadranka steered the car to the side of the road.

"A war profiteer from Herzegovina," she decided as the Mercedes sped away. Then she let out a stream of curses.

The exodus had begun. The Serbs were leaving Sarajevo.

On a cold overcast morning, troops from NATO's Implementation Force (IFOR) had stopped traffic from the airport to escort a Serbian convoy from the suburb of Ilidža to Republika Srpska and beyond. The cars, trucks, and wooden carts were stuffed with household goods and belongings, farm animals and furnishings, even coffins, since the Serbs were digging up their dead to ferry them away. The ruins of Dobrinja formed the backdrop to the latest instance of what Churchill once called "the disentanglement of populations." It was a strange and horrible sight, which irritated the impatient photographer giving me a lift into town. I asked him why he was not taking any pictures.

"I'm looking for soldiers, not refugees," he said, gunning the engine until IFOR gave him the signal to drive on. He did not speak again before he dropped me off by the Presidency building.

Soldiers were easy to find. Under the terms of the Dayton Peace Accords, 60,000 NATO troops, half of them American, were arriving in Bosnia, a Great Power occupation in the tradition of the Berlin Congress decision to let Austria take over the troubled country in 1878. Two days ago, on New Year's Eve, I had traveled to Županje, Croatia to watch the American forces enter Bosnia, and long before reaching the Sava I could hear the continuous roar of hundreds of tanks, Bradley armored vehicles, and Humvees idling on the levee, preparing to cross the largest pontoon bridge built by the U.S. military since the spanning of the Rhine in World War Two.

Bridging the Sava was no small task. One day the rising waters swamped an encampment; in the new lake next to the river were wooden crates, tents, and an overturned truck. More than a dozen speedboats held the spans in place, a stone's throw from the bridge blown up early in the war; MASH helicopters hovered overhead. And the rain had not let up, making the muddy road even slicker. Yet the commanding officer said he was having fun, and the townspeople who came to watch the convoy were in a festive mood. A Catholic priest from Newark, celebrating Mass under a girder of the destroyed bridge, enlisted me to do the first reading, from Sirach. Three wooden crates stacked one on top of another were his altar; his assistant chaplain carried an M-16. The priest shouted above the roaring engines, and still I lost my place in the text. *Pray for us* were his parting words to me.

This NATO operation was not without its critics. There was widespread fear of casualties, and despite talk of a new role for the alliance—Russian troops, for example, were serving for the first time alongside Americans—it was too soon to tell if NATO would succeed in disarming the warring parties. "Nothing happened, of course, until the Americans decided to play," a Canadian peacekeeper ruefully told me at Butmir Airport. He was a wiry young man gone prematurely grey—from bitterness, I decided. "And this is no peacekeeping force," he muttered. "The minute we use disproportionate force against one side we become a player. And there are already seven or eight players here."

I mentioned the firepower I had seen in Županje.

"They won't attack our strength," he said. "They know better than that."

In his opinion the international community had been hoodwinked in Dayton. Serbian and Croatian nationalists alike interpreted the peace agreement as Bosnia's formal partition: they would sabotage efforts to unify the country, which was split roughly evenly between the Bosnian-Croat Federation and Republika Srpska. It was only a matter of time before the Serbs and Croats merged with their motherlands to create Greater Serbia and Greater Croatia.

Perhaps the best the Bosnians could say of the peace agreement was that they were awarded Sarajevo because, as Milošević told them in Dayton, they deserved it, having held out for more than three years against what he called "the cowards" shelling them from the hills: his partners in crime, on whose behalf he negotiated the agreement. It was indeed because those "cowards" had fired one shell too many at Sarajevo that the West finally took robust action against them. Operations Blitz and Storm changed the balance of power; calls for intervention reached a crescendo with the revelation of the massacre at Srebrenica and then the deaths of three American diplomats forced to travel over Mount Igman (the Serbs refused to guarantee their safety at the airport); after a shell struck the Sarajevo Market in late August, killing thirty-seven, NATO launched waves of air raids against Serbian positions, destroying fuel depots, ammunition dumps, and communications centers; and when thirteen Tomahawk cruise missiles were fired at the command and control center near Banja Luka, Mladić agreed to withdraw his heavy weaponry from around Sarajevo, setting the stage for Dayton.

It was true that the Bosnian Army, working in tandem with Croatian forces, had retaken the Bihać pocket in northwest Bosnia and was advancing on the Serbian stronghold of Banja Luka when American officials warned Izetbegović not to go any further. So the Bosnians won back about as much land as they were then offered at Dayton. Some terms of the peace agreement negotiated by Assistant Secretary of State Richard Holbrooke, a dynamic figure who bullied the Balkan leaders into signing, were straightforward: armies would pull back from front line positions, refugees could return to their homes, the economic sanctions on Yugoslavia would be suspended, the arms embargo lifted, and so on. Less clear was the will of the international community to arrest war criminals, for example, or to settle the dispute between the Bosnians and Serbs over the disposition of land in the narrow Posavina corridor linking the two halves of Republika Srpska. The Bosnian Army would receive training from the U.S. military and weaponry to match Serbian forces, preparing to take matters into their hands, if and when NATO left. Which was only to be expected, a British officer told me in Sarajevo. "You have to admire these people," he said. "They didn't just lay down and die. They defended themselves well. After all, they had no choice."

Perhaps I will never understand Ferida Duraković's decision to return to Sarajevo. Last winter, the poet had gone to an artist's colony in Oregon, where she could have stayed for the duration of the war. Yet in May, at the height of the Bosnian offensive, when the airlift was suspended indefinitely, she took a bus from Zagreb as far up Mount Igman as Serbian sniping and shelling would allow, then hiked for more than two hours, with four heavy bags of luggage, through the forest until she came to a house in Hrasnica. There she waited until dark to risk going through the tunnel under the airport the day after Serbian gunners had shelled the entrance, killing ten. But first she and several others had to cross a cornfield and wait for an hour and a half outside the tunnel while Bosnian soldiers went through. It was nearing curfew when Ferida, the last civilian in line, claustrophobic and struggling to catch her breath (she suffered from tachycardia), started down the stairs—and almost fainted. She went outside and sat down. She considered spending the night the night in Hrasnica, but her fear of dying before delivering to her family the presents she had brought from America persuaded her to try again. This time in the tunnel a draft of air caught her by surprise. I realized I could breathe, she said. It was marvelous. The tunnel was only wide enough for one, and Ferida had to stoop lower and lower until she was nearly crawling. "God is so near, so hard to grasp." She was reciting Hölderlin to keep up her courage. "But where danger lies, salvation also grows." Then she was in another cornfield, lugging her bags to Dobrinja, where she found a taxi driver who took her partway home, claiming to be out of gas; before the night was over she would walk another kilometer then up fifteen flights of stairs to her brother's apartment. And within two weeks of her return, telling her boyfriend it was now or never (the Serbian shelling was continuous), Ferida was pregnant.

"Why did I do this?" she said. "I guess you could say it was the Grand Yes."

As for the peace accord?

"This could have happened two hundred thousand lives ago," she said.

How I regretted quarreling with Zvonko Radeljković. At dinner one night the professor accused his wife, a lawyer friend, and me of being sheep following Alija Izetbegović into Islamic hell. Nonsense, I replied. Zvonko flew into a rage. What triggered his anger was the president's letter, printed in *Oslobođenje,* condemning Sarajevan excesses on New's Year Eve—drinking, dancing, drugs, televised pornography, soldiers firing rounds off into the sky. No one has the right to celebrate after so much suffering, he wrote. Sarajevans called him a fool, attributing their frenzy to the release of pent-up anger, not joy over the city's liberation. One wit said the president was trying to become a comic writer. Zvonko called the letter proof of Izetbegović's determination to build an Islamic state.

"Why is the world standing by this Muslim fundamentalist responsible for two hundred thousand deaths," he said, opening another bottle of red wine, over the objections of his wife, Ivanka. "Why did I have to suffer for four years while Alija Izetbegović hid in a bank vault?"

He blamed Izetbegović for the warlords, corruption, and lack of gas, water, electricity, and food in Sarajevo. He said the ruling party had kept the war going to make money; that Bosnia could not become an Islamic state without Muslims dying in the tens of thousands; that Izetbegović was therefore the biggest murderer in history. When Ivanka contradicted him, he told her to shut up.

"Nothing is black and white," he said. "All sides are to blame."

But if not for those who first came to the city's defenses, I said, we might be having this conversation in the next world.

"You will learn not to argue with me," he growled. "I will not be beaten."

So it seemed. In the morning he blamed our argument on my innocence and bad Herzegovan wine. But before we set out to teach his class on Walt Whitman

Ivanka pulled me aside to apologize, saying the war had gone on for too long. "Our nerves are shot," she explained.

Walking through the snowy streets, Zvonko found another target for his ire: the American Embassy, which refused to support his new anthology of contemporary American writers or *Sarajevo Country Club*, the 115th program of which he would record that day. And he took no pleasure in my decision to begin his class, in the unheated library of the British Embassy, by reciting his favorite lines from Whitman's "Song of Myself": "Do I contradict myself? Very well, I contradict myself. I am large, I contain multitudes." We were barely speaking by the time we parted.

But he was all smiles when he showed up at Ferida's flat that afternoon. The three of us set out for the National Library. Zvonko gleefully told how its architect had committed suicide before it was finished; the sunlight did not refract through the windows the way he thought it would. The four-story structure, a mishmash of nineteenth-century architectural elements, part Austro-Hungarian, part Moorish revival, served as city hall until after World War Two, when it was converted into the National Library. In August 1992, when Serbian gunners attacked it with phosphorous shells, flames filled the luckless architect's windows. The fire burned for three days, destroying more than a million books. The writer James Carroll called the records, documents, and books contained in the libraries targeted by the Serbs (including thousands of Islamic and Jewish texts in the Oriental Institute and the vast holdings of the National Museum), "the memory of a precious exception—Serbs, Croats, Muslims, and Jews living as one people." That memory was fast receding.

It was very cold when we stood at dusk in the gutted building by the Miljačka River. A dog rooted through the rubble. I could just make out the six-panel wall hanging done by a Czech artist. In December, at a ceremony to unveil the artwork, the library director placed a book on one brick. It was, he said, the first to be deposited in the library since the fire.

There was an explosion on the hill above us.

"Land mine," said Zvonko.

"Too loud," said Ferida.

"Then an anti-tank land mine," said Zvonko.

Small arms fire began across the river.

"Snipers," said Ferida.

"Celebratory fire," Zvonko corrected her with a laugh.

They had grown closer during the siege, partly because of common literary interests, partly because much of the intelligentsia had fled the city. For now it did not seem to matter that she was Muslim and he was Croatian. These terra cotta ruins, they agreed, transcended ethnicity.

"When we heard the library was burning, my friends and I came here to cry," said Ferida. "It was impossible to imagine someone shelling the National Library."

"I have a mixed feeling about this building," said Zvonko. "My first job was as a tourist guide thirty years ago, and before they'd let me take foreigners around the city I had to attend a lecture on Sarajevo's cultural history. They said this building was ugly—a violation of the Islamic spirit. But now that it's gone all I can think is, what a terrible loss."

"This was the place to exchange opinions," said Ferida, "even if it turned out that I spent more time in the café with boys! But this was also where everything began for me as a poet."

Zvonko recalled the video his former best friend, Nikola Koljević, had made before the war, which opened with a shot of him reading his translation of Rebecca West's *Black Lamb and Grey Falcon* from the balcony of the library. "Then he destroyed it. Can you imagine?"

Ferida shook her head. "But what will happen now?" she asked.

Zvonko shrugged.

The library's future was uncertain. This year marked its 100th anniversary, a bitter celebration. There were rumors of the Austrian government rebuilding the graceful structure, and while some Bosnians wanted the ruins left as a monument to the death and destruction visited upon Sarajevo, others said the building should be returned to its original purpose as city hall. Indeed the country's future was being debated over just such questions as the disposition of these ruins. Ferida hoped they would not become a sightseeing attraction.

It was full dark when we left. Next door to the library was a building with a neon light shining above the doorway, a notorious prison during the early years of the Communist regime. Zvonko's neighbor, for example, was once locked up for the "crime" of working at the British Embassy. And anyone who spoke or studied English was suspect, Ferida added. The irony of housing political prisoners in sight of the National Library was not lost on her.

"It was a wonderful time," she said sarcastically.

"But look," said Zvonko. "Someone's opened a new café there. Shall we?"

—∿∿—

The night train to Budapest was late. After standing out in the cold for hours I could barely breathe: a case of flu had turned into pneumonia, though I was a week away from getting that diagnosis. When I found a seat in a crowded second-class compartment with two guest workers from Switzerland, an elderly couple who had lost their home in Knin during the Croatian offensive, and a middle-aged woman and her two grandchildren, I could not stop coughing. "My son," the grandmother said matter-of-factly, "is fighting in Bosnia." One guest worker asked how long I had stayed in Serbia.

"Three days," I said.

"Then you saw nothing," he sneered. "Like all journalists."

It was a sentiment bolstered this weekend by Peter Handke, who created an uproar in Germany with the newspaper serialization of his new book, *A Journey to the Rivers: Justice for Serbia*. Convinced the media had demonized the Serbs, the writer made a short tour of Serbia this fall, asking "What does a stranger know?" Very little, was his answer. But since he had not bothered to learn much about the Serbian war record his book offered a fine example of what Hubert Butler called "the modern indifference to evidence." Not only did Handke ignore the reports of Serbian atrocities compiled by Human Rights Watch and Amnesty International but also the evidence collected by investigators for the War Crimes Tribunal. Even the opinions of the Serbs he met did not seem to interest him, perhaps because he was afraid to surrender his illusions about them. In his eyes Serbia "seemed the huge room of an orphaned, yes, of an orphaned, abandoned child."

Handke's reflections on photographs from the concentration camps caused the biggest stir. He accused Muslim prisoners of assuming "the pose of suffering" to win the world's sympathy: "Who can tell me I am mistaken or even malicious when, looking at the pictures of the unrestrainedly crying face of a woman in close-up behind the bars of a prison camp, I see *also* the obedient following of directions given by the photographer of the international press agency outside the camp fence; and even in the way the woman clings to the wire I see something suggested by the picture merchant?" Malicious *and* mistaken, the novelist had secured a place for himself in the annals of infamy.

It did not take long for a middle-aged man drinking *rakija* to discover an American was on board, and for the next three hours he hovered over me, screaming about Bill Clinton and Warren Christopher. Do you know what NATO did to the Serbian people? he would shout, then draw his finger across my throat, egged on by the guest workers and the grandmother.

As it happened, I had seen some of the damage inflicted by NATO warplanes, having hitched a ride out of Sarajevo with a relief worker. He began to scowl when we drove into Ilidža, the Serbian-occupied suburb he called "The Twilight Zone." Fifty meters beyond the front it was impossible to tell that a vicious war had been waged nearby: these houses, bridges, and fields were unscarred. The Bosnian guns could only reach the confrontation line. What a change from the devastation in the city.

"These people thought the war was nothing, because they didn't suffer at all," muttered the relief worker, who accurately predicted that NATO would not stop the Serbs from setting fire to this suburb when they joined the exodus to Republika Srpska. Bearded soldiers milled in the streets, peasant women hawked cigarettes, the cafés were crowded with old men. Around a corner on the outskirts of town, just beyond a warehouse used by humanitarians, we slowed to look at a tangle of twisted metal: the remains of a communications center. The neighboring houses were intact.

"One smart bomb did this," said the relief worker. "It shows you what could have been done."

I carried that image with me to Split, where I caught a flight to Zagreb then took an overnight train to Budapest. During my layover in the Hungarian capital I went to a café with my old friend Aleš Debeljak, who was using his fellowship to the Central European University to write a new book of poems. He was also engaged in a polemic with the Serbian literary community, touched off by an interview he had given to the Belgrade weekly *Vreme,* the gist of which could be summed up in one question: Where were my Serbian friends when the JNA attacked Slovenia?

The gloves came off last month when three Serbian intellectuals published caustic replies in *Vreme*, accusing Aleš of being "a paid patriot." What most upset Aleš was that two of the writers were friends—Zoran Milutinović, the professor of Slovenian literature I knew from my first trip to Belgrade, and the novelist Dragan Velikić. They accused him of acting in bad faith, first for adopting the Joycean stance of *non serviam* toward Slovenia's independence movement in the 1980s, then for becoming a nationalist when war broke out. Aleš said they did not understand the difference between belonging to an attacking nation and to one under attack. He had a moral responsibility to defend his country. On the other hand, Zoran's silence during the war was morally despicable. Locked closets, he warned the professor, are usually empty. To Velikić, who insisted on separating literature and politics, he said his artistic model must be Mephisto, the hero of Klaus Mann's novel, an actor who thinks he can appear on stage in Nazi Germany without suffering any consequences. It's too easy to ignore the policies of a genocidal state, Aleš thundered. Too elegant to say, I'm only a writer, when writers were instrumental in orchestrating this war. There can be no reconciliation in the Balkans until Serbian writers repudiate what was done in the name of Greater Serbia, because for a man with a hammer the whole world looks like a nail. For Serbian writers the word *library* conjures an image of a cozy room filled with books and papers. For Bosnians the word means smoke: books burning in the libraries destroyed by the Serbs or in their stoves to cook a meal. How can we understand one another if the same word now has such different connotations?

To test his thesis, I called Dragan Velikić when I arrived in Belgrade eight hours later. But it was the first night of Orthodox Christmas—the novelist could not see me until after the holiday—and so on Sunday morning, aching and feverish, I walked through a driving rainstorm, from one church to another. I wanted to be with the faithful who were kissing icons and lighting candles in the incense-swirling dark. At the entrance to the new cathedral an old lame man and a Gypsy woman with two infants sprawled on a piece of cardboard were begging. I was about to give them some dinars when a priest ran them off. There was, of course, a gulf between the genocidal criminality of Croatian Catholic clerics in World War Two and the shameful behavior of Serbian Orthodox priests in this war. But the Serbian Patriarch still refused to acknowledge his Church's role in fomenting hatred and sanctioning crimes against humanity. "Serb religion, history, literature, art, and culture have been used to justify a crime," writes Michael A. Sells, "the true proportions of which the world has yet to grasp."

Late in the afternoon I returned to the hotel, took a long bath, and went to bed, with the television tuned to Studio B. The independent station was broadcasting a male choir singing the Orthodox liturgy interspersed with rock-and-roll videos when I dozed off; waking after midnight, drenched in sweat, I saw on the screen the novelist and leader of the Serbian Renewal Movement, Vuk Drašković, delivering a fiery speech, a warmup for protests that would begin on 17 November, the ninth anniversary of the start of the Velvet Revolution in Prague. Serbian students would take to the streets of Belgrade to protest Milošević's decision to annul the municipal elections; for more than three months they would march through the city, blowing whistles, pelting the state TV station with eggs, demanding the resignations of the rector and dean of the university. Drašković, Vesna Pešić of Civic Alliance, and Belgrade's new mayor, Zoran Djindjić, would form *Zajedno* (Together), an opposition coalition uniting disgruntled Serbs from all walks of life, and they would become familiar faces to television viewers around the world. Commentators would make analogies to the Velvet Revolution and the fall of Communism. Talk of Milošević being toppled—or worse—would fill the airwaves. But *Zajedno* would collapse within weeks of the last demonstration. And Milošević? Though he would use force against the protesters, which would not stop them, and eventually grant them their wishes, he would not only stay in power but even increase his stranglehold on Serbian politics. For Drašković was no Havel: one reason why Belgrade, once the most dynamic city of the Eastern Bloc, was now far removed in spirit and prosperity from Prague, which was fast acquiring the same magnetic draw for artists and writers as Paris in the 1920s.

In the morning I was sicker yet, and walking to the center of town to meet Dragan I knew I would have to curtail my stay in Serbia; if it was hard for Serbs to get medical treatment (a shortage of supplies had created an emergency), I could only imagine what sort of reception would await me at a hospital. In the Cool Jazz Café, I drank tea instead of *šljivovica* with Dragan, who was eager to talk about his new novel. *North Wall*, written in Vienna (courtesy of an Austrian government grant), juxtaposed the story of James Joyce in Trieste at the start of World War One with that of a contemporary Serbian couple in Vienna watching BBC reports on the siege of Sarajevo; what linked these characters were lines written by Gavrilo Princip on the north wall of the prison cell in which he died: *I wish I had lived another life. I wish I had sailed to Brazil. I wish I had been a wise man in the Himalayas. I wish I could be on a boat in a harbor, where the fog horns would wake me in the morning and the sun would reflect off the water. I wish I had a child . . .* To call a seventeen-year-old boy a revolutionary is stupid, said Dragan. He was just chosen by accident.

I told him I did not see the connection.

"Can you imagine?" he said. "Two sisters in Vienna were dead for seven years before they were discovered. I thought: how is this possible? Perhaps the wind blew out the stench. One was in a wheelchair, the other lay on the floor. The first one fell and couldn't get up, then the one in the wheelchair couldn't call for help, and she died of hunger. It was so funny."

"A metaphor for Serbia and Croatia?" I asked.

"Obsession has its own power," he said, "and when you are obsessed with a topic you touch some knowledge which is completely out of you."

"I'll say," I replied, and ordered more tea.

When I pressed him to talk about political matters—Srebrenica, the refugees overrunning Serbia, the war crimes tribunal—all he could say was, "Catastrophe." Greater Serbia? Just an idea that Milošević had exploited in his rise to power. The same went for the so-called threat to Serbs in Kosovo: the novelist thought the province might even have its autonomy restored. It was as if nothing had changed (after all, Slovenian and Serbian literati were mired in another dispute), except that countless Bosnians, Croats, and Serbs were dead or maimed, millions more were refugees or living in exile, hundreds of villages, towns, and cities lay in ruins, and perhaps as many as six million land mines covered the war zone. Yugoslavia, once the richest country in Eastern Europe, was now the third poorest, after Albania and Bosnia. Dragan was right. It was a catastrophe.

The drunk on the train was singing a popular song: "Who is saying—who is lying—that Serbia is small?" I was coughing, trying to catch my breath, recalling a sentence from the novel Dragan had urged his Slovenian friends to read during the Ten-Day War: "But it is the fate of the vanquished and the victor to merge into the flame of a single being."

"You were there three days, and with the language barrier, what did you see?" repeated the guest worker. "Nothing."

"Do you want some *rakija*?" the drunk demanded. I told him I did not want to make him sick, though in truth I was entertaining the notion of infecting him. His voice rose in volume. The old man from Knin whispered in my ear, "Sorry, sorry." The guest worker shook his head. The train was slowing down near the Hungarian border when the drunk pulled out a long knife and waved it at me, saying something I did not understand. Everyone laughed. But I was too sick to appreciate the joke: the knife was made in America. Even the passengers in the corridor were laughing.

21

Barcelona

April 1996

We were in the Hall of Chronicles, eating caviar. Up the Black Staircase, beyond a mural celebrating the traditional costumes of Catalonia, waiters in tuxedos greeted us with sparkling wine. A docent called attention to the paintings covering the walls and ceiling, depicting Roger de Flor's fourteenth-century expedition to the Orient: the defense of Gallipoli, the vengeance of the Catalans, the boarding of the ships in the Bosporus. The ceiling was given over to the defense of the tower of Adrianople, a grand optical illusion in which the defenders appeared to follow viewers around the room. The effect was more interesting than the subject, one Bosnian decided, even if it was a trick.

UNESCO had pulled out all the stops for the Barcelona Conference on Higher Education in Bosnia and Herzegovina. Sixty delegates from the Bosnian government, academic institutions, NGOs, and the EU had come to city hall for an evening reception hosted by the mayor, an affable man who welcomed us with a joke about Spain temporarily lacking a government.

It sounded familiar: a small nation within a nation agitating for autonomy. Catalans wanted more concessions from Madrid; with only 6 percent of Spain's territory and less than 14 percent of the population, they produced a quarter of the country's exports, and resented the taxes levied by the central government. Growing more strident about their identity, they had made Catalan their official language in schools, television, and government; Spanish was disappearing from public life. Castilians, meanwhile, were tired of Catalonian demands and suggestions that Barcelona was more "European" than the rest of Spain. Yugoslavia had fallen apart over the same kinds of issues.

"The day Dubrovnik was bombed," said the mayor, "I remembered what was said after Hiroshima: When a man dies, his memory travels among his friends and family, but when a city dies, the memory of it dies, too. And when Sarajevo was shelled, I realized the fate of all European cities was at stake. If we could not save Sarajevo, we knew no multicultural city could survive."

The mayor called for a grass-roots Marshall Plan—university to university, city to city, club to club—not one imposed from on high. At the conclusion of his remarks he invited the Bosnian minister of education to join him for a private meeting. The dean of Sarajevo's Fine Arts Academy winked at me. "One small step for Bosnia," he said, "one giant step for Europe."

"Do you want to see my first office?" said Kemal Bakaršić.

The exhibition on the ground floor of city hall, devoted to the National Library in Sarajevo, included a pair of photographs from before and after its destruction. The second-story window that caught Bakaršić's eye in one photograph was a gaping hole in the next. His first supervisor, he said, waving his hand over the ruins, had destroyed the library from his new home in Pale.

Kemal, a thin, grey-haired man of forty who looked fifty, radiated a kind of innocence, even with bags under his blue eyes. He had a reticent manner, which in others might have been mistaken for furtiveness, and a librarian's way of speaking: you had to lean forward to hear him. During the war he had taken a position in the ministry of education, and at this conference, while his colleagues pleaded their case, he stayed in the background. I had not noticed him until this afternoon, when he gave me a copy of his essay about the Sarajevo Haggadah, which I read with uncommon interest.

This fourteenth-century illuminated manuscript (one of the oldest extant Haggadahs, telling the Exodus story of the Jews rescued from slavery in Egypt) was the work of Sephardic calligraphers in Aragón. It belonged to a family in Barcelona, whose descendants, fleeing the Inquisition, brought the sacred text to Italy. Sometime in the sixteenth century the Haggadah arrived in Bosnia, where it survived not only the Nazi occupation of Sarajevo but also the Serbian siege, during which Kemal, chief librarian of the National Museum, pieced together its strange odyssey.

How did it end up in Sarajevo? There were two legends: in one, a Bosnian Jew studying in Padua fell in love with a beautiful girl and was given the Haggadah as a wedding gift; in the other, a Jewish merchant from Sarajevo, who saved an Italian companion from a bad business deal in Florence, received the Haggadah as a reward. These legends, which Kemal first came across as a student researching his thesis in the National Library, reminded him of stories his father had told him as a child. No story was more mysterious than that of the Haggadah's rescue during World War Two.

Soon after the Nazis occupied Sarajevo, in April 1941, the Old Jewish Temple was destroyed, along with its library and archives; Jewish property was confiscated; all but a handful of Jews were sent to concentration camps, including the entire

Cohen family, which had sold the Haggadah to the National Museum at the end of the nineteenth century. The Germans coveted the Haggadah, and the story goes that when high-ranking German and Croatian officers went to the museum to seize it the director told them a lieutenant had come to his office two hours earlier with the same demand.

"I gave it to him," said the director, feigning surprise.

"Who is this lieutenant?" shouted General Johann Fortner, the commander of German and Croatian troops in eastern Bosnia. "Give me his name! His unit! Who authorized him to do this?"

The director said he thought the lieutenant—he did not catch his name—was acting on the general's orders. As a matter of fact, the Haggadah was in his briefcase; and when the officers left he gave it to a librarian who, with the help of a Muslim cleric, hid it under the threshold of a house in a mountain village near Sarajevo, where it remained until the war was over.

Kemal recounted this legend at a conference in the Sarajevo Holiday Inn in September 1992 to commemorate the 500th anniversary of the Jewish expulsion from Spain. Alija Izetbegović opened the event by praising relations between Sarajevo's Jewish and Muslim communities: an extraordinary story of religious tolerance. "At the same moment we were watching the bombardment of the Jewish Cemetery," the librarian told me. Kemal was struck by parallels: that five hundred years after the Jews exiled from Spain settled in Bosnia, Bosnian Muslims were "ethnically cleansed" from their homeland; that fifty years after the Sarajevo Haggadah was saved from the Nazis, it was spared again, this time from Serbian gunners targeting the National Museum. The museum director, a Muslim guided by a *djinni*, rescued the Haggadah and hid it in the central bank vault.

"Guided by what?" I said.

"Every time you open the Haggadah a *djinni* appears and does good deeds, strange deeds," Kemal said with a straight face. "It was the joker in the war."

I nodded, inclined to credit the rumor that during the siege he had traded a computer, donated to him by a humanitarian, for LSD. But his elliptical manner of telling a story, he gave me to understand, was influenced by the *djinni*, not drugs. This *djinni* (or good ghost, as Kemal sometimes called it) was even responsible for the speculative nature of his writing and argument. Only with some difficulty had the librarian reconciled his training as a scientist with the *djinni's* tricks.

"I know how to write a scientific article, not one filled with maybes," he said. "But this was an exercise in history, confusion or confession, and parallels."

"Go on," I said.

"If you have not had the experience of war," he said, "you look for parallels to survive. For example, I quoted Hölderlin's lines to myself: 'Where danger lies, salvation also grows.' 'Patmos' is a beautiful poem, but the first parallel that came to mind was from *Woodstock*: 'There's always a little bit of Heaven in a disaster area.' Do you know what that means to a man who does not know if he will make it back to his home? We were brave about going out, because there was no reason to stay in the house. It was something we had to do. So if you woke up in the morning you said simple things, like *Fuck the rest, I'm the best*, to give you courage to walk through the snipers. You had to develop defense mechanisms, which for me meant finding parallels."

The most important parallel was the Haggadah, the book of remembrance and redemption. Kemal said the Sarajevo Haggadah was a Spanish, Jewish, and Bosnian manuscript, since most of Bosnia's ancient documents had been stolen or taken elsewhere during wars. Four times during the Passover Seder the child asks why this night is different from all other nights. The Haggadah (literally, the Telling) provides the answer in the story of the Jewish exodus, which revives the memory of Israel; in commentaries on the ten plagues, the crossing of the Red Sea, and the significance of the Passover lamb, unleavened bread, and bitter herbs; in blessings and hymns and a prayer for vengeance on the oppressing nations. The Haggadah represents the collective hope of the Jewish diaspora: the story of the Israelites' delivery from bondage offers a model for redemption. "Every oppressor is Pharaoh," writes one historian of the Haggadah, "and Egypt every exile." The last line is: "Next year in Jerusalem!" And Sarajevo, Kemal reminded me, is known as "Little Jerusalem."

"After every war someone has to tidy up," writes the Nobel laureate Wisława Szymborska. "Things won't pick themselves up, after all." We were in Barcelona to pick up the pieces of Bosnia's higher educational system. Sarajevo University had sustained more than $100 million worth of damage to its buildings, furniture, laboratory equipment, and libraries; none of its twenty-six schools was spared; ten were decimated. And while the regional universities established in the 1970s, in Tuzla, Mostar, and Banja Luka, had escaped devastation, their operations and academic development were curtailed when their students and faculty fled the country, joined the army, or were killed. The Dayton Peace Accords called for a decentralized approach to higher education, but the cantons lacked basic resources; and since the international community had yet to decide how to fund higher educational reconstruction efforts it was not clear how the universities, which everyone agreed were the best places to begin reunification, would survive. The rectors from Sarajevo, Mostar, and Tuzla emphasized how little help they had received, and how fast the new school year was approaching.

"The armies are going back to their barracks. Now the intellectuals must step out and do what they must do," said Srebren Dizdar, an English professor heading UNESCO's Bosnia Commission. "Our higher educational system is at a crossroads. Every postponement means a multiplication of the problems at the universities. There is no time to lose."

But what the conference revealed was how difficult it was to piece together something that, in Dizdar's words, did not exist, especially when the Bosnian Serbs had sent no representatives. Yet the meetings proceeded as if partition were only a distant possibility. The EU, the European Training Foundation, UNESCO, student groups, all offered to help bring together the Bosnian academic community and international partners. These were small steps, though. "We are not reconstructing a system that collapsed," said Dizdar. "What we are trying to reconstruct is life."

Nonetheless an Italian diplomat from the Council of Europe found a silver lining in this cloud. He described Bosnia as "a dream situation for countries attempting to reform their educational systems. In my own country there are any number of political theorists who dream of having the educational system destroyed so they can reform it. You have almost a tabula rasa," he told the Bosnians. "This is an opportunity. Diversify, don't isolate yourselves, and don't decentralize yourselves out of existence. Let us try to avoid the mistakes made in other countries."

The Bosnian delegation stared at him in stunned silence.

"So we meet in another exotic setting," Zvonko Radeljković said with a smile.

The professor took pleasure in discovering that the hotel in which the conference was held, an unfinished building in a newer section of Barcelona, was located around the corner from Karl Marx Place. But little else amused him. He was grieving over his mother's death on Easter Sunday. Nor had he forgotten our argument in Sarajevo about Bosnia's future.

"Do you see that everything I predicted has come true?"

Brigadier General Jovan Divjak, he said, the Serbian deputy commander who had defended Sarajevo, was now a parade general. If he was not the only Serbian or Croatian officer in the Bosnian Army to find himself marginalized, Divjak was the most visible. Unlike Muslim officers with no choice in the matter, Divjak had joined the Bosnian side, at great personal risk, when its military position had appeared hopeless. His former JNA colleagues branded him a traitor; had the Bosnians lost, the Serbs would have executed him. Izetbegović rewarded Divjak for his courage by trying to force him to retire; the general hung on to a ceremonial post, but other Serbs and Croats in the military were not so fortunate. The Bosnian Army was becoming an Islamic corps.

Meanwhile, the Dayton agreement guaranteed to the more than three million people uprooted in the war the right to return to their homes, but hardliners on all sides viewed this as an impediment to the creation of ethnically "pure" states. For example, bands of displaced Muslims from other parts of Bosnia had taken to threatening, beating, and expelling the few Serbs who stayed in the Serb-held suburbs of Sarajevo now under government control. Ethnic cleansing was continuing apace.

"So you see," said Zvonko.

"You win," I said.

"Hardly," he said.

Dissatisfaction with the peace agreement was not confined to Bosnia's warring parties. The lesson that Kosovar Albanians took from Dayton, where their dispute with Belgrade was ignored, was that the West respected only force; an armed campaign for independence was thus in the offing. Within the year, a new militia, the Kosova Liberation Army (KLA), would receive a windfall of weaponry from Albania, when failing pyramid investment schemes plunged that country into chaos, its arsenals emptied, its streets filled with the sounds of gunfire. "The Yugoslav crisis began in Kosovo," went the popular saying, "and it will end in Kosovo." And KLA guerrilla actions in 1998—assassinating Serbian policemen, kidnapping civilians—would provide Milošević with an excuse to ethically cleanse Kosovo, drawing NATO into its first military campaign. In the spring of 1999, the alliance would bomb Yugoslav targets for weeks to bring the Serbs to heel, though not before they had massacred thousands of Albanians and displaced a million more—crimes against humanity for which Milošević and his military leadership would be indicated by the War Crimes Tribunal. But Kosovars would return to their razed homes, protected by a NATO-led occupying force, even as Serbs fled the province, fearing revenge at the hands of the KLA. The flawed peace treaty would haunt the world into the next millennium.

And Zvonko's former friend, Nikola Koljević? The Shakespeare scholar and vice president of Republika Srpska was said to be depressed about his diminished role in the political process. Soon he would be found in his office with a fatal bullet wound to the head, unrepentant to the end.

"We are all vulnerable to barbarians," said Fahrudin Rizvanbegović. Bosnia's soft-spoken minister of education had adopted a harsh tone for his official remarks to the conference. "It is the world's obligation to help us rebuild Bosnia and Herzegovina, because the world was an accomplice to our tragedy. We have been attacked together, you and us, and therefore we must be defended together."

Barbarism was his new specialty. A professor of Yugoslav literature at Mostar University, he had spent six months in a Croatian concentration camp, where the guards (including his own students) tortured him, knocking out his teeth and breaking his ribs. He lost seventy-five pounds. He said that after four years of barbarism his country needed concrete assistance to rebuild, not verbal support.

The divided city of Mostar presented a special problem: now it had two universities instead of one. The original university was located in West Mostar, and when Muslim faculty and students were expelled from that side of the city, classes were moved first to the port city of Neum, then to Jablanica and Konjic. Eventually classes were held in East Mostar, in basements and public schools. At the same time, the new Croatian University of Mostar, housed in the original buildings, added a pedagogical faculty and an academy of fine arts. The separate universities stood for the continuing division between Muslims and Croats. The Bosnian-Croat Federation was united only on paper.

"Unfortunately, the Croats have a fascistic regime," Kemal Bakaršić told me. "Everyone is allowed to attend our universities. That is not the case in West Mostar. No one knows how to communicate or set up structures of interethnic cooperation."

And not only in Mostar. Classes were being divided up along ethnic lines throughout Bosnia, even in Sarajevo—Muslims here, Croats there; Serbs who wanted to hear lessons in something other than Muslim or Croatian history, literature, and language had to go to schools in the Serbian-occupied suburb of Lukavica. The chances of recovering a common usable past, a cultural thread strong enough to stitch together the war-torn country, diminished with each passing day. Auden was right about what all schoolchildren learn: "Those to whom evil is done/ Do evil in return."

Zvonko Radeljković summed it up. "You can fix anything, even Warsaw after World War Two, but is there any spirit? I am afraid my countrymen did not treat our new minister of education very well, and you will see what that will mean for Bosnia."

Remembrance and redemption.

The questions remained: how to remember a tragedy when, as Szymborska writes, "All the cameras have gone to other wars," and then, how to redeem it?

"We were the victims of aggression, but let us not insist on that," Kemal Bakaršić said one day in the hotel bar. "No one listens to the victims, which means they will unite in some kind of Hezbolla and teach radical traditions."

His main concern was reconstructing the National Library, the ruins of which he did not want to see turned into a monument. Determined to reproduce as much of the holdings as possible, Kemal and his bibliography students at the university were creating a record of what was lost.

He also had to restore his mother's apartment in Grbavica, the last of the Serbian-held parts of Sarajevo to be handed over to the Bosnian government. It was stripped of everything, even the windows, and from the balcony Kemal was surprised to see more space than he remembered: the Faculty of Forestry, where his father once lectured, had been burned to the ground. Like Ilidža, which had gone up in flames soon after my drive through it in the winter, Grbavica had been looted and torched before the Serbs fled to Republika Srpska and beyond.

"Here's an irony," said the librarian, ordering a *cappuccino*. "The foreign press used words like multicultural, multireligious, multiracial to define Bosnia. We described ourselves in other terms, like good neighbors, good behavior, home, education, respect for older people. For most Sarajevans religious holidays were a chance to visit relatives and friends: it was always mixed. Cakes, drinks, socializing. The competition of cuisines is all that counts!"

But the trust that tied together Sarajevo's different peoples was gone. The ethnic cleansers had won. The first war crimes trial in fifty years was about to begin in The Hague, but few believed the architects of these crimes against humanity would be brought to justice. Only that morning in my room I had watched Holocaust survivors telling their stories on television, and now the unthinkable had happened again. Here were three peoples set on the road by war criminals and true believers, three new versions of Exodus. Redemption? Not likely, at least not in the near term.

Kemal remained optimistic, though. "If we learned anything from our bad experience it is that you cannot stop life. Only brutal killing stops life, but you are always inventing something to feel normal. Finding parallels. Pascal said you can boil a liter of water with three hundred twenty five grams of paper: a ridiculous lie. If I trusted that I would have had to burn up my entire private library. But with practice you learn that two hundred grams is enough for a liter, which is enough for a cup of tea or coffee.

"We faced total isolation, we suffered so much we may just drop in our cage, like a turtle withdrawing into itself. No one can say what the future will be. But I'm looking forward to refurbishing my home, to fetch some glass for my windows, to redecorate my living room—a decent normal thing. I don't want anyone to remember me as a victim, because I don't want to be pathetic."

"Bar*the*lona!" cackled the Bosnians in the back of the bus, mimicking the tour guide's Catalonian accent. She pointed out the old prison, and they broke into laughter.

"The pri*th*on," they cried, "in Bar*the*lona!"

The conference was over, resolutions had been passed, and this afternoon we were being shepherded around the city. I sat next to a Serbian woman, Stamenka Uvalić-Trumbić, a dark-haired professor's daughter who had moved from Belgrade to Bucharest to work for UNESCO. "Of course the Serbs were invited," she was saying, "but they did not come."

When invitations were extended to Pale's minister of education and the rector of Banja Luka University, the political leadership decided to send instead the rector of the new Serbian University of Sarajevo, Vojislav Maksimović. But no one could determine if such an institution existed: the Serbs refused to provide any information about it. Nor would UNESCO permit Maksimović, a former literature professor in the Faculty of Philosophy, to attend the conference: he was on both the American and Bosnian lists of war criminals. His exclusion provided the Serbs with an excuse to stay away.

"They're working against their own interests," Stamenka said. "We asked them to reconsider their decision, because we hoped to start talking about a unified, autonomous higher educational system. The question is, Will Bosnia survive as a state or be partitioned? We would like to restore the tradition of four peoples living and working together."

The Serbian boycott of this conference and the international donors' conference in Brussels the week before offered more proof of their opposition to reunification. They were not allowing Muslim and Croat refugees to return to their homes in the Serbian enclave, and they would block the formation of a central bank, ignore arms control limits, and sign an economic and military cooperation agreement with Yugoslavia, all violations of the peace accord. Radovan Karadžić, who was amassing a fortune on the black market, would not step down as president until just before the September elections. Yet from behind the scenes, protected by the police, he would continue to control the economy, media, and political process. Republika Srpska was integrating with Serbia, not Bosnia.

Nevertheless Stamenka predicted change for the better if the extremists were removed from power. "If I were a student I would rather go to Sarajevo University than to Banja Luka or even to Belgrade, because they're so isolated," she said, though tens of thousands of university-age Serbs were already fleeing the destitute enclave, certain they had no future in the Balkans. "But maybe part of Bosnia will be Croat, part Serb, and then there will be the poor Muslims."

The scholar Bernard Lewis traces modern ideological racism's origins to the Christian reconquest and unification of Iberia in 1492, when Jews and Muslims were given the choice of conversion, exile, or death. The exiles found refuge in the Ottoman Empire, and five hundred years later Barcelona seemed an appropriate venue for a conference on promoting ethnic and religious harmony in the most enduring symbol of the Ottoman advance on Europe. But how to reconcile enemies or find justice for the war victims, and not only for the dead and wounded. Nearly 20,000 Bosnians were missing, 80 percent of them Muslim, almost all of them men. "Find out—record—reflect—but then move on. That is the least bad formula I know for truth *and* reconciliation," writes historian Timothy Garton Ash. But as we drove around Picasso's hometown I recalled the answer the artist gave to the Nazi officer who admired his *Guernica,* the painting inspired by the German saturation bombing of the Basques' ancient capital during the Spanish Civil War. Did you do that? said the officer. No, said Picasso. You did.

The tour was almost over, and the guide had yet to mention the Spanish Civil War. Sixty years ago, George Orwell returned to Barcelona, after months at the front, in time for the Uprising that spelled the end of cooperation among the forces allied against Franco and the Fascists. Here was a city with such a history of street fighting, Orwell noted, that one wag suggested numbering the paving-stones to save "trouble in building and demolishing barricades." But revolution was no longer in the air. I was thinking of the penultimate stanza of Szymborska's "The End and the Beginning":

> Those who knew
> what this was all about
> must make way for those
> who know little.
> And less than that.
> And at last nothing less than nothing.

To clean up the debris from a war requires a certain forgetfulness, without which it is impossible to rebuild houses, bridges, and train stations. How to commemorate the war in such a way that, years hence, when someone lies on the ground, staring up at the sky, he will not be moved, as Szymborska warns, to "dig up a rusted argument from underneath a bush" and use it to start another war?

Our last stop was the unfinished Expiatory Church of the Holy Family, the sculptor Antonio Gaudí's crowning achievement. Orwell said the Anarchists showed bad taste in not blowing up the church during the Rising, but I disagreed. In the lengthening shadows of a warm spring afternoon I reveled in Gaudí's idiosyncrasies. Here was an architectural journey from the Gothic to the fantastic, a spiritual autobiography in stone and tile and glass, in gargoyles and mosaics and four crenelated spires. The ecstatic façades of the Nativity, the Passion, the Glory, and the Assumption concealed a spare interior, like folds of flesh. Inspired by frescoes of martyrdoms (nails hammered into the eyes of the faithful waving goodbye as they sink into a cauldron of boiling oil, and so on), what Gaudí added to the light and clarity of the Gothic idea was a modern essay in exaggeration, expressed in broken surfaces and rubble and bright colors. The church was closed (the guide could not say when, if ever, construction might be completed), so we began to circle it.

I fell in with Kemal Bakaršić, who was musing on an unexpected consequence of the siege: how it had created among Sarajevans a strange will to survive. "If the Serbs had let us have an outlet, it would have been a catastrophe," he said. "No one would live there. Everyone would have made their escape plans. Instead, during the siege, there was fantastic movement among people in the city, a daily inner energy. This was something the Serbs did not predict would happen."

That movement, that energy, I realized with a sudden force, was what drew me to Sarajevo.

"This is all that counts in a dangerous time," Kemal said. "Good humor while you wait in the road to fetch water, a joke, wearing a nice tie, a clean shirt, perfume, or makeup. We were in such a small area, so I told my wife we cannot get angry at someone because in the years to come we will meet them again. This is something you have to think about: how to behave in a closed situation."

Our conversation turned to his next project. Like a tree with rings around the core—that was how he envisioned the facsimile edition of the Haggadah he planned to publish, with commentaries in the margins of the text. He believed its survival augured well for the future of his country. "When the Haggadah is opened, as it was last Passover, good things happen," Kemal said, slowing to a halt. "Perhaps this conference was one of the deeds of the *djinni*. Hölderlin was right: 'Where danger lies, salvation also grows.' I recited those lines throughout the war."

He looked hard at me. "Can you imagine? They saved my life."

Epilogue

14 September 1997

The road from Senj was empty. The hard rain that had slowed our drive down the coast, from Trieste to Rijeka and past the island of Krk, turned to a drizzle once we had climbed the Dinaric Alps. And it was on the harsh plains of Krajina, passing dynamited houses and barns, abandoned fields and pastures, that Krzysztof said the music seemed to change. Gone was the nostalgia he had felt along the windswept Adriatic, listening to tapes of Madredeus. The folk melodies of the Portuguese band now haunted the writer—like a lament, he said, speeding up. Except for a truckload of humanitarian supplies, the road was clear to the border; on the other side was a long line of cars carrying guest workers, refugees, and vacationing Bosnians waiting to cross over.

Krzysztof Czyżewski was a fine traveling companion, open-spirited, linguistically gifted, and in command of a new red Volkswagen. I admired his nomadic streak (twenty years ago, when he abandoned a promising soccer career to join a troupe performing medieval religious dramas in Poland and abroad, he made a point of visiting the Continent's remoter outposts) and his independence. During the period of martial law, he founded an underground cultural journal, with the blessings of his father, a high-ranking military officer who could ill afford any scandal. With Communism's fall, Krzysztof gave up the journal to create one of Poland's first nonprofit organizations, a center for arts, cultures, and nations. Now in his essays he was sketching a vision of Central Europe rooted in the cultural power of the peripheries. A small bearded man, he was himself a source of cultural riches. After the last reading at Vilenica one year he had sung for a group of us in the empty cave, his rich baritone filling the tectonic cathedral with hymns from the Kaddish. I was happy to see him again at the literary festival this year, not least because he accepted without hesitation my proposal to drive to northwest Bosnia when it was over. A reckless plan, some writers said. Krzysztof was undeterred, having made several wartime trips to Mostar and Sarajevo. We set out before daybreak.

This was a momentous weekend in the Balkans. Even as Slovenians celebrated the fiftieth anniversary of the partitioning of lands around Trieste, which they called the rightful return of Istria, Bosnians voting in their first municipal elections since 1990 were effectively ratifying their country's partition into ethnic enclaves. Very few refugees had been allowed to return to their homes (especially in areas under Croatian and Serbian control), and the officials they elected, by absentee ballots or after being bused, under heavy guard, to polling stations in ethnically cleansed towns like Srebrenica and Žepa, might never set foot in their precincts. A dozen governments-in-exile could issue from this democratic exercise, and more exiles. Witness the traffic on the Bosnian side of the border.

Entering Bosnia was just as time-consuming, though there was no one ahead of us. First we had to come up with 92 deutsche marks for Krzysztof's visa, then we were told that his car insurance was invalid in Bosnia. A policeman sent us to a row of kiosks next to the border crossing, where a clerk demanded 64 marks for two weeks of coverage—twice what the writer paid per year.

"But my card is good for all of Europe," Krzysztof protested.

"But Bosnia is not Europe," the clerk replied nonchalantly.

So it seemed. And we needed the coverage, since the roads were in abysmal shape, as we discovered driving into Bihać, the "safe area" with the strangest war record of all. The Bihać pocket was a good place to study the senselessness of the war. Besieged by Serbian forces in May 1992, no less than five separate armies would fight over it, including rival Muslim parties caught up in a civil war, before it was liberated in August 1995. Civilians suffered the brunt of the attacks.

It was an old Balkan tale—the cult of personality leading gullible people into ruin. Before the war, the main employer in this predominately Muslim region was Agrokomerc, a food distribution company under the direction of Fikret Abdić, a corrupt figure who nevertheless received more votes than Alija Izetbegović in the 1990 election to Bosnia's collective presidency. The northern half of the Bihać pocket stayed loyal to Abdić when the war began, and when he broke with Izetbegović in 1993 to create his own fiefdom his people followed him into battle with the Bosnian government. By the winter of 1994 thousands were living as refugees in Agrokomerc chicken coops in Krajina; the next summer, when Abdić founded the Republic of Western Bosnia, they had the distinction of pledging allegiance to the shortest-lived state to emerge from the former Yugoslavia: it was overrun one week later. Though he had been supported by Bosnian and Croatian Serbs, Abdić sought asylum in Zagreb (the Bosnian government had charged him with treason), perhaps because the Croats planned to install him as the leader of northwest Bosnia, if they ever incorporated it into Croatia proper.

"*Stećci*," Krzysztof cried, excitedly waving his hand.

He turned onto a side street to get a better look at the stone monuments marching up a nearby hill, the medieval gravestones, often decorated with carvings of human figures, horses, spirals, and crosses, once thought to belong to the Bogomils. However, the slabs and standing blocks, thousands of which remain in Bosnia, Herzegovina, and Dalmatia (and, to a lesser extent, in Croatia, Serbia, and Montenegro), were built for the nobility of all faiths, Bosnian, Catholic, and Orthodox. Little else is known about them, except that they are peculiar to these lands. A mystery on the same order as Stonehenge, the *stećci* figure prominently in *Stone Sleeper*, the masterpiece of Mak Dizdar (1917–1971), Bosnia's greatest modern poet. *Stone Sleeper* is at once a historical work on the enigmatic Bosnian Church in the face of persecution and a meditation on Bosnian identity. Which is to say: its theme is the enduring mystery of the human condition. Dizdar's "Kolo of Sorrow," for example, enacts a *kolo* or round dance of the South Slavs, the steps of which have not changed since medieval times, according to carvings on the *stećci*. Nor has the tragic nature of life in the Balkans, as Francis Jones's translation makes clear:

How long the kolo from hollow to hollow
How long the sorrow from hollow to hollow

How long the dread from stead to stead
How long the tombs from coomb to coomb

How long the blood we are judged to pay
How long the death till the judgment day

How long the kolo from hollow to hollow
How long the sorrow from kolo to kolo

Kolo to kolo from sorrow to sorrow

The circular sorrow of this poem brought to mind a conversation with a Croatian journalist who claimed the inscriptions on the *stećci* were proof that Bosnia was once part of Croatia. "They're weird, like Bosnia itself," she said with assurance. "That's why I like them. The *stećci* are me. I'm talking about myself when I talk about them. But I wouldn't want to have them as neighbors."

But the misnamed Bogomil tombstones did not make the same impression on the young man we stopped to ask for directions to them. He did not even know what they were.

"I only finished grammar school," he said, and walked off toward the *stećci*.

"He must be a holy fool," said Krzysztof, rejoining the traffic streaming into Bihać.

Bosnia was flush with foreigner observers—diplomats, election monitors, and journalists—as well as 3,000 fresh NATO troops providing extra security during the voting (the size of the original force was cut in half when the mission was extended into a second year). And the first person we met at the post office was a perky American woman married to a refugee from Banja Luka. They came here twice a semester from their home in Germany, once for her husband to register for classes at the university, once for exams. He was completing his mechanical engineering degree, begun before the war, through the mail. Crazy, he said. His absentee ballot he called useless. Like his degree, said his wife. Probably we'll have to start over in America. But they appreciated the humor of the OSCE decision to give everyone pencils with which to mark their ballots. No wonder 70 percent of Bosnians would vote in these elections, including Serbs, who had threatened to boycott them.

"It will take time," said the refugee, "but Muslims and Serbs will live together again."

"Is that right?" I said, surprised.

"But never with Croats," he insisted. "They want everything separate—their own schools, media, everything. Do you know they're building villages along the border to join with Croatia?"

From the post office we went to a bar to change money with a pudgy war profiteer. Then a stroll through the city; compared to the surrounding countryside, it did not seem badly damaged. The buildings we saw were intact and trees stood in the park; the underground airport destroyed by the Serbs (along with $50 million worth of sophisticated equipment) was not on our itinerary. Our lunch, we decided, returning to the car, would be taken in Republika Srpska.

But Krzysztof was stopped for speeding even before we had reached the outskirts of Bihać. "What an expensive trip this is," he sighed. Then he was spared a ticket. The policemen did not have to tell him to slow down: the road along the Una River hugged a cliff from which hundreds of rocks had fallen during the storm; potholes stretched from shoulder to shoulder; more landslides were imminent. The villages on both banks of the river lay in ruins. The railroad tracks were overgrown with grass. In Bosanska Krupa, a fiercely contested, devastated town central to Serbian designs before it was retaken by the Bosnians, Krzysztof spoke of a Muslim woman he knew who had lost her father and two brothers in the fighting. She no longer lived here.

"More Madredeus?" he said.

Just past the temporary bridge in Otoka (guarded by two NATO soldiers in a truck), we came to a curious sight. On either side of a wooden hut marking the border between the Federation and Republika Srpska were camp trailers, plastic chairs and tables with umbrellas, cars, and cabs. Muslims and Serbs had created a market of sorts, where you could buy, trade, or barter for almost anything—gas, liquor, cigarettes, cars, sheep. Men were shaking hands and laughing in the drizzle. Cab drivers called out to those who had traded away their cars and needed rides home. No policemen or soldiers anywhere.

On to the grimy city of Bosanski Novi, which Serbian authorities had renamed Novi Grad—New City. There was nothing new about the place, aside from the red signs proclaiming, in Serbian, that life was better under the Communists and the posters of Radovan Karadžić plastered to lampposts and walls, which read, in English, "He Means Peace"—a veiled threat to NATO officers contemplating his arrest. The pro-Karadžić posters, which dated from a summer NATO operation to arrest two Bosnian Serbs charged with war crimes (one of whom, Prijedor's former police chief, had been killed in a shootout), gave the false impression of Bosnian Serb unity. Civil war was brewing in Republika Srpska.

Karadžić and his protégée, Biljana Plavšić, who replaced him as president when he was forced out of politics in the summer of 1996 (a concession won from Milošević during an emergency trip that Richard Holbrooke had made to the region to salvage the peace agreement), were fighting for the soul of the Bosnian Serbs. Plavšić's nationalism was no less fanatical than Karadžić's. If his beliefs grew out of poetry, hers were rooted in biology: the former professor from Sarajevo had even defended ethnic cleansing as "a natural phenomenon." And for much of her presidency she was little more than Karadžić's puppet until she turned on him this summer, accusing him and his ally, Momčilo Krajišnik, the Serbian member of the Bosnian collective presidency, of mafia-style corruption. They monopolized imports and paid no duties or taxes, leaving the government unable to meet the salaries of doctors, teachers, and civil servants, or to deliver checks to pensioners; with unemployment running at 90 percent and thousands of refugees from neighboring Krajina living in squalor, Plavšić found a ready audience for her attacks on the men she said were impoverishing the Serbs.

Karadžić had his own support, including a large contingent of well-paid bodyguards; hence there were two functioning governments in the Serbian entity, Plavšić's elected seat in Banja Luka and Karadžić's shadow government in Pale, each with its own police force and media; with Ratko Mladić still wielding power behind the scenes, the army had yet to decide whom to support. Plavšić refused to travel to Pale, with good reason. NATO had already foiled one coup attempt, and Serbian police had recently detained her at the airport in Belgrade. The international community was giving her military and economic aid because she seemed willing to consider honoring some of the terms of the Dayton Peace Accords. (Not the right of return, however, the most important provision; of the 70,000 Muslims and Croats expelled from Banja Luka, for example, fewer than 200 had been allowed to go home, and only because their houses were too badly damaged for Serbs to live in.) What the West offered, though, was inadequate. Plavšić had some protection, but it was not until Pale TV, which routinely spewed vile propaganda against her and NATO, likened the Western troops to Nazis that NATO shut down its television translators. To prevent mob violence against Western election observers, the OSCE had not disqualified those Serbian candidates whose campaign posters included photographs of Karadžić, in flagrant violation of the peace agreement. Indeed Karadžić operated with impunity, though conventional wisdom among Serbs and Muslims alike was that his days were numbered: either he would be killed in a NATO raid (and thus enshrined in the pantheon of Serbian martyrs), or he would commit suicide. The chances he would stand trial for war crimes at The Hague were nil. Nor was he likely to collect the millions of dollars he had deposited into Swiss numbered accounts, a detail certain to be overlooked by his future *guslers* and hagiographers.

Republika Srpska was, in many respects, worse off than even the hardest-hit parts of the Federation, where Western aid was flowing in and rebuilding had begun. Perhaps only the western half of the Serbian entity would remain in Bosnia. Those in the eastern enclave seemed determined to join with Serbia: one reason why the Bosnian government was rearming, preparing to overrun Serbian territory if and when the NATO mission ended. The Bosnians would make short work of the bankrupt, dispirited Bosnian Serb forces, according to some observers. No telling how Serbia would react, especially since the elections set for the next weekend would bring Vojislav Šešelj, the murderous paramilitary leader, into a power-sharing arrangement with Milošević's party. Last winter's "pro-democracy" protests in the Serbian capital had been for naught.

Belgrade was even more miserable than I remembered, Dušan Kovačević assured me at lunch. The thick-fingered carpenter invited us to join him at one of the two tables in a café on the main street of Novi Grad, and his company was better than the *ćevapčići* we ordered. He punctuated each phrase with a swipe of his hand, whether he was telling jokes about Slovenians (he hated their music) or describing his wartime losses (cars, houses, friends). A long scar ran down his throat, and he spoke in a husky voice, part of his vocal cords having been removed in one of the seven operations he had undergone after being wounded by a grenade. His monthly disability check amounted to the price of a beer. It cost him three months' pension to tell us his story.

"Do you know the difference between the man who drinks and the man who doesn't drink?" he asked when three young men entered the café and took their seats at the other table.

Krzysztof shook his head.

"The man who drinks will die. And the man who doesn't drink will follow him into death."

Dušan let out a hearty laugh. The impresion he gave was of an innocent aging man caught up in a meaningless war: the first two years he had spent hiding in his house in Bosanska Krupa. Then he enlisted in the medical corps as a stretcher-bearer; while carrying a wounded soldier across a bridge he was struck by shrapnel from a grenade, which killed the soldier and four friends. Dušan was recovering from another operation when he became a refugee. His recent trip to Belgrade left him even more embittered: for two months of work he had netted only two hundred marks. He planned to move to Tampa, where his brother lived, as soon as the paperwork came through. He could count to ten in English, and I taught him the word *shit,* which on his tongue came out *sheet.*

"You'll have no problem," I told him.

Only once did Dušan lower his voice, and that was in answer to a question about the elections.

"I voted here, in Novi Grad," he whispered sheepishly, eyeing the young men at the other table. "For Mrs. Plavšić. She is our president." He raised his voice to insist, wrongly, that the Muslims had started the fighting in Bosanska Krupa. Just as quickly he was speaking with genuine warmth about his Muslim friends there. Like the refugee we had met in Bihać, Dušan believed that Muslims and Serbs would live together again someday, but never with Croats. After all, he knew the Muslim family living in his house, as well as the Muslims whose house he now occupied in Novi Grad (they were refugees in Sanski Most). Before we parted he asked us to take a message to his old neighbors telling them his family was fine—a message, needless to say, we did not deliver.

But we did drive back and forth three times between Muslim and Serbian territory, first because we missed the sign to Croatia and then, when we had returned to Novi Grad, because the Serbian soldiers guarding the bridge said the Croats on the other side would not let us cross the river. So we retraced our route to Otoka. With business still brisk at the border market between the Muslim and Serbian entities, I joked that someone was bound to make an offer on Krzysztof's car. Across the bridge in Otoka we turned north and meandered through several Muslim villages, in each of which new mosques were under construction; the minarets, much taller than those destroyed in the war, rose from the hills like giant beanstalks. (In Republika Srpska, where every mosque had been razed, the authorities prevented the muftis from undertaking any reconstruction.) Through the gloaming we drove, past horse-drawn carriages and milling crowds, boys playing soccer and an invalid hobbling down the road. Everywhere were signs for Agrokomerc.

It was dark when we arrived in Velika Kladuša, Abdić's former stronghold, and I sighed with relief once we had cleared the Croatian border. Too soon, as it turned out. The border guard directed us to a dirt road with so many craters and potholes that Krzysztof had to drive at a walking pace. Every house and barn in the next ten kilometers had been dynamited; the fields were ravaged and, for all we knew, mined. We did not see a light or another car for more than an hour. A broken axle, a skid off the road—any number of grim scenarios passed through my mind. Thousands had trudged this way, fleeing from one army or another. What remained were skeletal buildings looming in the dark. Krzysztof turned up the music. We barely spoke until we came to a paved road. Then: What a hollow victory for Croatia. And: Who would want to live here? Tuđman had even invited Australian Croats to return and repopulate towns in Krajina like Cetingrad, a desperate unlit place. The main street was all but empty, and the man who gave us directions to the highway sent us down another crumbling road that, to our surprise and chagrin, took us to another border crossing. The lone guard was equally astonished to see us appear out of the dark. The hut on the Bosnian side was empty. No, Krzysztof told the guard, we do not wish to cross. Where is the highway? The guard laughed and, leaning on the crossbar, proceeded to give him detailed directions to Karlovac. There was really no reason to return to Bosnia. Beyond the hut was pitch darkness, and we were very tired.

Glossary of Names and Terms

Abdić, Fikret: Bosnian Muslim businessman and renegade warlord
Açka, Flutura: Albanian poet
Akashi, Yasushi: UN Special Representative (1993–1995)
Andrić, Ivo (1892–1975): Yugoslav novelist, diplomat, and Nobel laureate
APC: Armored Personnel Carrier
Arkan (Željko Ražnjatović): Serbian paramilitary leader
Arnautalić, Ismet (Nuño): Bosnian musician and co-founder of SAGA
Bakaršić, Kemal: Bosnian librarian
Basha, Eqrem: Kosovar Albanian poet
Berisha, Sali: Albanian cardiologist and president (1992–1997)
Black Hand/Union of Death: Secret society dedicated to Serbian expansion
Boban, Mate: Leader of the Bosnian Croats
Bogdanović Bogdan: Serbian architect and writer
Boutros-Ghali, Boutros: UN Secretary General (1991–1996)
Branković, Vuk (d. 1397): Serbian nobleman who in Serbian epic poetry is said to have betrayed his father-in-law, Prince Lazar, at the Battle of Kosovo
Butler, Hubert (1900–1991): Irish writer and author of *Independent Spirit: Essays*
Cankar, Ivan (1876–1918): Slovenian novelist
Carrington, Lord Peter: First EC peace envoy
Çavdarbasha, Agim: Kosovar Albanian sculptor
Četnik: Serbian militia
Ćosić, Dobrica: Serbian novelist and president of Yugoslavia (1992–1993)
CSCE/OSCE: Conference/Organization on Security and Cooperation in Europe
Czyżewski, Krzysztof: Polish essayist
Debeljak, Aleš: Slovenian poet and essayist
Divjak, Jovan: Deputy commander of the Bosnian army

Dizdar, Mak (1917–1971): Bosnian poet and author of *Stone Sleeper*
Djilas, Milovan (1911–1995): Montenegrin writer and Partisan leader
Djordjević, Milan: Serbian poet
Drakulić, Slavenka: Croatian writer and journalist
Drasković, Vuk: Serbian journalist, novelist, and opposition leader
Duraković, Ferida: Bosnian poet
Dušan, Stefan (1300–1355): Tsar of the Serbs, Greeks, Bulgarians, and Albanians
EC/EU: European Community/European Union
Elytis, Odysseas (1911–1996): Greek poet and Nobel laureate
Gaj, Ljudevit (1814–1872): Croatian linguist and founder of the Illyrian Movement
Gjuzel, Bogomil: Macedonian theater director and translator of Shakespeare
Gligorov, Kiro: President of Macedonia
Handke, Peter: Austrian novelist and playwright
Havel, Václav: Czech playwright and president
HDZ/Croatian Democratic Community: Radical nationalist party
Herceg-Bosna: Self-proclaimed Bosnian Croat state
Holbrooke, Richard: Assistant U.S. Secretary of State
Hoxha, Enver: Communist leader of Albania (1945–1985)
Hudhri, Bujar: Albanian publisher
HVO/Croatian Defense Council: Croatian militia
Ibrišimović, Nedžad: Bosnian Muslim novelist and president of the Bosnian Writers' Association
ICRC: International Committee of the Red Cross
IFOR: NATO's Implementation Force in Bosnia
IMF: International Monetary Fund
IRC: International Rescue Committee
Ismajli, Rexhep: Kosovar Albanian social linguist and vice president of LDK
Izetbegović, Alija: President of Bosnia-Herzegovina
Keeley, Edmund (Mike): American novelist and translator of Greek poetry
Janša, Janez: Slovenian Minister of Defense
JNA: Yugoslav National Army
Jelačić von Buzim, Baron Joseph (1801–1959): Croatian military hero
Jósef, Attila (1905–1937): Hungarian poet
Kadaré, Ismaïl: Albanian novelist living in Paris
Karadjordje (1768?–1816): Black George, Serbian patriot and founder of the Karadjordjević dynasty
Karadžić, Radovan: Bosnian Serb poet, leader, and indicted war criminal
Karadžić, Vuk (1787–1864): Serbian philologist and folklorist
Kardelj, Edvard (1910–1979): Yugoslavia's leading ideologist and author of *The Slovene Nationality Question*

Kenović, Ademir: Bosnian film maker and co-founder of SAGA
Kiš, Danilo (1935–1989): Serbian novelist and author of *Garden, Ashes*
Kocbek, Edvard (1904–1981): Slovenian poet and essayist
Koljević, Nikola (1936–1997): Bosnian Serb Shakespeare scholar and and vice-president of Republika Srpska
Konrád, George: Hungarian novelist, politician, and president of PEN International
Kosovo Polje: The Field of Blackbirds, where Ottoman forces defeated Prince Lazar's army in the Battle of Kosovo, 28 June 1389
Krašovec, Metka: Slovenian painter
Kučan, Milan: President of Slovenia
Kulenović, Tvrtko: Bosnian writer
Kundera, Milan: Czech novelist living in Paris
Lalić, Ivan (1931–1997): Serbian poet and translator
Lazar, Prince (1329–1389): Serbian leader killed at the Battle of Kosovo
LDK: Democratic League of Kosova
Maković, Zvonimir (Zvonko): Croatian poet and art critic
Marković, Ante: Last Yugoslav Prime Minister
Matevski, Mateja: Macedonian poet
Meštrović, Ivan (1883–1962): Croatian sculptor
Milarakis, John: Greek journalist
Milošević, Slobodan: President of Serbia
Miłosz, Czesław: Polish poet and Nobel laureate
Milutinović, Zoran: Serbian professor of Slovenian literature
Mladić, Ratko: Commander of the Bosnian Serb military forces
Momirovski, Tomé: Macedonian novelist
Mujahedeen: Islamic holy warriors
Novak, Boris: Slovenian poet and president of Slovenian PEN Club
NATO: North Atlantic Treaty Organization
NDH: *Nezavisna Država Hrvatska*—Independent State of Croatia (1941–1945)
NGO: Non-governmental organization
Obilić, Miloš: Legendary Serbian knight who in Serbian epic poetry is said to have killed Sultan Murad at the Battle of Kosovo and then was beheaded by the Turks
Obrenović, Miloš (1780–1860): founder of Obrenović dynasty and father of modern Serbia
Open Society Institute: Humanitarian organization funded by George Soros
Oslobođenje: Sarajevo's daily newspaper
Owen, Lord David: European Community Special Envoy to the former Yugoslavia

Panić, Milan: Serbian-American millionaire and prime minister of Yugoslavia (1992)
Pašović, Haris: Bosnian theater director and co-organizer of the Sarajevo Film Festival
Pavelić, Ante (1889–1957): Ustaša leader
Pavlović, Miodrag: Serbian poet and novelist
Perović, Slavko: Montenegrin poet and leader of the Liberal Alliance
Perse, St.-John (1887–1975): French poet, diplomat, and Nobel laureate
Pešić, Vesna: Serbian opposition leader
Petrović-Njegoš, Petar I (1747–1830): Montenegrin metropolitan (Saint Peter of Cetinje)
Petrović-Njegoš, Petar II (1812–1851): Montenegrin metropolitan, poet, and author of *The Mountain Wreath*
Petrović-Njegoš, Nikola I (1841–1921): Montenegrin king and "father-in-law of Europe"
Pintarić, Jadranka: Croatian art critic
Plavšić, Biljana: Bosnian Serb biologist and president of Republika Srpska (1997–1998)
Plečnik, Jože (1872–1957): Slovenian architect
Podrimja, Ali: Kosovar Albanian poet
Popa, Vasko (1922–1991): Serbian poet and author of *Earth Erect*
Prešeren, France (1800–1849): Slovenia's national poet
Princip, Gavrilo (1895–1918): Serbian political agitator and assassin of Archduke Franz Ferdinand
Purivatra, Miro: Director of Obala, an art gallery/cultural center in Sarajevo
Razović, Maja: Croatian journalist
Radeljković, Zvonimir (Zvonko): Bosnian professor of American literature
Republika Srpska: Self-proclaimed Bosnian Serb republic, which after the Dayton Peace Accords became an internationally recognized entity, making up 49 percent of Bosnia-Herzegovina
Ritsos, Yannis (1909–1990): Greek poet
Rotberg, Dana: Mexican film maker and co-organizer of the Sarajevo Film Festival
Rugova, Ibrahim: President of Kosovar Albanians
Rupel, Dimitrij: Slovenian novelist and foreign minister
Şahin, Haluk: Turkish journalist and novelist
SAGA: Sarajevo Group of Auteurs
Šalamun, Tomaž: Slovenian poet
Schneider, Jana: American photojournalist wounded in Sarajevo (1992)
Seferis, George (1900–1971): Greek poet, diplomat, and Nobel laureate

Šešelj, Vojislav: Serbian unltra-nationalist politician and paramilitary leader
Simic, Charles: Serbian-American poet and translator
Simić, Goran: Bosnian poet living in Toronto
Stinger: U.S.-made, shoulder-held surface-to-air missile
Strela: Soviet-made, shoulder-held surface-to-air missile
Stepinac, Alojzije (1898–1960): Archbishop of Zagreb during the NDH
Strossmayer, Josip (1815–1905): Croatian bishop and founder of the first Yugoslav Academy
Stup: Front-line position in Sarajevo
Szymborska, Wisława: Polish poet and Nobel laureate
Taufer, Veno: Slovenian poet and organizer of the Vilenica Literary Festival
TDU: Territorial Defense Unit
Tito, Marshal (Josip Broz): President of Yugoslavia (1945–1980)
Tuđman, Franjo: Partisan leader, historian, and president of Croatia
UNHCR: United Nations High Commissioner for Refugees
UNPROFOR: United Nations Protection Forces
Ustaša: Croatian Fascist regime (1941–1945)
Vance, Cyrus: UN Special Envoy to the former Yugoslavia
Velikić, Dragan: Serbian novelist
West, Dame Rebecca (1892–1983): English writer and author of *Black Lamb and Grey Falcoln*
Zadruga: Extended family
Zagoričnik, Nina: Slovenian journalist
Zimmerman, Warren: Last U.S. Ambassador to Yugoslavia
Zupan, Uroš: Slovenian poet
Zupančič, Boštjan: Slovenian constitutional lawyer and writer

Select Bibliography

Agee, Chris (editor). *Scar on the Stone: Contemporary Poetry from Bosnia.* Newcastle upon Tyne: Bloodaxe Books, 1998.

Akhmatova, Anna. *The Complete Poems of Anna Akhmatova.* Edited by Roberta F. Reeder. Translated by Judith Hemschemeyer. Somerville, MA: Zephyr Press, 1990.

Ali, Rabia, and Lawrence Lifschultz (editors). *Why Bosnia? Writings on the Balkan* War. Stony Creek, CT: The Pamphleteer's Press, 1993.

Allen, Beverly. *Rape Warfare: The Hidden Genocide in Bosnia-Herzegovina and Croatia.* Minneapolis: University of Minnesota Press, 1996.

Anderson, Benedict. *Imagined Communities: Reflections on the Origin and Spread of Nationalism.* London: Verso, 1983.

Andrić, Ivo. *The Bridge on the Drina.* Translated by Lovett F. Edwards. Chicago: University of Chicago Press, 1977.

———. *Bosnian Chronicle.* Translated by Joseph Hitrec. New York: Alfred A. Knopf, 1963.

———. "A Letter from 1920." Translated by Lenore Grenoble. In *Three Stories about Bosnia: 1908, 1946, 1992.* Belgrade: Association of Yugoslav Publishers and Booksellers, 1995.

Appelfeld, Aharon. *Beyond Despair: Three Lectures and a Conversation with Philip Roth.* Translated by Jeffrey M. Green. New York: Fromm International, 1994.

Ash, Timothy Garton. *The File: A Personal History.* New York: Random House, 1997.

———. "In the Serbian Soup." *New York Review of Books* (April 24, 1997): 25–30.

Bakaršić, Kemal. "The Story of the Sarajevo Haggada." *Judaica Librarianship,* Volume 9 (Spring 1994/Winter 1995): 135–143.

Balkan War Reports. Bulletins of the Institute for War and Peace Reporting, London.

Banac, Ivo. *The National Question in Yugoslavia: Origins, History, Politics.* Ithaca, NY: Cornell University Press, 1984.

Benderly, Jill, and Evan Kraft (editors). *Independent Slovenia: Origins, Movements, Prospects.* New York: St. Martin's Press, 1994.

Blinkhorn, Martin, and Thanos Veremis (editors). *Modern Greece: Nationalism & Nationality.* Athens: Eliamep, 1990.

Brodsky, Joseph. *Watermark.* New York: Farrar, Straus & Giroux, 1992.

Burry, Mark. *Expiatory Church of the Sagrada Família: Antoni Gaudí.* London: Phaidon Press Limited, 1993.

Butler, Hubert. *Independent Spirit: Essays.* New York: Farrar, Straus and Giroux, 1996.

Calvino, Italo. *Invisible Cities*. Translated by William Weaver. New York: Harcourt Brace, 1974.

Carnegie Endowment for International Peace. *Report of the International Commission to Inquire into the Causes and Conduct of the Balkan Wars*. Washington, 1914.

Char, René. *Leaves of Hypnos*. Translated by Cid Corman. New York: Grossman Publishers, 1973.

Chatwin, Bruce. *The Songlines*. New York: Viking, 1987.

Cohen, Lenard J. *Broken Bonds: The Disintegration of Yugoslavia*. Boulder, CO: Westview Press, 1993.

Ćosić, Dobrica. *This Land, This Time* (including *Into the Battle, A Time of Death, Reach to Eternity*, and *South to Destiny*). Translated by Muriel Heppell. New York: Harcourt Brace Jovanovich, 1983.

Crnobrnja, Mihailo. *The Yugoslav Drama*. Montreal and Kingston: McGill-Queen's University Press, 1994.

Cviic, Christopher. *Remaking the Balkans*. New York: Council on Foreign Relations Press, 1991.

Danforth, Loring M. *The Macedonian Conflict: Ethnic Nationalism in a Transnational World*. Princeton, NJ: Princeton University Press, 1995.

Danner, Mark. "The U.S. and the Yugoslav Catastrophe." *New York Review of Books* (November 20, 1997): 56–64.

———. "America and the Bosnia Genocide." *New York Review of Books* (December 4, 1997): 55–65.

———. "Clinton, the UN, and the Bosnian Disaster." *New York Review of Books* (December 18, 1997): 65-81.

———. "Bosnia: The Turning Point." *New York Review of Books* (February 5, 1998): 34–41.

———. "Bosnia: Breaking the Machine." *New York Review of Books* (February 19, 1998): 41–45.

Debeljak, Aleš. *Anxious Moments*. Translated by Christopher Merrill and the author. Fredonia, NY: White Pine Press, 1994.

———. *Twilight of the Idols: Recollections of a Lost Yugoslavia*. Translated by Michael Biggins. Fredonia, NY: White Pine Press, 1994.

Dizdarević, Zlatko. *Sarajevo: A War Journal*. Translated from the French by Anselm Hollo. Edited from the original Serbo-Croatian by Ammiel Alcalay. New York: Fromm International, 1993.

———. *Portraits of Sarajevo*. Translated by Midhat Ridjanović. Edited by Ammiel Alcalay. New York: Fromm International, 1994.

Djilas, Aleksa. *The Contested Country: Yugoslav Unity and Communist Revolution, 1919–1953*. Cambridge, MA: Harvard University Press, 1991.

Djilas, Milovan. *Njegoš: Poet, Prince, Bishop*. Translated by Michael B. Petrović. New York: Harcourt Brace and World, 1966.

———. *Wartime*. Translated by Michael B. Petrović. New York: Harcourt Brace Jovanovich, 1977.

Donia, Robert J., and John V. A. Fine Jr. *Bosnia and Hercegovina: A Tradition Betrayed*. New York: Columbia University Press, 1994.

Drakulić, Slavenka. *The Balkan Express: Fragments from the Other Side of the War*. New York: W.W. Norton, 1993.

———. *Café Europa: Life after Communism*. New York: W.W. Norton, 1997.
———. *How We Survived Communism and Even Laughed*. New York: W.W. Norton, 1992.
Ehrenreich, Barbara. *Blood Rites: Origins and History of the Passions of War*. New York: Henry Holt, 1997.
Elytis, Odysseas. *Open Papers: Selected Essays*. Translated by Olga Broumas and T Begley. Port Townsend, WA: Copper Canyon Press, 1995.
———. *Selected Poems*. Chosen and introduced by Edmund Keeley and Philip Sherrard. New York: Viking Penguin 1981.
Fine, John V. A., Jr. *The Early Medieval Balkans: A Critical Survey from the Sixth Century to the Late Twelfth Century*. Ann Arbor, MI: University of Michigan Press, 1983.
———. *The Late Medieval Balkans: A Critical Survey from the Late Twelfth Century to the Ottoman Conquest*. Ann Arbor, MI: University of Michigan Press, 1987.
Gibbon, Edward. *The History of the Decline and Fall of the Roman Empire*. London, 1813.
Gjelten, Tom. *Sarajevo Daily: A City and Its Newspaper under Siege*. New York: HarperCollins, 1995.
Glenny, Misha. *The Fall of Yugoslavia: The Third Balkan War*. London: Penguin, 1996.
Green, Peter. *Alexander of Macedon, 356–323 B.C.: A Historical Biography*. Berkeley and Los Angeles: University of California Press, 1991.
Greenfeld, Liah. *Nationalism: Five Roads to Modernity*. Cambridge, MA: Harvard University Press, 1992.
Gutman, Roy. *A Witness to Genocide*. New York: Macmillan, 1993.
Handke, Peter. *Repetition*. Translated by Ralph Manheim. New York: Farrar, Straus and Giroux, 1988.
———. *A Journey to the Rivers: Justice for Serbia*. Translated by Scott Abbott. New York: Viking Penguin, 1997.
Havel, Václav. *Open Letters: Selected Writings, 1965–1990*. Selected and edited by Paul Wilson. New York: Alfred A. Knopf, 1991.
———. *Summer Meditations*. Translated by Paul Wilson. New York: Alfred A. Knopf, 1992.
———. *The Art of the Impossible: Politics as Morality in Practice*. Translated by Paul Wilson and others. New York: Alfred A. Knopf, 1997.
Helsinki Watch. *War Crimes in Bosnia-Herzegovina: A Helsinki Watch Report*. New York: Human Rights Watch, August 1992.
———. *Open Wounds: Human Rights Abuses in Kosovo*. New York: Human Rights Watch, March 1993.
———. *War Crimes in Bosnia-Herzegovina: A Helsinki Watch Report, Volume II*. New York: Human Rights Watch, April 1993.
———. *The Macedonians of Greece*. New York: Human Rights Watch, April 1994.
———. *Civil and Political Rights in Croatia*. New York: Human Rights Watch, October 1995.
———. *A Threat to "Stability": Human Rights Violations in Macedonia*. New York: Human Rights Watch, June 1996.
———. *Human Rights Watch/Helsinki Reports*. Bulletins of Human Rights Watch, New York.
Hobsbawm, Eric. *The Age of Extremes: A History of the World, 1914–1991*. New York: Vintage Books, 1994.
Holbrooke, Richard. *To End a War*. New York: Random House, 1998.
Honig, Jan Willem, and Norbert Both. *Srebrenica: Record of a War Crime*. London: Penguin, 1996.

Howard, Michael. *The Causes of Wars.* Cambridge: Harvard University Press, 1983.
Hukanović, Rezak. *The Tenth Circle of Hell: A Memoir of Life in the Death Camps of Bosnia.* New York: BasicBooks, 1996.
Izetbegović, Alija Ali. *Islam between East and West.* Indianapolis, IN: American Trust Publications, 1984, 1989.
Jelavich, Barbara. *The History of the Balkans, Volumes I and II.* Cambridge: Cambridge University Press, 1983.
Jones, Lloyd. *Biographi: A Traveller's Tale.* New York: Harcourt Brace, 1994.
József, Attila. *Winter Night.* Translated by John Bákti. Oberlin, OH: Oberlin College Press, 1997.
Judah, Tim. *The Serbs: History, Myth and the Destruction of Yugoslavia.* New Haven, CT: Yale University Press, 1997.
Kadaré, Ismaïl. *Chronicle in Stone.* Translated by Al Saqi Books. New York: New Amsterdam Books, 1987.
———. *The File on H.* Translated by David Bellos and Jusuf Vrioni. London: Haverill Press, 1997.
———. *Palace of Dreams.* Translated by Barbara Bray and Jusuf Vrioni. New York: William Morrow, 1993.
Kapuściński, Ryszard. *Imperium.* Translated by Klara Glowczewska. New York: Vintage International, 1995.
Karahasan, Dzevad. *Sarajevo, Exodus of a City.* Translated by Slobodan Drakulić. New York: Kodansha International, 1994.
Keegan, John. *A History of Warfare.* New York: Alfred A. Knopf, 1993.
Kiš, Danilo. *Garden, Ashes.* Translated by William J. Hannaher. New York: Harcourt Brace Jovanovich, 1975.
———. *A Tomb for Boris Davidovich.* Translated by Duiška Mikić-Mitchell. New York: Harcourt Brace Jovanovich, 1978.
———. *Hourglass.* Translated by Ralph Manheim. New York: Farrar, Straus and Giroux, 1990.
———. *Homo Poeticus: Essays and Interviews.* Edited by Susan Sontag. New York: Farrar, Straus and Giroux, 1995.
Konrád, George. *The City Builder.* Translated by Ivan Sanders. New York: Harcourt Brace Jovanovich, 1977.
———. *Antipolitics.* Translated by Richard E. Allen. New York: Harcourt Brace Jovanovich, 1984.
———. *The Melancholy of Rebirth: Essays from Post-Communist Central Europe, 1989–1994.* Selected and translated by Michael Henry Heim. New York: Harcourt Brace, 1995.
Krleža, Miroslav. *On the Edge of Reason.* Translated by Zora Depolo. London: Quartet Books Limited, 1989.
Kundera, Milan. *The Art of the Novel.* Translated by Linda Asher. New York: HarperCollins, 1989.
———. "The Tragedy of Central Europe." Translated by Edmund White. *New York Review of Books.* Volume 31 (April 26, 1984): 33–38.
Labon, Joanna (editor). *Balkan Blues: Writing Out of Yugoslavia.* Evanston, IL: Northwestern University Press, 1994.
Lalić, Ivan V. *The Passionate Measure.* Translated by Francis R. Jones. Dublin: The Dedalus Press, 1989.
———. *Roll Call of Mirrors: Selected Poems of Ivan V. Lalić.* Translated by Charles Simic. Middleton CT: Wesleyan University Press, 1988.

Lindsay, Franklin. *Beacons in the Night: With the OSS and Tito's Partisans in Wartime Yugoslavia.* Stanford, CA: Stanford University Press, 1993.
Maass, Peter. *Love Thy Neighbor: A Story of War.* New York: Alfred A. Knopf, 1996.
Maclean, Fitzroy. *Eastern Approaches.* London: Penguin, 1991.
Magaš, Branka. *The Destruction of Yugoslavia: Tracking the Break-up 1980–1992.* London and New York: Verso, 1993.
Malaparte, Curzio. *Kaputt.* Translated by Cesare Foligno. Evanston, IL: Northwestern University Press, 1995.
Malcolm, Noel. *Bosnia: A Short History.* New York: New York University Press, 1994.
———. *Kosovo: A Short History.* New York: New York University Press, 1998.
Mandelstam, Nadezhda. *Hope Abandoned.* Translated by Max Hayward. New York: Atheneum, 1974.
———. *Hope Against Hope: A Memoir.* Translated by Max Hayward. New York: Atheneum, 1970.
Mandelstam, Osip. *Complete Critical Prose.* Edited by Jane Gary Harris. Translated by Jane Gary Harris and Constance Link. Dana Point, CA: Ardis, 1997.
Mann, Thomas. *Death in Venice and Seven Other Stories.* Translated by H.T. Lowe-Porter. New York: Vintage Books, 1954.
Matevski, Mateja. *Footprints of the Wind: Selected Poems.* Translated by Ewald Osers. London and Boston: Forest Books, 1988.
Matthias, John, and Vladeta Vučković (translators and editors). *The Battle of Kosovo.* Athens, OH: Swallow Press Books, 1987.
Mehmedinović, Semezdin. *Sarajevo Blues.* Translated by Ammiel Alcalay. San Francisco: City Lights Books, 1998.
Miłosz, Czesław. *The Captive Mind.* Translated by Jane Zielonko. New York: Penguin Books, 1980.
———. *The Collected Poems 1931–1987.* New York: The Ecco Press, 1988.
———. *Beginning with My Streets: Essays and Recollections.* Translated by Madeline G. Levine. New York: Farrar, Straus and Giroux, 1991.
Morgenthau, Henry. *Ambassador Morgenthau's Story.* Garden City, NY: Doubleday, 1918.
Moynihan, Daniel Patrick. *Pandaemonium: Ethnicity in International Politics.* Oxford: Oxford University Press, 1993.
Musil, Robert. *The Man without Qualities.* Translated by Sophie Wilkins and Burton Pike. New York: Alfred A. Knopf, 1995.
Newhouse, John. "Dodging the Problem." *The New Yorker* (August 24, 1992): 60–71.
———. "No Exit, No Entrance." *The New Yorker* (June 28, 1993): 44-51.
Njegoš, Petar II Petrović. *The Mountain Wreath.* Translated and edited by Vasa D. Mihailovich. Irvine, CA: Charles Schlacks, Jr., 1986.
Norwich, John Julius. *Byzantium: The Early Centuries.* New York: Alfred A. Knopf, 1988.
———. *Byzantium: The Apogee.* New York: Alfred A. Knopf, 1991.
———. *Byzantium: The Decline and Fall.* New York: Alfred A. Knopf, 1995.
Orwell, George. *Homage to Catalonia.* New York: Harcourt Brace, 1952.
Osers, Ewald (editor and translator). *Contemporary Macedonian* Poetry. London and Boston: Kultura/Forest Books, 1991.
Pavlović, Miodrag. *The Slavs beneath Parnassus: Selected Poems.* Translated by Bernard Johnson. London: Angel Books, 1985.
Paz, Octavio. *The Collected Poems of Octavio Paz, 1957–1987.* Edited and translated by Eliot Weinberger. New York: New Directions, 1987.

———. *The Other Voice: Essays on Modern Poetry*. Translated by Helen Lane. New York: Harcourt Brace Jovanovich, 1990.

Perry, Duncan M. *The Politics of Terror: The Macedonian Liberation Movements. 1893–1903*. Durham, NC: Duke University Press, 1983.

———. "Macedonia: A Balkan Problem and a European Dilemma." *RFE/RL Reasearch Report*. Volume 1, number 25 (19 June 1992): 35–45.

Pfaff, William. *The Wrath of Nations: Civilizations and the Furies of Nationalism*. New York: Simon and Schuster, 1993.

Popa, Vasko. *Collected Poems*. Translated by Anne Pennington. Revised and expanded by Francis R. Jones. London: Anvil Press, 1997.

———. *Homage to the Lame Wolf: Selected Poems*. Translated by Charles Simic. Oberlin, OH: Oberlin College Press, 1987.

———. *The Golden Apple: A Round of Stories, Songs, Spells, Proverbs and Riddles*. Chosen and translated by Andrew Harvey and Anne Pennington. London: Anvil Press Poetry, 1980.

Poulton, Hugh. *The Balkans: Minorities and States in Conflict*. London: Minority Rights Group, 1991.

———. *Who Are the Macedonians?* Bloomington, IN: Indiana University Press, 1995.

Ritsos, Yannis: *Exile and Return: Selected Poems, 1968-74*. Edited and translated by Edmund Keeley. New York: The Ecco Press, 1985.

Runciman, Sir Steven. *The Fall of Constantinople 1453*. Cambridge: Cambridge University Press, 1965.

Šalamun, Tomaž. *The Four Questions of Melancholy: New and Selected Poems of Tomaž Šalamun*. Edited by Christopher Merrill. Fredonia, NY: White Pine Press, 1996.

———. *The Selected Poems of Tomaž Šalamun*, edited by Charles Simic. New York: The Ecco Press, 1988.

Samary, Catherine. *Yugoslavia Dismembered*. Translated by Peter Drucker. New York: Monthly Review Press, 1995.

Sells, Michael A. *The Bridge Betrayed: Religion and Genocide in Bosnia*. Berkeley and Los Angeles: University of California Press, 1996.

Seferis, George. *A Poet's Journal: Days of 1945–1951*. Translated by Athan Anagnostopoulos. Cambridge, MA: Harvard University Press, 1974.

———. *Collected Poems*. Edited and translated by Edmund Keeley and Philip Sherrard. Princeton, NJ: Princeton University Press, 1995.

Sherrard, Philip. *The Wound of Greece: Studies in Neo-Hellenism*. London: Rex Collings, 1978.

Silber, Laura, and Allan Little. *Yugoslavia: Death of a Nation*. New York: TV Books, 1995.

Simic, Charles. *The Unemployed Fortune-Teller: Essays and Memoirs*. Ann Arbor, MI: University of Michigan Press, 1994.

Simić, Goran. *Sprinting from the Graveyard*. Translated by David Harsent. Oxford: Oxford University Press, 1997.

Suško, Mario (editor). *Contemporary Poetry of Bosnia and Herzegovina*. Sarajevo: International Peace Center, the Writers' Association of Bosnia and Herzegovina, 1993.

Szymborska, Wisława. *Poems New and Collected, 1957–1997*. Translated by Clare Cavanagh and Stanisław Barańczak. New York: Harcourt Brace and Company, 1998.

Tanner, Marcus. *Croatia: A Nation Forged in War*. New Haven, CT: Yale University Press, 1997.

Thompson, Mark. *A Paper House: The Ending of Yugoslavia*. New York: Pantheon, 1992.

———. *Forging War: The Media in Serbia, Croatia and Bosnia-Hercegovina*. London: Article 19, 1994.

Van Creveld, Martin. *The Transformation of War.* New York: Free Press, 1991.
Vickers, Miranda. *The Albanians: A Modern History.* London: I. B. Tauris, 1995, 1997.
Vulliamy, Ed. *Seasons in Hell.* London: Simon and Schuster, 1994.
Vreme. News Digest Agency. Belgrade.
Wachtel, Andrew. *Making a Nation, Breaking a Nation: Literature and Cultural Politics in Yugoslavia.* Stanford, CA: Stanford University Press, 1998.
West, Rebecca. *Black Lamb and Grey Falcon.* London: Viking Press, 1941.
West, Richard. *Tito and the Rise and Fall of Yugoslavia.* New York: Carroll and Graf, 1994.
Wheatcroft, Andrew. *The Ottomans.* London: Viking Press, 1993.
Wilson, A. N. *Paul: The Life of the Apostle.* New York: W.W. Norton, 1997.
Wolff, Larry. *Inventing Eastern Europe: The Map of Civilization on the Mind of the Enlightenment.* Stanford, CA: Stanford University Press, 1994.
Woodward, Susan L. *Balkan Tragedy: Chaos and Dissolution after the Cold War.* Washington, DC: The Brookings Institution, 1995.
Yerushalmi, Yosef Hayim. *Haggadah and History.* Philadelphia and Jerusalem: The Jewish Publication Society, 1997.
Zimmerman, Warren. *Origins of a Catastrophe.* New York: Random House, 1996.

Index

About the Author

Christopher Merrill's books include three collections of poetry, *Workbook, Fevers & Tides,* and *Watch Fire*; translations of Aleš Debeljak's *Anxious Moments* and *The City and the Child*; several edited volumes; and two works of nonfiction, *The Grass of Another Country: A Journey Through the World of Soccer,* and *The Old Bridge: The Third Balkan War and the Age of the Refugee*. He holds the William H. Jenks Chair in Contemporary Letters at the College of the Holy Cross. He and his wife, violinist Lisa Gowdy-Merrill, are the parents of a daughter, Hannah.

Merrill,
Christopher.

Only the nails
remain.

DATE			